Diel Vertical Migration of Zooplankton in Lakes and Oceans

Joop Ringelberg

Diel Vertical Migration of Zooplankton in Lakes and Oceans

Causal Explanations and Adaptive Significances

Springer

Joop Ringelberg
Stationsstraat 90
3881 AG Putten
Netherlands
j.ringelberg@hccnet.nl

ISBN 978-90-481-3092-4 e-ISBN 978-90-481-3093-1
DOI 10.1007/978-90-481-3093-1
Springer Dordrecht Heidelberg London New York

Library of Congress Control Number: 2009936304

Printed on acid-free paper

Springer is part of Springer Science+Business Media (www.springer.com)

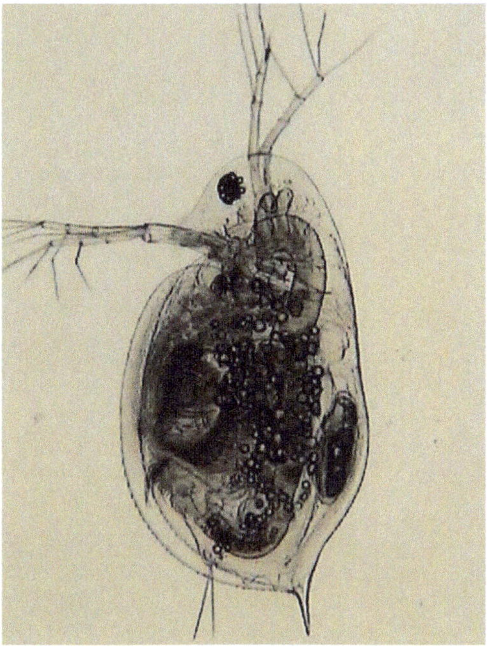

Plate 1 *Daphnia galeata x hyaline* from Lake Maarsseveen (note that the compound eye has been turned in a ventral direction). The hybrids played an important role in the study of DVM in the lake and in the laboratory experiments of the Department of Aquatic Ecology, University of Amsterdam. (Photograph by Michaela Brehm, Centre for Limnology, Nieuwersluis, the Netherlands.)

Preface

Whatever theory may be advanced to explain diurnal migration, the underlying reactions involved must be demonstrated conclusively in the laboratory before the explanation can be finally accepted

George L. Clarke 1933 p. 434

In oceans and lakes, zooplankton often make diel vertical migrations (DVM), descending at dawn and coming up again in late afternoon and evening. The small animals cover distances of 10–40 m in lakes or even a few hundred metres in the open oceans. Although not as spectacular as migrations of birds or the massive movements of large mammals over the African savannas, the numbers involved are very large and the biomass exceed the bulk of the African herds. For example, in the Antarctic oceans swarms of "Krill" may cover kilometres across, with thousands of individuals per cubic metre. These Euphausiids are food for whales, the most bulky animals on earth.

Zooplankton are key species in the pelagic food web, intermediary between algae and fish, and thus essential for the functioning of the pelagic community. Prey for many, they have evolved diverse strategies of survival and DVM is the most important one. Most fish are visually hunting predators and need a high light intensity to detect the often transparent animals. By moving down, the well-lit surface layers are avoided but they have to come up again at night to feed on algae. This description is too simple, however, because a trade-off between a decreased predation and less food is present. Thus benefits may be smaller than costs. That depends on circumstances, of course, which implies that the individual has to "decide" when to migrate and how deep. During evolutionary time, a physiological–behavioural mechanism must have been selected for with information from the environment as decisive input. This topic is hardly touched upon in the hundreds of articles that appeared in more than a century of research. For decennia the simple hypothesis that zooplankton followed a "preferred" light intensity when moving up and down with the setting and rising of the sun reigned. In the last 30 years the aspect of predator avoidance got most attention as if by that the final explanation was given. And few authors are concerned with causal mechanisms and decision-making.

In this book, the physiological–behavioural mechanisms are central topics. Light intensity changes trigger complex photobehaviour, guiding the animals into a timely descent and a prudent ascent. Photobehaviour has no obvious relation with adaptive goals and in the opinion of some is, therefore, not important. However, without light intensity changes DVM would not occur. Causal mechanisms were predominantly studied in *Daphnia*, but the results can be applied to many migrating species in lakes as well as oceans. The crucial light changes of dawn and dusk are essentially the same, everywhere and always (yes, I know of seasons and latitudes). About the causal explanations presented in this book, the last word is not said. It never is in science. But they are presently the best explanations. Over the years of study, the mechanisms have become complicated and cannot be comprehended by reading summaries of the original articles only. I try to give a comprehensive picture now. Ultimate aspects are not shunned but as long as speculations are a large part of the literature, mentioning of these aspects had to be modest. Clarke's dictum also holds for the ultimate level of investigations.

I do not treat the literature exhaustively. There are excellent reviews where the many results and opinions are presented objectively, without much comment. I had to make choices from the large number of articles because I tried to scrutinise what was implied and to compare with the principles and philosophy of the book. Did I violate the principle of being objective? It is up to critical readers to say so. Compared to other topics in biology, I think that DVM research has not made much progress and, after the stimulating impulse of realising how important predators are, is slowing down again.

The set-up of the book is perhaps a bit unusual and needs explanation. Traditionally, descriptions of ecological phenomena and relevant environmental factors precede the presentation of mechanisms and explanations. Because most readers will be familiar with DVM, I start with the underlying physiology of stimulus perception and the resulting behaviour (Chapters 1, 2, 3, 4 and 5) as obtained with laboratory experiments. That opened a way to think about the role of essential "light" and "temperature" as "biological factors" (Chapter 6). I am aware of the independent existence of physical factors but I also recognise that each individual perceives and processes abiotic factors in a way that depends on a species-specific physiology. Processing occurs predominantly in the central nervous system and the factor gets new biological properties.

The book continues with DVM, as encountered in the field (Chapters 8, 9 and 10). I looked for studies with extensive and detailed data. Thereafter, in Chapter 11, the conclusions from the experiments were integrated with field data in a final validation of ideas.

Diel vertical migration is multifaceted with, amongst others, consequences for population dynamics, food web structure and community composition. In Chapter 12, the development of a *Daphnia* population during a period of migration was presented and in the last chapter the information flow as superimposed on the material flow in a tritrophic food chain was presented.

DVM cannot be studied without dedicated colleagues. Ten years of unique field work in Lake Maarsseveen would have been impossible without the organisation and control of Ben Flik and the talent for being able to fix everything by Dick Lindenaar. Koen Royackers participated in the first years and Erik van Gool in the last years. Niels Daan helped when the size and distribution of the 0^+ fish population had to be estimated. An intensive acoustic study in the lake by Annie Duncan, Brian Quiliiam (Royal Holloway University of London) and Jan Kubecka (Hydrobiological Institute, Ceske Budejovice, Czech Republic) contributed much to our knowledge of the juvenile fish population in the lake. I did not always have the time to participate in the field study because of duties at the university. Moreover, my attention was also focused on photobehaviour experiments, in the last years together with Erik van Gool. We had long discussions before we understood (do we?) "decision-making" in *Daphnia*. DVM starts long before sunrise and occurs predominantly after sunset. In early summer, nights are short and the need to sleep was not always felt on these weekly days at the lake. We were among friends, I think.

Drafts of some chapters were read by Jan Verheijen, Erik van Gool, Koos Vijverberg, Niels Daan and Otto Siebeck for critical comments on several chapters.

For 30 years Ben Flik played a central role in the organisation of our Department of Aquatic Ecology, University of Amsterdam. His friendship was often of great help to me.

Finally, I thank my son and daughters for support and unremitting love.

Putten, the Netherlands Joop Ringelberg

Contents

Chapter 1
Windows: An Introduction

Contents

1.1 Introduction

We study animals from an anthropomorphic perspective. The sense organs of animals differ from ours and, of course, also their central nervous system. A bee, with a spectral sensitivity extended in the ultraviolet light, sees a meadow with (for us) white daisies but reflecting UV quite differently. Yet the biotope is familiar to us.

If the environment is unfamiliar too, as the pelagic water masses of oceans are, the "world of plankton" is difficult to imagine. Henschel, at the end of the nineteenth century, coined the name "plankton" for what he thought were passive particles, floating in the vast expanse of the ocean. Like dust in the air. We now consider the passivity of zooplankton a matter of scale. Certainly, the oceanic circulations carry plankton over large distances even to localities where they are unable to breed (Van der Spoel and Heyman, 1983). On a much smaller scale, however, copepods and cladocerans are not lost in space and the "weak locomotor powers" in the definition of plankton do not hold. Individuals actively swim around, searching for food, for mates, or avoiding one another if necessary. How large must a volume of water be in order to be considered the "Lebensraum" of a copepod? That depends on the action radius of the animal. Of even more importance is the distance over which information about the environment can reach the animal. Like all animals, zooplankton have a behavioural programme to process relevant information from the environment. That programme is very different from ours but both evolved so as to be adequate for survival. We will always look from the outside at the behaviour and the

J. Ringelberg, *Diel Vertical Migration of Zooplankton in Lakes and Oceans*,
DOI 10.1007/978-90-481-3093-1_1, © Springer Science+Business Media B.V. 2010

environment of animals. Our studies create a window only and different views create as many windows. Each window produces its own picture of a transient reality. How is it to be a copepod in the Atlantic Ocean?

1.2 Individuals

Individuals are structured as open systems: matter, energy and information are continuously needed from the environment to keep the physiological state of the body in dynamic equilibrium. Food items are diverse but before entering the organism (the digestive tract is considered the outer world) digestion has reduced it into simple, biochemical substances. Energy, generated in the cells, is even simpler. Of matter and energy especially quantity is important. Therefore, it is "easy" to study the flow of matter in an organism, in an ecosystem. However, "one cannot live by bread alone". That also holds for zooplankton. Information is needed and information is very diverse. It must be obtained from physical, chemical and biological properties of the environment. These properties get their specific information content only after perception and processing by the individual organism and, therefore, are species-specific. Matter and energy are "consumed" and the original source disappears. The available amounts are often limited and the animal dies when there is not enough. Information is not consumed and is not depleted. The source of potential information is very rich. Ultimately, biodiversity is based on information rather than on matter or energy. The continuous stream of potential information impinging on the individual organism must be restricted to relevant information. A first step is filtering by the sense organs. Of the broad band of radiation in nature, only a small part is admitted due to the properties of the pigments in the eye. Successive selection follows in the nervous system. Finally, from the environment only a small, relevant part is retained for processing. This species-specific part was long ago called "Umwelt"[1] by Von Uexküll (1921). His terminology is largely forgotten but, nevertheless, remains relevant to understand how an organism deals with the environment.

Some factors from the "Umwelt" are important for vegetative functions such as metabolic rates, respiration, osmotic balance, and this part Von Uexküll called the " Wirkwelt". For a proper physiological functioning, the configuration and the values of the "Wirkwelt" factors must be suitable. It should be realised that most "Wirkwelt" factors cannot be detected directly. When the internal, physiological state of the body is threatened because of inadequate values of Wirkwelt factors, the animal is informed by interoceptors.[2] A correcting action should improve the situation but because "Wirkwelt" factors cannot be perceived problems arise. Now another part of the "Umwelt", which can be perceived directly by sense organs (exteroceptors (see Note 2)), comes to the rescue. This part of the environment Von Uexküll called the "Merkwelt". Perception and processing of these factors is fast and the information, obtained by this route, is rapidly translated into commands and adaptive behaviour. In life-threatening situations, "Merkwelt" factors are of primary importance. For example, copepods and cladocerans avoid low oxygen concentrations in nature as well as in laboratory cylinders (Nauman, 1921; Ubrig, 1952). Yet, no exteroceptor sense organ is available to detect low oxygen levels and no cue is available to direct the animal towards higher concentrations. A gradient in the water column might be present, but the slow-motion trial-and-error method is not optimal to reach higher concentrations when time is pressing. The solution is an association between "Wirkwelt" and "Merkwelt" factors. A low oxygen concentration in the blood (or a high carbon dioxide concentration) is coupled to a positively phototactic reaction. A *Daphnia* in low oxygen stress swims in the direction of the highest intensity of incident light, which is normally coming from above, and in that way reaches rapidly the water surface and in most cases also a higher oxygen concentration. The "Merkwelt" factor "light" has nothing to do with the problem of oxygen stress, but is nevertheless of life-saving importance. I will discuss at length another example of this principle of association of "Merkwelt" and "Wirkwelt" factors. Diel vertical migration (DVM) is about avoidance of predators and a chemical associated with visually hunting fish is very important as an early warning. This chemical is present in the water day and night and provides no cue for a diel reaction or to realise a timely retreat from the danger zone. Again, an association with the "Merkwelt" factor light is available and is used to initiate and to direct DVM.

Some "Wirkwelt" and "Merkwelt" factors seem to be identical but differ in effective modality. For example, temperature influences the rate of many vegetative functions like metabolic rate and development time and thus it is a "Wirkwelt" factor. On the other hand, a sufficiently large increase or decrease in temperature per unit time can be perceived directly and may evoke an immediate change in swimming behaviour as in zooplankton migrating through a thermocline. Each temperature modality must be considered a different "Umwelt" factor.

These nearly century old and forgotten German terms draw attention to various aspects of the environment as encountered by animals. If the principle is kept in mind, the role of light in DVM is more easily recognised.

1.3 Sensory Physiology and Behaviour

Many ecologists are now predominantly focused on the adaptive significance of behaviour. How this result is achieved hardly gets attention. However, speculations on the optimality of behavioural strategies run the risk of implicitly using performances, which the animal is unable to carry out. Therefore, a few remarks on general aspects at the base of animal behaviour might be useful. In the first chapters of this book, first principles of animal physiology play an important role. It is another window to look at animals.

From the moment of perception of a physical or chemical factor, transcription starts. The fate of a perceived signal is schematised in a sensor-actor chain, as presented in Fig. 1.1. In the first place, the modality and the quantitative and qualitative range of a perceived "Merkwelt" factor are limited by the properties of the

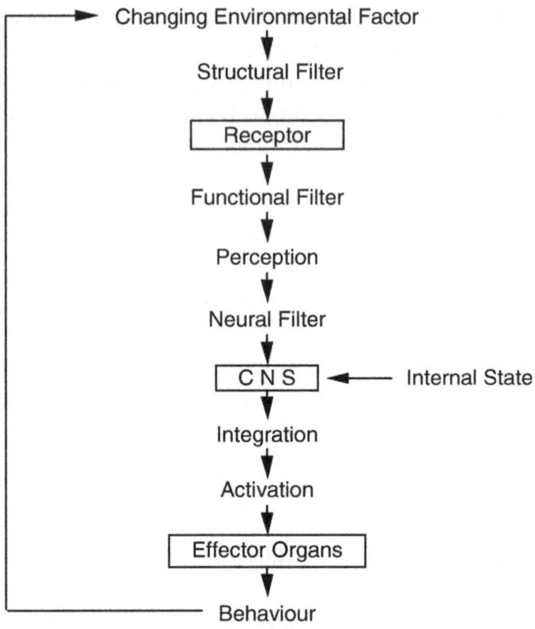

Fig. 1.1 A general scheme of a sensor-actor chain. The feedback *arrow* indicates that the result of a reaction is perceived and influences the course of the reaction (modified after Greve, 1995)

sense organ. For example, not all eyes are able to measure the plane of polarisation of the radiation from the sky. Moreover, eyes always absorb only a small part of the radiation spectrum. Once absorbed, the signal is transcribed in a form that can be transported by afferent nerves. At several levels of the nervous system, filtering, enhancement and inhibition takes place until the most relevant information is left. The mechanisms that realise these modifications came into existence by an evolutionary selection process in direct relation with the environment. In the central nervous system, also signals from interoceptors about the internal state of the organism exert their influence. Thus, before behaviour starts, information of the environment has changed in a species-specific way. In Fig. 1.1, a line is drawn from the end of the chain back to the beginning. This line suggests that the change in the initial value of the factor resulting from the response behaviour determines continuously this behaviour, which makes the simple reflex chain a closed system looking like a cybernetic system as in many technical control mechanisms. Of importance then are target or norm values inherent to the system. The input must be reformulated as the current difference between the ambient factor value and the internal norm value. The response diminishes the difference and continues to do so until a minimal difference depending on the sensitivity of the system is reached. Several vegetative functions can be described as negative feedback systems, for example, blood sugar control, osmoregulation and thermoregulation. Behavioural steering is another example. In that case the norm is a target, a prey to be caught or a course to be followed. Driving a motorcar in a straight line occurs through a continuous process of small corrections of which the magnitude depends on the perceived difference between actual course and the straight-line course. The paradigmatic explanation of diel vertical migration of zooplankton has long been the supposition that the animals follow an optimum light intensity. If true, this intensity must be considered a norm that is fixed in the central nervous system (CNS), and swimming is determined by the difference between the ambient light intensity and this norm intensity. Research was aimed, if not at the cybernetic character of reactions to changes in light intensity, in any case to show that an optimum light intensity was maintained (Clarke, 1932; Harris and Wolfe, 1956; Harris and Mason, 1956). The experiments failed to demonstrate this because the upward or downward movements were nearly always smaller than the theoretically expected. In the chapters to follow, I return to this topic.

As mentioned above, the sensor-actor chain was changed from an open chain into a closed one by means of a feedback. The difference between an open, causal chain type of reaction and a feedback mechanism is illustrated in Fig. 1.2a, b. Typical open chain reactions are simple reflexes, such as involuntary responses like sneezing or the Babinski toe reflex in human babies. Analysis of simple behavioural patterns might start as an open chain description, thus as a causal reaction to a stimulus. In subsequent steps analysis might lead to more complicated behavioural models.

The organism is not a response-machine, waiting for stimuli from the "milieu exterieur"; behaviour is also initiated "spontaneously" from the "milieu interieur" (nice terminology of the French physiologist Claude Bernard, 1813–1878). For example, "hunger" is an internal state that is registered by pressure receptors in the gut (strictly speaking the gut is not "milieu interieur") or by chemical receptors that signal low concentrations of sugars in the blood (the true "milieu interieur"), leading to feeding behaviour. Moreover, a nervous system is an active system where spontaneous behaviour is initiated. In that case the sensor-actor chain still holds but the beginning is at the level of the CNS but again many new stimuli from the "milieu interieur" and the "milieu exterieur" arise. For example, faster swimming in *Daphnia* involves changes in muscle tension and a different posture that is registered by proprioceptors. Swimming in a vertical direction also induces light intensity changes that are perceived (Ringelberg, 1964). These self-induced light changes must be distinguished from externally imposed light changes.

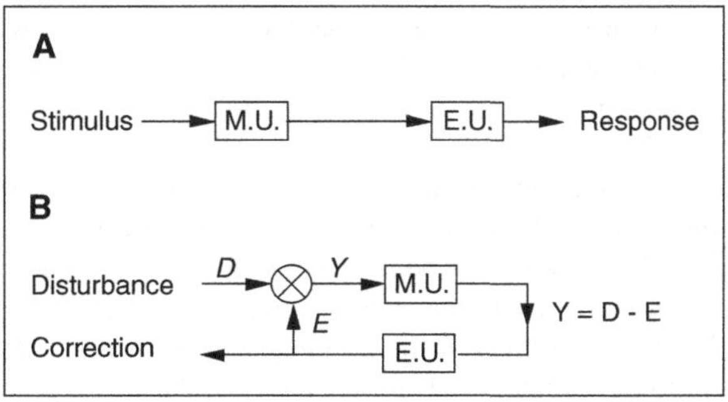

Fig. 1.2 The information flow chart of an open, causal chain (**a**) and of a closed, control system (**b**). M.U. and E.U. indicate the measuring unit and the effector unit, respectively (modified after Ringelberg, 1964)

If this distinction would not be made, the self-induced light intensity changes would result in counteractions, thus making spontaneous swimming impossible. On the other hand, if a *Daphnia* is passively displaced vertically, for example, by upwelling water movements, reactive downward swimming follows indeed as Schröder (1962) observed in Lake Constance. If swimming is initiated by the animal, the CNS obviously "expects" environmental changes and does not react to these self-induced changes. These aspects of animal behaviour were worked out in a brilliant paper by Von Holst and Mittelstaedt (1950). They called the central expectation "Efferenzkopie", which is a sensory input by a pertinent factor from the environment, caused by the execution of a command. A stream of information from the sense organ to CNS is called the "Afferenz" and, in the case of a self-induced environmental change, the "Reafferenz". According to the "Reafferenzprinzip" of Von Holst and Mittelstaedt (1950), a "Reafferenz" is compared with the "Efferenzkopie" that goes with it and if both match, spontaneous behaviour continues as intended. In this way, a stimulus from the environment (called "Exafferenz") can be distinguished from self-imposed stimuli during spontaneous action. The hypothesis was illustrated by several examples of ingenious experiments. They must be studied in the original before the principle can be fully appreciated. The paper was published in German, but an English translation is available (Dodwell, 1971). Nevertheless, the impact has been less than it deserves. It was thought identical to cybernetic control theory. Although elements are certainly the same, the "Reafferenzprinzip" is richer in the biological context.[3] It provides an important way of thinking about animal behaviour and is another window to look through.

1.4 The Ecological Context

Individual organisms live in a complex, undivided environment and they evolve under pressure of a combination of physical and chemical factors. This has led to the idea of "wholeness", which means that the influence of factors does not withstand analysis without the loss of essential features. Many (system) ecologists tend to have a more or less holistic view of nature and eschew a strong reductionist approach. Wholes or "holons" [4] are problematic in scientific studies, however. We can only grasp parts or aspects and what we experience as a whole can only be described in terms of smaller items (Popper, 1960). Also behaviour, which is evidently a whole, must be made into smaller units. Recognising this has been the success of ethology (Tinbergen, 1951). What we call an explanation is a renewed description of (parts of) the phenomenon but it is now in terms of relations between components and individual causal factors. The procedure may be found unsatisfactory even if a final synthesis, a model, consisting of an integration of the components is made because characteristics may be absent that emerge in the original, undivided whole. The discussion about these emergent properties is an old one in biology and psychology and not very fruitful as long as the origin of these properties is the blue sky.

Emergent properties arise from the organisation of the parts. Isolation of the parts makes emergency invisible.

Several authors have tried to do justice to the intuitive thought of an organism as embedded in the integral set of its environment or habitat. For example, Hutchinson proposed the multidimensional niche[5] consisting of *n* factors with *n* corresponding responses. Therefore, niche is a response entity and as such different from habitat, which comprises the physical part inhabited by a species. Species may live in the same or nearly the same habitat as the pelagic zooplankton do, but then the physiological and behavioural responses of the various species must be sufficiently different to prevent exclusion by competition. This makes niche separation an important feature of the pelagic community. The food niche of the herbivorous zooplankton must overlap and it will be interesting to see how competitive exclusion is prevented. This is a complicated problem because the species is not a homogenous unit. Individuals change during ontogenetic development, also with regard to migration behaviour. The nauplier stages of calanoid copepods do not migrate but with successive stages of the copepodites, migration starts and the amplitude gradually increases. In addition, a species population is composed of genetically different individuals. If reproduction is by parthenogenesis, the population is composed of various clones with different although strongly overlapping niches. In Lake Maarsseveen, a particular subgroup of the *Daphnia* population characterised by a different allozyme became dominant in a few weeks during summer (Chapter 12). These large shifts in dominance may be ascribed to strong competition. Clonal dominance shifts from year to year as well as within a year (Spaak and Ringelberg, 1997). Subtle niche differences may be responsible but we lack the data and tools to compose complete niche descriptions. However, a space of a few dimensions mostly suffices to separate taxons in the ecological context. The Hutchinsonian niche is a way of thinking about organisms in nature. It is again a window to look through at the life of animals.

Attention in ecology has shifted from proximate mechanisms to ultimate aspects such as adaptations. The study of causation in individual animal behaviour has declined rapidly over the last 20–30 years at the expense of the study of function (Hogan, 2005). The adaptive significance of behaviour is as important for understanding animals in nature as the physiological basis is. Zooplankton migrate daily to greater depth and back again and the current function is the prevention of mortality by predation of visual predators. This notion has strongly stimulated research. Zooplankton detect the presence of fish by chemical substances that are associated with these fish (see Larsson and Dodson, 1993 for review). The dominant role into the functional aspect of predator evasion has led occasionally to the erroneous idea that these predators are also the direct cause of DVM. Ultimate cause and proximate cause are sometimes confused. Recently, Bolhuis (2005, p. 466) explicitly stated that "there are no ultimate causes". He continued by saying that natural selection might be thought off as the causal factor in the historical process of evolution. However, we now talk about another level of concepts than that of the physiological mechanisms with the causal factors that trigger everyday swimming downwards and upwards again. The attention for ultimate, adaptive aspects has sometimes given rise to undue speculation about the evolutionary origins of migration behaviour. The evolution of a trait is a historical phenomenon that refuses to be analysed in an empirical way. One way out of the dilemma is to study present-day proximate aspects of behaviour in combination with the factors and circumstances that determine costs and benefits, although Cuthill (2005) warns that current function may not equal past function. Physiological and behavioural mechanisms change by selection. The costs–benefit balances are the final arguments in this selection process. Diel vertical migration is a daily shift from a food-rich, though sometimes dangerous habitat, to a food-poor habitat with a greatly reduced risk of predation. The vertical distance individuals swim depends on circumstances and individual animals must "choose" the best trade-off solution between food requirements and predation risk; therefore, it must have a kind of "decision-making" mechanism requiring that essential factors must be assessed and that the CNS has to process the obtained information before behaviour starts. In this way, proximate mechanisms serve adaptive needs. That makes the proximate analysis of behaviour necessary and highly interesting (Ringelberg, 1999a).

1.5 Finally

Windows provide different views on the life of animals in their environment. Our outlook on nature is determined for a large part by the window through which we look. Research programmes are determined, a priori, by notions of windows (or by the "Zeitgeist" or "fashion").

Presently, evolutionary thinking is a fashionable window. Although the individual (or even less) is central, in DVM research individuals are almost hidden. Many models were made to explore the adaptive meaning of DVM. All finally lean on assumptions because knowledge about underlying mechanism by which individuals operate in nature is insufficient. This has to be repaired. Gradually, the low number of studies in causation, recognised by Hogan (2005), will return to a necessary level because otherwise progress is choked in assumptions (Krebs and Davies, 1993, p. 385; Lima, 1998).

Notes

1. The notion "Umwelt" seems superficially to be identical to "habitat". However, the first indicates a set of environmental factors that somehow interacts directly with an individual. It is more associated with the physiology of an organism while habitat is an ecological term to denote the physical part of the environment where one can come across a particular species. "Umwelt" is closer to "niche", which is a species-specific, multidimensional combination of factors and responses. The latter is again a typical ecological notion and subdivided in an ecological way as pre- and post-interactive niche, or the (larger) part before biotic factors like predators and competitors have exerted an influence. The division of "Umwelt" factors (see text) is linked with the physiological division of body functions. "Wirkwelt" factors are connected with vegetative functions and "Merkwelt" factors with animal functions. Vegetative

functions are autonomic functions as digestion, blood circulation, endocrine secretions by glands, etc. To the animal functions belong sensory perceptions and muscle contraction.

2. Interoceptors are sensory nerve endings, specialised for the registration of the physiological state in the body. To the exteroceptors belong the well-known sense organs as eyes, ears, taste buds, etc.
3. In the English literature, the Reafferenzprinzip is seldom mentioned. Bullock (1988) considered it a useful tool in behavioural systems analysis, but, nevertheless, rejected it because no neural grounds would be present for notions like "Efferenzkopie" and "Reafferenz". On the other hand, he cites with approval a model published by Sokolov (1960, 1975) in which "expectation cells", "sensory reporting cells" and "comparator cells" were introduced. The latter cells fire whenever there is a discrepancy between the expectation cells and the sensory-reporting cells. According to Bullock (1988), cells of each kind have actually been found, so neural equivalents seem to be present after all. What is more, he introduces the term "expectation" to mean "an inferred state of the nervous system accounting for any of a wide variety of behavioural or physiological signs that a more or less specific sensory input is anticipated..." (Bullock, 1988, p. 260). In one breath, different phenomena like "innate reactions to environmental features", "temporal fluctuations in readiness to respond" as due to endogenous rhythms, "stimulus acquisition" and "innate reactions to one's own movements" are given as examples. Terms that apply to such a broad scale of different phenomena are vague. "Efferenzkopie" is judged inappropriate for studies of the physiology of the brain and replaced by "corollary discharge", defined as "a message created by a voluntary command (to effector organs) and left behind in the brain to influence the evaluation of consequent input" (Bullock, 1988, p. 274). It seems that "corollary discharge" has been copied from "Efferenzkopie". The precisely defined "Reafferenzprinzip" was replaced by rather vague notions that are inspired by it.
4. The term "Holon" seems to have been coined by Koestler (1964).
5. See Chapter 5 in Hutchinson (1978) for a development of the notion.

Chapter 2
Swimming in a Strange Biotope

Contents

2.1 A Strange Environment

The vast expanse of the open ocean is a strange biotope to us. Imagine a blue nothingness, a total absence of the variety of optical cues that are so characteristic on land. In woods, savannahs, tundras, mountains and even deserts, the multitude of forms and colours provide an overwhelming array of optical stimuli. Physiological selection prevents overloading and only the species-specific crucial items that provide the cues for orientation, food finding, and so on are experienced. Optical over-stimulation in the pelagic environment cannot easily occur. However, the emptiness also means that places to hide, places to build a home, or special places to meet mates are absent. As Hamner (1996) puts it, "the absence of cover in the epipelagic ocean is without question the single most important issue for pelagic ecology and evolution" (p. 25). Several authors (Zaret, 1975; Hamner, 1996; Isaacs, 1977) have drawn attention to the absence of refuges but no one has discussed the consequences of an optically empty environment for sensory physiology and sensory ecology. We can expect traits in zooplankton that are absent in terrestrial animals. Niches must be different in character, and community structure will reflect this strangeness of the environment. Thus if the "Merkwelt" of zooplankton is exceptional, the problems that arise must be formulated accordingly. For example, notwithstanding the optical emptiness, zooplankton in lakes and the upper 1,000 m of oceans have often well-developed eyes. To synchronise physiological processes to diel rhythms, perception of diel light changes is necessary, but the

J. Ringelberg, *Diel Vertical Migration of Zooplankton in Lakes and Oceans*,
DOI 10.1007/978-90-481-3093-1_2,

elaborate structure of many eyes, even of the considered simple ones of copepods, is far too complicated to this end. Being a filter feeder, *Daphnia*, with 22 ommatidia in the compound eye, a thick bundle of optic nerves and a well-developed optical ganglion, does not need this elaborate structure to collect algae. For *Daphnia*, the complexity concerns body orientation and vertical direction finding with regard to the underwater light distribution.

It might seem strange that gravity is not used for the orientation of body positions. Gravity perception is present in the majority of terrestrial invertebrates and vertebrates but only few pelagic invertebrates have statocysts. For example, *Mysis* has pits on the endopodites of the uropods. The walls are lined with sensory hairs and in the hollow a calcareous body, the statolith, is situated (Borradaile et al., 1955). Also Decapod crustaceans have statocysts but, tellingly, not as long as they are planktonic larvae. Not until the last, or Megalopa stage, when bottom life is approaching, are these structures developed. The relative unimportance of gravity in marine, pelagic crustaceans also becomes clear from studies on pressure sensitivity (Rice, 1964, 1965). The copepods *Temora, Calanus*, *Acartia* and the euphausiid *Meganyctiphanes* reacted to large changes in pressure by orientations to incident light only. On the other hand, several species of Mysidacea, some of them pelagic, always swam upwards or downwards, independently of the light direction. Zoaea larva of the brachyuran Decapoda orient to light; however, as megalopa they showed a geotactic response. Perhaps the absence of gravity receptors in planktonic animals has to do with the specific gravity of the zooplankton body, which is the same or slightly higher than water. Otoliths, comparable structures to statocysts and common in vertebrates, including humans, do not function properly at microgravity, as astronauts, floating in space, experience. In that situation, vision takes over in spatial orientation (Grigorova and Kornilova, 1996). If gravity cannot function sufficiently in zooplankton either, the absence of gravity receptors can be understood. Strickler (1982) found that a copepod, made neutrally buoyant artificially in a suited density fluid, is unable to orient its body position. Normally, hairs on the first antennae function as mechanoreceptors to regulate body posture. Strickler (1982) distinguished between a feeding and a cruising mode of swimming. In the latter mode, animals might swim with different body positions but this author does not tell if and in what way the mechanoreceptors on the first antennae then operate in orientation. I propose that optical orientation takes over. Also *Daphnia* has two modes of orientation: optically, when light intensity is above a certain value, and by "gravity", when optical body orientation is impossible, as it is in darkness. In Chapter 7, the discussion about orientation in the pelagic environment will be continued.

Scaled to the swimming abilities of zooplankton, the pelagic of oceans and lakes is homogenous in horizontal directions. Stable microhabitats, an important element in structuring the terrestrial biotope, are rare or absent. The action radius of the individual planktonic rotifer, cladoceran or copepod is limited and a different configuration of environmental factors can mostly not be reached in a lifetime. Therefore, swimming in horizontal directions is without sense and mechanisms for horizontal plane orientation, for example sun compass orientation, which is common in terrestrial invertebrates and vertebrates, have not been found.[1] On the other hand, the vertical plane is of paramount importance. The oceanic and freshwater community is structured vertically and over the vertical dimension extensive seasonal and diel shifts in composition occur. Thus we should expect mechanisms for orientation in the vertical plane in order to maintain or change depth. Pressure perception is an obvious suggestion and several researchers have studied the effect of pressure changes (Rice, 1964; Lincoln, 1971; Knight-Jones and Morgan, 1966; Baylor and Smith, 1957). These had to be large, however, and evidence of a sensitivity to be of use is not convincing. In the future it will be known whether Bullock (1996) is right that "a hitherto unknown sense organ, indeed a new class of sense organs is awaiting discovery" (p. 3). The best evidence for responses to changes in hydrostatic pressure is found in Megalopae of Decapod crustaceans (Rice, 1964; Bainbridge, 1961). Experiments by Forward et al. (1996) with this larval stage of the blue crab (*Callinectes sapidus*) have demonstrated that swimming responses to changes in hydrostatic pressure occur, but these authors also argued that the determined thresholds were too high to be of use as a cue for ascent into the water column during a rising tide. Hydrostatic pressure perception, like gravity perception, seems to become a factor of importance only when pelagic life is finished and bottom life commences.

Pelagic biodiversity of oceans is low (Angel, 2000). This has certainly to do with the horizontal homogeneity and the absence of stable microhabitats. If barriers like latitudinal temperature differences would not prevent this (Angel, 1993, 2000), the turnover of the major oceans in 500 years (Stuiver et al., 1983) would realise homogenous distributions of the pelagic species. Now, latitudinal differences exist in several taxa. For example, in the North-East Atlantic, the number of fish species, Ostracoda, Decapoda and euphausiids decreases between 10 and 70°N (Angel, 1993). However, the difference is over very large distances and do not undermine the idea of horizontal homogeneity and the absence of allopatric speciation. Biodiversity is higher in the vertical plane. Several authors have distinguished layers with a specific species composition, for example, Legand et al. (1972) and Angel (1977). Habitat differentiation due to physical and chemical factors is responsible in the first place. But also over the relatively short distance of the first 1,000 m species are not mixed randomly. We will meet an example of a vertical separation of euphausiid species and it will be suggested that this is due to niche differentiation and maintained by competition. DVM is of importance too.

During summer stratification in lakes, a vertical habitat differentiation of zooplankton can also been encountered (Nauwerck, 1963, 1993; Siebeck, 1960). The steep temperature gradient is a strong barrier that realises vertical space differences. Even within the confines of the epilimnion, the vertical daytime distribution of cladocera and copepoda may differ, although not sufficiently stable over time to speak of habitat differentiation, and differences in migration amplitude are often responsible (Fig. 2.1) (Siebeck, 1960; Einsle, 1969).

Compared to terrestrial and oceanic environments, lakes are simple research systems. The easy access and the cheap logistic demands have made extensive research in lakes possible. Similar studies are difficult and costly to realise in the oceanic environment. Fortunately, the pelagial of oceans and the open water of lakes have several properties in common and DVM is to a large extent comparable.

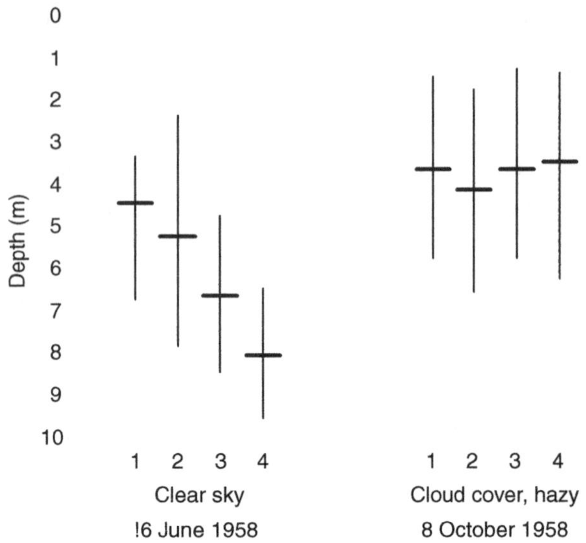

Fig. 2.1 Vertical distribution of four species in the Lunzer Untersee. Females of 1. *Eudiaptomus gracilis,* 2. *Cyclops tatricus,* 3. *Daphnia longispina,* 4. *Bosmina longispina* (data from Siebeck, 1960). The figure suggests the influence of light intensity on day depth. In 1958, the stimulating effect of fish on DVM was unknown. It is therefore plausible that in October not light intensity but the absence of predating fish caused the equal vertical distribution

2.2 Small-Scale, "Normal" Swimming

It is not easy to define unequivocally what must be understood by small-scale, "normal" swimming. A small distance for species A may be a large distance for species B and "normal" is an elastic notion. "Normal" swimming is swimming not directly the result of an external factor. Unmistakably, diel vertical migrations do not belong to this category, whether a 10 m vertical distance of *Daphnia* is involved or 400 m, as in the marine copepod *Pleuromamma xiphias* (Buskey et al., 1989). Vertical migration is reactive swimming to changes in light intensity of dawn and dusk. Doall et al. (1998) observed that males of *Temora* made sinuous routes until chemical signals (pheromones) in the wake of a female were tracked. This cruising behaviour also does not belong to the "normal" swimming I have in mind. Ninety percent of the accounts of zooplankton swimming in the literature refer to feeding behaviour, predator avoidance or mate seeking. Species-specific differences get much attention but not the swimming patterns they have in common. Because they all have adapted to the same unstructured biotope and had to cope with the same specific problems, they must have features of swimming in common. This is the case, indeed.

Because the marine and lake environment is unstructured in all directions, swimming of planktonic animals has been considered a movement in random directions. For example, modelling encounter rates between predator and prey was done by assuming that the animals were dots in three-dimensional space, moving randomly (Gerritsen and Strickler, 1977; Giguere et al., 1982). Strickler (1998) wrote that this mathematical approach has met with considerable success. However, the "life at random" paradigm has a problem: randomly moving animals tend to disperse. If swimming of males and females in the pelagic were in random directions, encounter of the sexes would probably be too incidental to sustain a population. Therefore, when empty space is large, and density becomes too low, problems of reproduction may arise.[2] Strickler (1998) thought of two mechanisms to overcome this problem of dispersal and, consequently, of a too low encounter rate of the sexes: 1. diel vertical migration and 2. one sex swimming horizontally and the other one vertically (Gerritsen, 1980). Vertical swimming during DVM concentrates animals indeed. As soon as migration is over, dispersal sets in again, as is illustrated in Fig. 2.2. During the stationary nighttime depth, variance (standard deviation/mean, calculated over eight sampling depths) of the vertical distribution is high. The concentrating effect of DVM is present, but, probably, a by-product and no real solution for the threat of drifting apart. In the presented case of *Daphnia*, the positive effect on mate finding is not involved since sexual reproduction occurs in autumn and not during the early summer period of vertical migration.

With the development of three-dimensional video recording and computerised picture analysis (Ramcharan and Sprules, 1989; Strickler, 1998), the "life at random" paradigm has been rejected (Wong and Sprules, 1986; Van Duren and Videler, 1996).

These analyses were predominantly made with marine calanoid copepods and were focused on food catching, predator evasion and mating. Physically oriented researchers are interested in the specific hydrodynamic disturbances made by different species (Fields and Yen, 1996; Doall et al., 1998; Van Duren et al., 1998). The

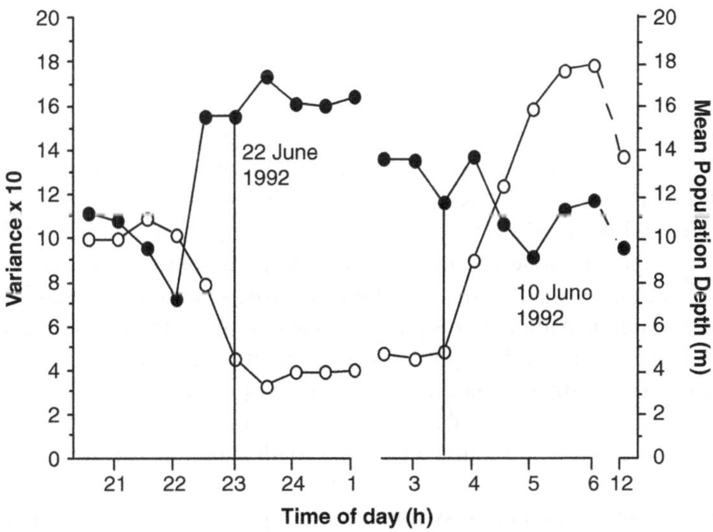

Fig. 2.2 Variance calculated over eight depth distributions of adult females without eggs of *D. galeata x hyalina* in Lake Maarsseveen during the evening ascent migration of 22 June 1992 and the morning descent migration of 10 June 1992. Variance (*black dots*) was multiplied by 10 to fit the mean population depth (*open dots*). Between the vertical lines the period of the stationary night period is located

applied techniques made it necessary to confine the animals in small vessels and to observe them often in the dark using infrared light. However, orientation and swimming in the dark might be different from swimming in the specific angular light distribution of the open water. Copepods have relatively simple eyes, which might easily lead to the view that they can only perceive light or darkness but are of no use in gathering detailed information about the environment (Van Duren, 2000). The elucidating experiments of Siebeck (1980) with the freshwater calanoid copepod *Mixodiaptomus laciniatus* have shown, however, that body axis orientation while swimming depends on de angular light distribution.

Several authors have described discontinuous, intermittent swimming with active upward and passive downward displacements (Wong and Sprules, 1986; Williamson, 1986; Bainbridge, 1952; Hardy and Bainbridge, 1954; Strickler, 1970). In calanoid copepods, the upward movements are often smooth glidings, for example, in *Senecella calanoides*, *Limnocalanus macrurus* and *Epischura lacustris*, or in a "hop-and-sink" motion (Wong and Sprules, 1986). Bainbridge (1952) observed *Calanus finmarchicus* in nature and saw that these animals all swam in an upward position and in a "hop-and-sink" way but seldom horizontally. In zooplankton the term "hop-and-sink" swimming is in use to indicate intermittent swimming, but not all authors have the same thing in mind. Dodson and Ramcharan (1991), studying *Daphnia,* called the unit of one stroke of the second antennae, plus the succeeding short sinking with outspread antennae, "hop-and-sink". Williamson (1986) used the term in a similar way for *Mesocyclops*. Several strong strokes – with short sinking phases in between – may alternate with a period of less vigorous antennal beats or even continuous sinking. In both cases an undulating displacement in the vertical plane results. These often quite regular vertical displacements were also called "hop-and-sink". For example, Hardy and Bainbridge (1954) used this term when marine calanoid copepods, such as *Calanus*, *Labidocera* and *Temora*, swam with short bursts upwards and then sank for some time. Mills (1981) speaks of "hop-and-sink" to indicate the oscillating swimming pattern of some medusae, covering amplitudes of 1–2 m. A succession of pulses of the swimming bell propels the animal upwards, while the downward phase consists of passive sinking. In *Daphnia*, oscillations might likewise be of a relatively large amplitude and the downward phase may even consists of active, head downwards, swimming (Viaud, 1938; personal observations). Although zooplankton swimming is more variable than two categories can encompass, I propose to call "hop-and-sink" swimming the one single upward thrust with the succeeding short phase of passivity and "oscillating swimming" a more or less regular pattern of upward displacement, consisting of several "hop-and-sink"-like units, alternating with a period during which descent predominates.

Under constant illumination, swimming of *Daphnia* consists of "hop-and-sink" units. The behaviour was analysed by Dodson and Ramcharan (1991). A "hop" consists of a power stroke of the second antennae, propelling the animal upwards. The frequency is about 3–4 beats/s and seems to be independent of body size. The latter fact suggests an internal origin of the rhythm because it is independent of growth and metabolism-determining factors like food and temperature.

Of course, larger daphnids have stronger strokes and the average speed per hop is larger. A hop is followed by a phase of sinking, lasting 0.1–0.6s, with average sinking rates between 0.8 and 2.7 mms^{-1}. The sinking rates depend on body size, but do not follow Stoke's Law for physical particles, as is the case in anaesthetised animals (Gorski and Dodson, 1996). Temperature is of influence only in anaesthetised animals. Therefore, sinking is under biological control, for example, through changing the body surface area or changing the posture and configuration of the parachute-like second antennae. Sinking in *S. calanoides* might also be controlled through changing the orientation of the appendages and setae (Wong and Sprules, 1986). This indicates that the downward phase is not merely an inhibition of the active upward phase but can be directed by the animal for adaptive purposes. Consequently, net vertical displacements by hops and sinks can be varied, resulting in vertical drifts. Endogenous and exogenous factors may induce these slow drifts. For example, in the presence of fish kairomones, a downward drift occurs (Dodson, 1988; Ringelberg, 1991). In Fig. 2.3, drifts of *D. galeata x hyalina* over a period of nearly 2 days are presented. The origin of these changes in average depth is unknown. An endogenous, circadian rhythm might be involved but the observation time under constant light intensity is too short to warrant such conclusion.[3]

Under constant conditions, oscillating swimming is often quite regular. In *D. "longispina"*[4] an average duration of both phases of 60.4s (s.d. = 16.54, n = 98) with an average vertical displacement of 2.9 cm (s.d. = 1.39, n = 98) was determined (Ringelberg, 1995a). Swimming paths of euphausiids, observed in the Dalhousie Plankton Towers, consisted of oblique upward swimming, alternating with periodic intervals of sinking bouts, lasting 2–17s (Price, 1989). In small pelagic *Daphnia*, the amplitude might be very small. Then, the animals are very sensitive to changes in light intensity (Ringelberg, 1987b, 1999a). On the other hand, large amplitude oscillations of 50–150 cm can be also observed. Depending on size, the animals then move from one end of the experimental cylinder towards the other end

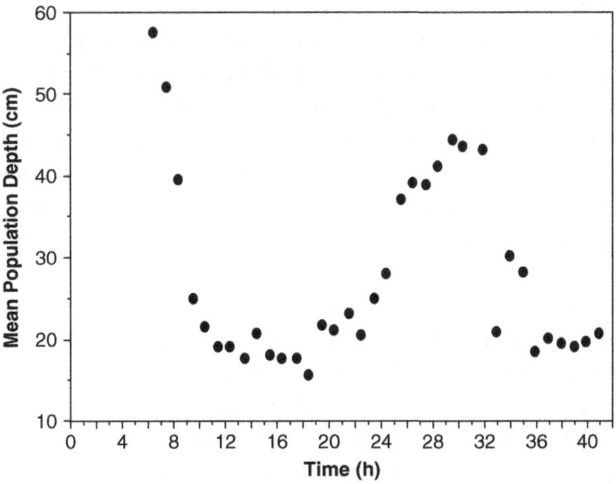

Fig. 2.3 The mean population depth of 30 hybrids of *D. galeata x hyalina* in a cylinder of 1 m height during nearly 2 days. Light intensity and temperature were constant during the period. Slow drifts leading to gradually changing depths are visible. The origin of these changes in depth position is unknown. An endogenous, circadian rhythm might be involved, but cannot be concluded from this registration alone

and going down is often the result of active, head down swimming. Period time is considerably increased. Amplitude is probably larger than the dimension of the cylinders and then swimming is hampered by the confines of surface and bottom. Consequently, the daphnids are stationary for some time at either end. Clarke (1932) called this kind of behaviour primary phototaxis to distinguish it from secondary phototaxis or reactions elicited by changes in light intensity. Oscillating swimming also occurs in horizontally placed troughs, illuminated from one side, thus swimming is optically oriented in both horizontal directions (Clarke, 1932). In this mode, the daphnids are unresponsive to changes in light intensity. Adverse environmental circumstances seem to cause the extremely oscillatory swimming behaviour. For example, I have often observed that on entering the laboratory in the morning, the animals swam up and down in the absence of food. Some time after the addition of algae to the water, "hop-and-sink" swimming was resumed.

The peculiar large amplitude oscillations have been observed by several authors and in several planktonic species (Baumbach, 1922; Rose, 1925; Ewald, 1910c; Clarke, 1932; Viaud, 1938; Rimet, 1960a, b). In an extensive narrative, as characteristic of the literature of the first half of the last century, Ewald (1910a, c) described the regular upward and downward swimming over the length of a 35-cm-long tube for *D. magna, D. quadrangula, Leptodora hyalina* and *Bythotrephes longimanus*. Period length differed between observations and species, ranging from 4 to 10 min, to give an order of magnitude derived from the presented figures. No precise measurements were made because the prevalent paradigm made that unnecessary. This paradigm prescribed the existence of an optimum light intensity and Ewald (1910b) interpreted the oscillations as alternating between the lower and higher boundary of such optimum, albeit he was well aware of the fact that the alternations remained in full force when the intensity of the overhead illumination was changed ("vielmehr adaptieren die Cladoceren an alle mittleren Lichtintensitäten", Ewald, 1910b, p. 392). Therefore, a relative optimum was proposed, as Clarke (1932) would do more than 20 years later.[5] From Clarke's figure a periodicity in *D. magna* of 4–5 min can be estimated and for *D. pulex*, in Viaud (1938), a period length of about 8 min. These are all rough estimates but the basic behaviour is evident. Amplitude cannot be determined because bottom and surface prevented continued swimming whether tube length is 35 cm (Ewald, 1910a, b, c), 50 cm (Clarke, 1930) or 80 cm (Viaud, 1938). *Daphnia* may stay for half an hour at either end as in Fig. 4 in Clarke (1930), which suggests a period length of about an hour. It seems that a continuous range of period lengths is possible in *Daphnia*. We might even speculate that the consistently positively or negatively phototactic *Daphnia*, as described by De Meester in several papers (De Meester, 1991, 1993a), are extremes, blocked in either phase of the oscillation rhythm. More or less regular upward and downward swimming has also been observed in the pelagic amphipod *Mysis* (Bauer, 1908), the marine calanoid copepods *Acartia tonsa* and *A. clausii* (Esterly, 1917) and in the freshwater calanoid *Eudiaptomus gracilis* (Van Gool, personal communication, see Fig. 5.7). Autocorrelation of the signal in the latter species revealed a period length of 56 min and an amplitude that nearly covered the length of the 100-cm tube. The pattern is highly comparable with that of *Daphnia*. Although the large amplitude oscillations occur when experimental circumstances in the laboratory are adverse, the observations by Bainbridge (1952) of free swimming *C. finmarchicus* in the sea indicate that this kind of behaviour also occurs in nature. At some distance from the water surface, *Calanus* appeared to swim vertically upwards and then downwards again over a distance of 1.50–1.80 m. Going downwards was with head down, thus active. The literature is loaded with anecdotal data, but these are difficult to bring into a coherent picture.

Nevertheless, I conclude that oscillating swimming is probably widespread in zooplankton and part of the behavioural repertoire. Two questions come to mind: 1. What function does it have? 2. What physiological mechanism is at the base?

2.2.1 The Function of Oscillating Swimming

Small-scale, oscillating "normal" swimming will have more than one function. Indeed, the first question can be met with suggestions about 1a. optimal energy expenditure, 1b. feeding aspects, 1c. predator–prey escape behaviour and 1d. prevention of dispersal.

1a. *Optimal energy expenditure*. Intermittent swimming might be advantageous because energy costs are less than in continuous swimming (Haury and Weihs, 1976). This has been found for diving seals and whales too (Williams et al., 2000). Intermittent swimming is more effective in terms of oxygen use. Also in humans, contracting muscles are able to cope with subsequent contractions better, if short periods of rest are present.[6]

1b. *Feeding aspects*. The 1–2 m oscillating swimming is obvious for feeding in the hydromedusa *Stomotoca atra* (Mills, 1981). The two long tentacles drag through a vertical plane when the animal performs the repeated swim-sink cycle, trying to catch other medusae on which it feeds. Also the oscillations performed by *Phialidium gregarium* are seen as part of the feeding behaviour but the food collecting mechanism is different. In the large observation tank, animals sink upside down for 1–2 min, then turn over and swim upward again over distances up to 2 m before turning again upside down to start sinking again. Each time the medusa turns over, a pocket of water is established in the dead space behind the bell. With the tentacles within this dead space, *P. gregarium* clears this volume from its prey. Each time, the intermittent upward excursions establish new feeding spaces. Collecting food in the dilute environment of the pelagic biotope might necessarily be a continuous process. Filter-feeding zooplankton feed all the time while swimming. The direction of the upward thrust of a hop in *Daphnia* is between 0 and 30° (Schröder, 1959, 1962) and they also turn often around the median body axis. Dodson and Ramcharan (1991) quantified this turning during "hop-and-sink", using Buskey's (1984) "net to gross displacement ratio (NGDR)".[7] The average value of NGDR was 0.48 and was independent of body size, speed and hops per second. The values of NGDR seem to be clustered halfway the extremes of 0 and 1, although suggestions were made of the influence of food, the opposite sex or predators. For example, Buskey (1984) reported a value of 0.46 for *Pseudocalanus minutus* and Wong (1988) an average value of 0.47 for *Metridia pacifica* when not feeding. In the latter species NGDR dropped to 0.23–0.24 when feeding on *Coscinodiscus* but, rather strange, the same value was found in water without algae. In *Temora longicornis*, Van Duren and Videler (1996) found slightly more turnings in the presence of fish kairomones (0.482–0.604 against 0.614–0.681), although the difference was only significant in starved females. Also the difference in NGDR for females and males in the absence and presence of the smell of the opposite sex was not significant and the overall values ranged from 0.466 to 0.561. For females of the same species Doall et al. (1998) found an average NGDR of 0.68. Therefore, the frequency of turnings seems to be a rather fixed part of swimming behaviour in cladocerans as well as calanoid copepods and only slightly influenced by environmental circumstances.

When the direction of the horizontal component is changed regularly during hops, each time a different volume of water can be searched for food. Without these turnings, the nearby volume of water would be rapidly empty of algae. In combination with oscillatory swimming, new pockets of water are reached continuously. Occasionally, the regularity of the oscillation is disturbed by a sudden and larger vertical displacement (Ringelberg, 1995). These irregular displacements are probably part of the oscillating mechanism and, perhaps, not without adaptive value since the food collecting animal is propelled far from the original feeding spot. So one answer to the question "what function has oscillating behaviour" is that it might be an essential part of feeding behaviour, not only in filter feeders but also in animals that catch food items individually like the mentioned medusa and copepods.

1c. *Predator–prey behaviour.* Although the oscillating pattern can be found in many planktonic animals, swimming itself has a species-specific character. The wake of swimming differs from species to species (see Boxshall, 1998, for overview) and can be recognised by predator and prey alike. If it is true that the sensitivity of mechanoreceptors that pick up the hydromechanical disturbances is reduced when the animal is swimming (Strickler, 1975), it is advantageous to stop once in a while to "listen" for disturbances in the environment. There are more indications in the literature that animals stop moving for sensory reasons (Wong et al., 1986; McLaughlin, see Pennisi, 2000). Predators should try to detect wake disturbances of potential prey and prey animals should swim intermittently to confuse these "listening" predators (Wong and Sprules, 1986). *Diaptomus* is only attacked by *Limnocalanus macrurus* and *Mysis relicta* when it jumps (Wong et al., 1986; Ramcharan and Sprules, 1989). So the prolonged sinking phase of "hop-and-sink" in the downward phase of oscillating swimming might also serve sensory perception (Gerritsen, 1978). If prey density is high enough to stimulate the receptors of predators from many directions, the first step in the chain of prey capture behaviour, sensory fixation, may be hampered if the signal of a fixed prey suddenly falls still and is replaced by stimuli from other prey animals. Then the process of fixation must start all over again. This kind of prey capture interference has been demonstrated for juvenile pike (*Esox lucius*) in the presence of a dense swarm of big *D. magna.* A visually fixed *Daphnia* is lost due to the swirling multitude within the visual field. Another prey is fixed again but before the next step in the prey catching sequence can take place, the fixated daphnid is lost again. Finally, the result might be that the juvenile pike flees into a corner of the aquarium.[8] Predators should also swim intermittently to make them less detectable for herbivorous copepods because these prey animals are also capable of detecting hydrodynamic disturbances (Kerfoot et al., 1980; Williamson, 1986).

1d. *Dispersal.* Last but not least, oscillating swimming might be a solution for the problem of dispersion in an unstructured environment. Species are often found within certain layers of the water column. Swimming in random directions will move them gradually apart, resulting in a more diffuse distribution. If all individuals perform in the same way, moving on average over an equal distance upwards and downwards, dispersal will be less.

A depth control mechanism was also suggested by De Meester (1993b). It was supposed that a genotype-specific optimal light intensity[9] was involved. This was defined as the light intensity "at which the threshold for relative changes in light intensity is minimal" (p. 174). Evidence for such thresholds exists in *D. magna* (Ringelberg et al., 1967) and the calanoid copepod *A. tonsa* (Stearns and Forward, 1984b). See Chapter 3. No evidence is present that these animals can perceive and keep to a certain absolute light intensity, however. Contrarily so, in the laboratory *Daphnia* is more often than not distributed over nearly the whole length of a cylinder, independent of the intensity of the overhead illumination.[10] (See, however, Fig. 2.1.) This is due to a large capacity of acclimatisation to light. Leibold et al. (1994) lowered long translucent tubes in a lake and concluded that "the magnitude of the acclimatisation effect seems large enough that almost any genotype could be found at almost any depth if it were to become sufficiently acclimatised to that depth" (p. 1329). Light intensity as such seems not to be a good cue for maintenance of a particular depth range.

2.2.2 The Mechanism at the Base of Oscillating Swimming

In general, locomotion, whether walking, swimming or flying, is a rhythmic phenomenon in vertebrates (Grillner, 1975) as well as invertebrates (Friesen, 1989; Golubitsky et al., 1999). It has been a point of discussion whether the central nervous system is intrinsically capable of generating the rhythms for locomotion or that locomotion is under peripheral control through the use of sensory feedback from the moving parts of the body. The principle of a central pattern generation is now well established (Delcomyn, 1980). A central pattern generator or neural oscillator can be imagined as a network of neurons capable of generating a cyclic pattern of spikes, notwithstanding a constant and continuous input. The origin of the rhythm must be looked for in the coupling of two or more separate units which, if operating alone, would not work in a cyclic, repetitive way (Friesen and Block, 1984). Of essential importance is the presence of a time lag in the mutual influence of the individual neural groups. Rhythmic performances are widespread in organisms. Well-known examples are pacemaker cells in the vertebrate heart and the respiration centre in the medulla. The importance of central pattern generation for periodic biological behaviour does not exclude peripheral influences. For *Daphnia*, a light intensity change, generated by swimming in the vertical plane, is perceived and of influence. It was demonstrated that the duration of a positively phototactic reaction, caused by a sudden decrease in light intensity, is also determined by the light increase experienced while swimming upwards (Ringelberg, 1964). To what extent sensory feedback modifies small-scale, normal swimming is unknown.

For *Daphnia*, a range of frequencies and amplitudes of oscillating swimming can be derived from the literature (Clarke, 1932; Viaud, 1938), although these authors never thought in terms of oscillations. I used different oscillations in *Daphnia* as a starting point for building a model with a coupled oscillator as a physiological basis for intermittent, oscillating displacements. This model has been extended into a model of photobehaviour and diel vertical migration in *Daphnia*. They are discussed in Section 5.2. An example of a simulation of "normal" swimming is presented in Fig. 5.4 with an actual swimming track for comparison.

Intermittent locomotion has several advantages over continuous walking or swimming: "in nature, animals that stop and start win the race" is the heading of an editorial article in Science (Pennisi, 2000).

2.3 Summary

Attention was drawn to the exceptional properties of the pelagial biotope as compared to terrestrial environments. One such property is the optical emptiness which led to the question why zooplankton often have well-developed eyes. The suggested answer was that body-axis orientation in the vertical plane was with regard to the typical angular light distribution under water. That raised the question why gravity was not used, as in terrestrial invertebrates and vertebrates. Most zooplankton do not have statocysts, however, and the few that have, as the larvae of Decapod crustaceans, only attain them in the last developmental stage, when bottom life soon is to begin. Gravity receptors do not function perhaps when specific gravity of the body nearly equals that of water.

The pelagial of oceans and lakes is homogenous in horizontal directions and lacks the microhabitat diversity of the terrestrial biotopes. Therefore swimming in horizontal directions is without much sense and a mechanism for horizontal plane orientation cannot be expected. In a vertical plane, diversification does exist and the zooplankton community is often structured along the vertical axis, especially in oceans. Cues to determine a vertical position are hard to find, however. It has never been demonstrated convincingly that the obvious factor, hydrostatic pressure, plays a role.

In the second paragraph, small-scale, normal swimming was discussed. Although zooplankton is a taxonomically diverse group, with a different morphology, feeding mechanism, mating, reproduction and locomotion apparatus, all have evolved within the same controlling environment and constrains are shared. Dispersal in an unstructured, three-dimensional space is a general problem. Mate finding and predator avoidance is another one. The wake of swimming is species-specific, and the hydrodynamic disturbances are perceived by mechanoreceptors. Acuity of perception is probably hampered if the perceiving animal is swimming and therefore disturbing the environment at the same time. Intermittent swimming might be a solution and a common feature of locomotion. Once in a while, prey could then "listen" for predators and the latter for prey. Alternating short periods of upward displacements with periods of sinking might make this possible. Regular frequencies and amplitudes have been described. This has led to the construction of a mechanistic model of oscillating swimming, presented in Chapter 5.

Notes

1. Sun compass orientation is the ability to use the position of the sun for horizontal plane orientation. Since the sun's position changes during the day, a circadian clock mechanism is needed to compensate for the movement of the sun across the sky (Dusenbery, 1992).
   Another mechanism of horizontal plane orientation makes use of the polarisation pattern of the sky. It has been demonstrated for several terrestrial animals and for two aquatic ones: *D. magna* (Jander and Waterman, 1960) and *Mysidium gracile* (Waterman, 1966a; Bainbridge and Waterman, 1958). It is, perhaps, not by chance that these two species possess a polarotaxis because the habitat of both is not the pelagic but the littoral.
2. This aspect of population dynamics has got no attention, as far as I know.
3. With a longer observation period, an eventual free running rhythm should be possible to determine. If the periodicity would remain the same, an external synchroniser (Zeitgeber) should be present.
4. Probably, these *Daphnia*, obtained in 1972 in Lake Maarsseveen, the Netherlands, belonged to *D. galeata* or the hybrid *D. galeata x hyalina*. In 1989 and 1990, these two taxa were nearly exclusively present in the lake. In 1990, the hybrid dominated (Spaak and Hoekstra, 1993). This continued to be so until the last allozyme analysis was made in 1998 at least.
5. The idea of an optimum or preferred light intensity keeps returning in the literature, also in more recent times. For example, De Meester (1993c) has proposed that *Daphnia*, staying continuously near either end of an experimental tube, are searching for a higher and lower optimum light intensity, respectively. In turn, his ideas are based on a paper by Hailman and Jaeger (1996). In due time, the strange behaviour of these daphnids will be discussed as the final expression of one mechanism that determines small scale oscillations of "normal" swimming behaviour and the oscillations of different amplitudes and period times.
6. Glycolysis of glycogen ends in pyruvic acid. Pyruvate dehydrogenase (PDH) plays a role in the aerobic oxidation of pyruvic acid, leading finally to ATP for muscle contraction. If PDH is not rapidly regenerated, pyruvic acid is shunted into the anaerobic pathway that creates lactic acid. This causes muscle fatigue. With intermittent contractions, PDH is regenerated faster, the aerobic pathway is maintained and muscle fatigue prevented (Pennisi, 2000).

7. Buskey et al. (1993) defined the mean net to gross displacement ratio (NGDR) as the linear distance between starting and ending points of a path (net displacement) and the total distance covered by the path (gross displacement). For completely linear paths, NGDR = 1; for convoluted paths the ratio approaches zero. Dodson and Ramcharan (1991) added displacements per 1/30s for 5s by the linear distance between beginning and end of the 5s observation period.
8. The described behaviour was part of a course in animal physiology and behaviour and has been observed many times.
9. The hypothetical "optimal light intensity" has for a long time been the paradigmatic explanation of diel vertical migration.
10. De Meester (1993b) suggested that differences of *D. magna* in depth distribution of a few centimetres in a small tube of 25 cm are indications of optical microhabitat preferences. However, the depth distributions were determined some few minutes after light was switched on. Since full light acclimatisation cannot have occurred in such a short time, the distribution likely reflected a transient situation.

Chapter 3
Light-Induced, Reactive Swimming

... very little is known about what characteristic of changing light field triggers the migration

(Forward, 1988, still cited by Frank and Widder, 1996, p. 191)

Contents

3.1 Introduction

Nowadays, ecologists are not very interested in the physiological–behavioural mechanisms that underlie ecological phenomena. Attention is focused on ultimate, evolutionary aspects. This also holds for investigators of diel vertical migration (DVM). Since we know that many migrations are strategies to avoid predation by visually hunting planktivores fish, research was stimulated and a large number of papers appeared. However, without a mechanistic approach deep understanding of migration is not possible. Adaptive goals have to be realised by mechanisms. Fish predation has become hype and the way in which migration offers a solution is sometimes presented as a mix of ultimate and proximate explanations. To say it bluntly: what evolved in evolutionary history were proximate mechanisms.

J. Ringelberg, *Diel Vertical Migration of Zooplankton in Lakes and Oceans*, DOI 10.1007/978-90-481-3093-1_3, © Springer Science+Business Media B.V. 2010

The adaptive significances provided criteria for selection of the physiological–behavioural mechanisms. With a short fairy tale I will illustrate the role of causal mechanisms in the evolution of diel migration behaviour.

Once upon a time, there were many *Daphnia* in a lake. They were happy because no planktivorous fish were present. The water was clear and on cloudless days ultraviolet radiation penetrated sufficiently deep to damage embryos in the first few metres. Photosynthesis was inhibited too and maximum primary productivity occurred lower down in the water column. Then a "mutation" occurred which made individuals sink over a few metres to the increasing light intensity of dawn. Photobehaviour was born. These happy few also found more algae to eat at the day depth and their numbers increased. However, they also went down on cloudy days, when UV was not damaging and maximum primary productivity was higher up in the water column. Therefore, their relative fitness was not always superior and the original genotypes were also maintained in the population.

On a bad early summer day, visually predating fish entered the lake and started to eat *Daphnia*. It was a pity that these individuals knew that only after they were eaten. Therefore, fish predation could not be a selecting factor. What lacked was a kind of "early warning system", alarming the daphnids that fish were present. Then a "mutation" occurred again: they became able to smell fish. This was not of much help either because behaviour was not altered. Only a kind of general sensitisation resulted.

After some time, a neural connection in the central nervous system grew between fish smell centres and the centres responsible for sinking upon the perception of changes in light intensity. The sensitisation made the individuals that possessed photobehaviour to swim actively down before sunrise, out of reach of the fish. Fish predation had finally become a selecting factor. However, the migrating individuals were also out of the food zone for most of the day and also at low temperatures. Death rate had dropped, but the birth rate too. At low predation rates, the non-migrating animals were sometimes better off. Costs and benefits determined the relative fitness of the diverse groups. Consequently, an ever-changing ratio between migrating and non-migrating *Daphnia* resulted, up to present times.

It must be evident that physiological–behavioural mechanisms have to be evolved and are the objectives of selection in migration behaviour. In Chapters 3, 4 and 5, these mechanisms are discussed as far as we understand them.

3.2 The Stimulus Concept

The term stimulus is often used in a general way but must be defined precisely. Fellow scientists must repeat experiments and, thus, experimental circumstances must be described properly. If from field observations correlations are made between a phenomenon and an environmental factor, both must be specified. It sounds obvious and elementary but it is often not done. For example, for decades, the paradigmatic explanation of DVM has been that zooplankton follow an optimal light intensity (see reviews by Cushing, 1951; Longhurst, 1976; Segal, 1970). Many authors presented this "explanation", although few had actually measured light intensity. On what ground did they offer this then? It is true that light meters could not be bought at the local hardware store, but simple selenium cells for measuring light intensity in oceans and lakes were used and described as early as 1933 (Atkins and Poole, 1933). Obviously, the need for precise measurements of this dominant factor was not felt, nor was it suggested what property was optimal. Since DVM occurred at dawn and dusk it was obviously sufficient to make "light" a factor of importance. But what aspect of "light"? Different modalities may figure as a stimulus, for example, absolute intensity, spectral composition, the plane of polarisation, etc. A change in light intensity may be thought essential, but is the absolute or the relative difference considered as stimulus? Reading the literature, I think the discussion has to start at an even more fundamental level: what is the general nature of a stimulus and when can an environmental factor be called a causal stimulus?

If a steady stream of solar radiation falls on a physical object like a rock, part of this radiation will be absorbed and the temperature of the stone will rise. Energy is neither gained nor lost. The body temperature of a lizard, lying on that rock, will also rise and, again, total energy will not alter. In both cases solar radiation is not called a stimulus. If suddenly a shadow covers the rock, the stone starts loosing heat and that is all, but the lizard gets excited and retreats rapidly. The animal will expend more energy than it had absorbed earlier. The change in radiation can be called a stimulus for the lizard, evoking the escape reaction. Several differences between a mere physical force and a stimulus can be listed. A stimulus is always related to an excitable system, which thus can be incited or roused into action. Essential is that a change in the environmental factor has to occur.[1] This change represents a small amount of energy (or material in case of a chemical stimulus) compared to the far more energetic outburst that mostly follows. This additional amount of energy was gained earlier from metabolism, and is not part of the sequence: stimulus – response. Thus, *reinforcement* is a characteristic feature of the result of stimulation. A stimulus is a carrier of *information*: in the example of the lizard of possible danger. The significant information has become connected with the particular factor-change during a process of evolution in the receiver organism. The process may have been one-sided if the "sender" is inanimate, as in the example of light and the lizard. If the sender is another organism, and if sending of the message is of interest to the sender too, a stimulus pattern might have evolved that optimises information content and transfer. The response can be of a fixed quantity, independently of the intensity of the stimulus. This is called an "all-or-nothing" reaction. But in most cases, the extent of the reaction is a function of the strength of the stimulus. A response to a stimulus is of a *temporary character* and fades away some time after the stimulation ends.

The relation between stimulus and reaction might be considered a simple causal chain (Fig. 1.2; see also the sensor–actor chain in Fig. 1.1). For example, when a receptor neuron is stimulated by an adequate environmental factor, a change in membrane polarisation in the generator region of the neuron occurs. The amount of depolarisation (or hyperpolarisation) depends on the strength of the stimulus. If the new value of the factor persists, *accommodation* occurs, that is, the change in membrane polarisation diminishes with time until a steady state is reached. With a slow rate of change in factor quantity, accommodation might be fast enough to prevent a response of the receptor neuron. If accommodation is complete and the factor does not change again, the membrane potential is in a new steady state, or has returned to the resting potential. If the generator potential of the neuronal membrane increases above a certain level, an *action potential* or *spike potential* is generated, which is conducted along the length of the axon of the receptor neuron. The frequency, with which the spikes travel along the axon, is directly proportional to the level of the generator potential, and, thus, to stimulus strength. After accommodation, spike frequency becomes nil, or has a low frequency, depending on the kind of receptor neuron. In either case the reaction is terminated.

After several synaptic connections have been passed, the spike train may reach, or may not reach, the central nervous system. Integration of information from other extero- and interoceptors takes place that modifies the outcome. Modifications happen before excitation of the efferent nerves occur. Finally, the effector organs, such as muscles, are eventually excited and behaviour is initiated. It is evident that between peripheral perception and peripheral action, filtering, facilitation and inhibition occur. Consequently, results from electrophysiology cannot be applied unconditionally to behaviour. For example, the threshold value of a stimulus that permits, if surpassed, a frequency spike train in afferent nerves is probably lower (never higher) than the one responsible for the resulting final behavioural response. Although the initiation of behaviour is a complicated process, the simple and well-established picture of excitation can be a guideline for the analysis of relatively simple reactions of zooplankton to environmental stimulation. This will be illustrated with the effects of light intensity changes on *Daphnia*.

When a light intensity (I) changes with an instantaneous step (ΔI), a swimming response occurs if ΔI exceeds the threshold. Intuitively we understand that this "just noticeable difference" has no constant value but depends on acclimation intensity. In 1832, Weber proposed that ΔI would be a certain fraction or percentage of I, but later it was found that this only holds for the middle range of the stimulating, environmental factor. Much discussion followed about the precise formulation of the function $\Delta \mathrm{I} = \mathrm{f}\ (I)$, especially in psychophysics where the discrimination of sensations in humans were of interest (see Blackwell, 1971 for a historical introduction). We cannot measure sensations in zooplankton, even if they have them; therefore, the psychophysics discussion is of no concern for zooplanktologists. As a *first approximation*, a relative change in light intensity, as suggested by Weber, will suffice for sensory and behavioural ecology. Thus a stimulus can be defined as $S = \mathrm{d}I/I\mathrm{d}t = \mathrm{dLn}I/\mathrm{d}t$. As Hartline and Ratliff (1972) argued, light intensity I can be best described in logarithmic units. This is convenient because the natural range in light intensities wherein eyes function is large and, consequently, eyes seem to operate logarithmically. Whether a definition of a stimulus satisfies depends on the kind of relation between the studied parameter of behaviour and the thus formulated change in environmental factor. If this relation has the simple form of a linear function, the definition will do for the time being. When analysis continues, it is possible that modifications have to be made or that additions are necessary. A stimulus definition is a working tool, not an ironclad rule.

Stimulus thresholds may change with time of day. Barlow et al. (1988) found for the lateral eyes of *Limulus polyphemus* that sensitivity for test flashes might increase by a factor 10^5 during the night. An endogenous rhythm was involved because the periodicity persisted in continuous darkness, although the rhythm drifted out of phase with the solar day. This is characteristic for circadian, endogenous rhythms with a free running period slightly different from 24 h. The origin of the particular rhythm is in the CNS, which is interesting because it means that the sensitivity of the eye is under efferent control. Changes in retinal pigment distribution are at the base of the phenomenon. During the night, pigment in the cells around the basis of the lens moves away to leave a larger aperture for light to enter the retinula cells (see Chapter 7). Efferent influence that changes sensory sensitivity is present in many species of invertebrates and vertebrates. Although the lateral eyes of *Limulus* are complicated structures compared to the nauplier eye of many zooplankton species, also in simple pigment cup eyes of echinoderms, a daily cycle was found (Cronin, 1988). An endogenous, circadian rhythm in response to an instantaneous decrease in light intensity is probably present in *D. magna* (Ringelberg and Servaas, 1971) because the percentage of reacting animals was less during the night-time hours, a pattern that persisted for some days in continuous light.

The general properties of excitable systems are to a large extent independent of the kind of system or species and are mutually comparable. However, internal and external circumstances easily modify the result of what might be essentially a simple excitable assemblage. This makes it sometimes difficult to recognise the basic stimulus–response pattern in complex behaviour. For this reason, the less variable sensory neuron was discussed as a model to make the analysis of photobehaviour in zooplankton easier.

3.3 Visual Thresholds

One important function of the eyes of copepods and cladocerans is the detection of changes in light intensity at dawn and dusk. At what lowest light level is this possible? Long before sunrise, downward migration of zooplankton starts. In early summer in Lake Maarsseveen, a change in depth of *Daphnia* became noticeable an hour and a half before the sun was above the horizon. At that time, and at the mean late-night-time-depth of the population, around 2 m, light intensity was about 3×10^{-5} μmol m^{-2} s^{-1}. At that low intensity, *Daphnia* seems to perceive changes in light intensity. This indirectly and roughly estimated value compares reasonably well with the photosensitivity thresholds determined for other zooplankton species by Forward and co-workers (Table 3.1). Dark acclimated animals were placed in the centre of a horizontally placed trough with illumination from one side. Within a few minutes after light was switched on, the distribution of the animals was determined. The lowest light intensity at which significant displacements occurred was

Table 3.1 Experimentally derived lower visual thresholds in dark acclimated zooplankton species

Species	Visual threshold		Wavelength (nm)
	Original unit	Converted unit	
Mesocyclops edax[a]	5×10^{-8} W m^{-2}	2.3×10^{-7} μmolm^{-2} s^{-1}	540
Sagitta hispida[b]	5×10^{13} photons m^{-2} s^{-1}	8.30×10^{-5} μmolm^{-2} s^{-1}	500
Pleuromamma gracilis[c]	9.5×10^{12} photons m^{-2} s^{-1}	1.58×10^{-5} μmolm^{-2} s^{-1}	480
Pleuromamma xiphias[c]	7.7×10^{10} photons m^{-2} s^{-1}	1.28×10^{-7} μmolm^{-2} s^{-1}	480
Rhithropanopeus harrisi[d]	1×10^{-7} W m^{-2}	4.18×10^{-7} μmolm^{-2} s^{-1}	500
Chaoborus punctipennis fourth instar[e]	3.3×10^{-6} W m^{-2}	1.52×10^{-5} μmolm^{-2} s^{-1}	
Chaoborus punctipennis second instar[f]	2×10^{-7} W m^{-2}	9.2×10^{-7} μmolm^{-2} s^{-1}	420
Acartia tonsa[g]	2.8×10^{11} photons m^{-2} s^{-1}	4.65×10^{-7} μmolm^{-2} s^{-1}	453–620

[a]Swift and Forward (1983), [b]Sweat and Forward (1985), [c]Buskey et al. (1989), [d]Forward et al. (1984), [e]Swift and Forward (1988), [f]Swift and Forward (1982), [g]Stearns and Forward (1984a).

in the range of 10^{-7}–10^{-5} μmol m^{-2} s^{-1}. The fourth instar of *Chaoborus punctipennis* (Diptera) seems to be less sensitive. Swift and Forward (1988) mentioned that the estimate is conservative, thus the actual threshold might be lower. A different method was applied. The threshold for upward migration of *C. punctipennis*, as determined from field sampling by Chaston (1969), was estimated at 2.5 lux (about 4.88×10^{-2} μE m^{-2} s^{-1}), thus much higher. However, estimates made from field observations are difficult to compare with those made in the laboratory with specifically designed apparatus. This especially holds for phantom larvae, burying in the mud.[2] This complicating circumstance was absent for *C. flavicans* migrating downwards after dark acclimation in three prealpine lakes in southern Bavaria (Germany). Wagner-Döbler (1988) reported estimates of 0.43 mlux on average for the three lakes. Converted[3] into 8.4×10^{-6} μmol m^{-2} s^{-1}, this value is within the range of the other animals in Table 3.1. For the chaetognath *Sagitta hispida*, a threshold was not explicitly given but I estimated a value of 5×10^{13} photons m^{-2} s^{-1} (= 8.3×10^{-5} μmol m^{-2} s^{-1}) from Fig. 1 in Sweat and Forward (1985) (30% positive phototaxis at the maximum spectral sensitivity of 500 nm).

A noticeable phenomenon in these dark acclimated animals is that a positive phototaxis followed upon an increase in light intensity. Considering the morning descent in nature and the results of most experiments on photobehaviour, a negative phototaxis should be expected. The phenomenon compares with the dawn ascent sometimes found in advance of descent migration (Cushing, 1951). With increasing acclimatisation light intensity, the percentage moving towards light decreased and the usual negative phototaxis took over.

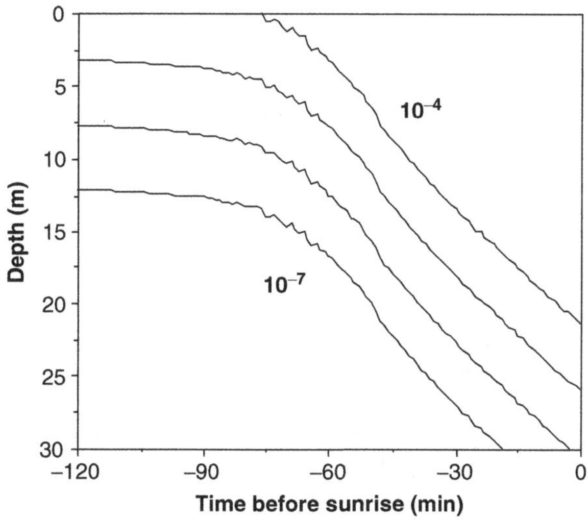

Fig. 3.1 Isopleths (μmol m^{-2} s^{-1}) at dawn in Lake Maarsseveen on 11 June 1992. Absolute visual thresholds for zooplankton are between 10^{-7} and 10^{-5}, thus vision seems possible in the upper 10 m from 1.5 h before sunrise. Complete darkness was absent because of reflexion of the night sky by the illumination from nearby town of Utrecht

The unexpected positive phototaxis, reminds one of the so-called dawnrise, sometimes performed by zooplankton at early dawn just before morning descent begins (Cushing, 1951).

The sensitivity to light in these various zooplankton species is high. In Fig. 3.1, isopleths within the range of thresholds is depicted for dawn on 11 June 1992 in Lake Maarsseveen. Species with thresholds within the range are able to perceive light of the sky during the whole night. Consequently, species-specific differences in visual threshold are of no importance for the timing of downward migration in this lake.

Still another aspect of threshold light sensitivity must be mentioned. Predation by the visual predator, 0^+ perch (*Perca fluviatilis*), is possible at a light intensity of 2×10^{-4} μmol m^{-2} s^{-1} (Flik et al., 1997). On 11 June 1992, at a depth of about 2 m in Lake Maarsseveen, this light intensity was not realised earlier than 1 h before sunrise (Fig. 3.1). Therefore, to prevent predation, *Daphnia* and *Eudiaptomus gracilis* have to go down before this time, which they did indeed.

3.4 Reactions to Instantaneous Changes in Light Intensity

If light intensity is decreased in one step, *Daphnia* reacts with swimming in the direction of the light; upon intensity increases swimming is away from light. Clarke (1930) demonstrated that swimming was in the direction of the light rays indeed, and was not a reaction directed to attain the original light intensity again. At the time – and for many years to come – the paradigmatic explanation of DVM was that zooplankton followed an optimal light intensity. Therefore, it was important to demonstrate whether or not absolute light intensity played a decisive role in phototactic reactions. Clarke (1930) observed *Daphnia* in a convergent bundle of light. Upon a light intensity increase, the animals swam away from the incident light and, consequently, moved into ever increasing intensity instead of into lower intensities. If the reaction had been to restore an optimal, absolute light intensity, swimming would have been in the opposite direction, of course. Clarke (1930) concluded that absolute light intensity, whether in the shape of an optimum or not, was not part of the reaction mechanism.

These reactions to instantaneous changes in light intensity look like reflexes: the intensity change triggers a response and that seems to be all. This is not so. Ringelberg (1964) demonstrated experimentally that a swimming animal perceives the self-generated change in light intensity. After a decrease in light intensity, the upward swimming *Daphnia* was kept at the same distance to the light source, thereby preventing a "restoration" of the acclimation intensity.[4] In case of a simple reflex chain, this should be without effect. However, the animals continued swimming for a longer time: the duration of the reaction was on average 1.58 times that of the control. The result demonstrates that the reaction is not a simple reflex, but an "optimum" is not involved either. It looks like a feedback system because the effect of the reaction, thus the self-induced change in light intensity, influences the response. However, if a real cybernetic feedback system was involved, swimming should have continued endlessly, or anyhow, until the end of the experimental tube had been reached. In a true cybernetic system, the discrepancy of the "norm" (in our case, the optimum or the acclimation intensity) would determine the reaction. In Fig. 1.2, the information flow chart of the open reflex chain reaction is compared with the flow chart of a closed loop cybernetic control system. The phototactic reaction of *Daphnia* looks like a kind of hybrid between the two. This hybrid character is expressed in Fig. 3.2. The self-generated change in light intensity is a feedback

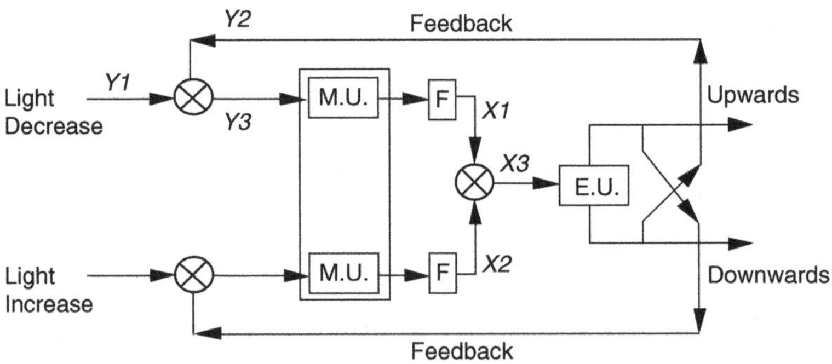

Fig. 3.2 The information flowchart of the phototactic reaction of *Daphnia magna*. Upward and downward swimming is indicated. The input of the measuring units (MU) is Y3 = Y1–Y2 (only indicated for the light decrease but also holding for the light increase). The feedback Y2 is the light intensity change generated by the swimming response. *F* indicates a "calculation centre" and represent the effect of the light change or stimulus *S* and the disintegration of this effect with time *t* (see text): $F = Y = (c_1S/c_2)(1 - e^{c_2t})$. The input for the effector unit, leading to upward or downward swimming is x3 = |x1–x2| (Modified after Ringelberg, 1964)

that influences the reaction, but a "norm" is absent. The value of the feedback must be a function of the swimming velocity and of the turbidity of the water. The higher turbidity is, the faster light intensity increases while swimming upwards, and thus the shorter the reaction time must be. If swimming velocity is not affected by the feedback, which was found (Ringelberg, 1964), the result must be smaller swimming distances if turbidity of the water increases. If the acclimation intensity is of influence, a smaller swimming distance should also be found with increasing turbidity. Therefore, it must be checked whether the original light intensity was reached or not. In Fig. 3.3, the ratio is presented between the empirically found swimming distance and the distance *D. magna* had to swim to reach the original light intensity again. At an attenuation coefficient of $k = 0.0217\ \text{cm}^{-1}$, the swimming distance is such that the acclimated light intensity is reached at the end of the reaction. At lower coefficients, thus in more clear water, swimming distance falls short; in more turbid water it is too large. In this experiment, an instantaneous decrease in light intensity of 30% was given. With a different percentage, perhaps the crucial attenuation coefficient would have been different. If so, the present one has no particular meaning and is an accidental outcome of the reaction mechanism. From data compiled from the literature, Dodson (1990) found that the amplitude of *Daphnia* migrations increased with increasing Secchi depth.[5] In other words, larger distances were swum in clear lakes than in more turbid lakes. With increasing clarity, visually hunting fish are able to predate deeper in the water column. Therefore, *Daphnia* has to migrate deeper in order to decrease the chance of being eaten. The results of the above-discussed experiments indicate what physiological mechanism is available to realise this adaptive goal.

Heberdey and Kupka (1941) determined what difference in light intensity two lamps,[6] placed at either end of a horizontal trough, must have to be discriminated by *D. pulex*. At an acclimation light intensity of 400 lux, the relation was minimal. For lower and higher light intensities, the perceived difference increased. Although both

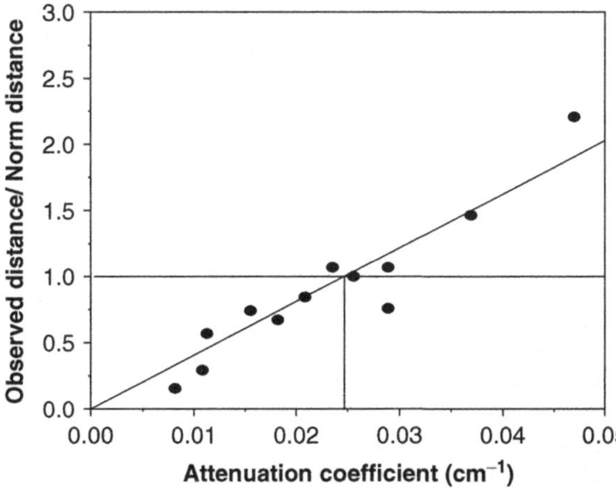

Fig. 3.3 Comparison of the swimming distance at the end of a positively phototactic reaction of *D. magna* and the distance to be swum in order to reach the acclimation light intensity again. Instantaneous changes in light intensity of 30% were applied at different turbidity in the experimental tubes. The best fit line is described by the function $y = 0.127 + 40.23\,x$; $R^2 = 0.899$ (Data from Ringelberg, 1964)

lamps were simultaneously present, and thus discrimination of a spatial difference was involved, it is possible that both lamps could not be seen simultaneously and that, in fact, a temporal difference was involved. Whether this is true or not, experiments with temporal changes led to results, comparable to those of Heberdey and Kupka (1941). In Fig. 3.4a and b, thresholds for instantaneous, relative decreases in light intensity, as a function of the acclimation intensity, are presented for *D. magna* and the marine calanoid copepod, *Acartia tonsa.* The resemblance is striking! A minimal threshold is for both species at more or less the same acclimation intensity, 0.2 and 0.07 μmol m^{-2} s^{-1}, respectively. At both sides threshold values rapidly increase. Referring to Heberdey and Kupka (1941), Waterman (1961) suggested that two different optical systems might be involved. The dual aspect of the relation in Fig. 3.4 supports this hypothesis for a copepod and a cladoceran. At the higher light intensities, the compound eye of *D. magna* is certainly involved but probably not at the lower light range because 50% of *D. magna* individuals were unable to orient the body axis by optical means at a light intensity of 0.14 μmol m^{-2} s^{-1} (Ringelberg, 1964).

Daphnia still has another light sensitive organ: the nauplius eye. Although the exact function of this pigment cup is unknown, it might be involved in the discrimination at low light intensities. Ewald (1910a) extirpated this eye but saw no difference in behaviour for various kinds of light stimuli as long as the compound eye was present. However, light intensity might have been too high to expect an

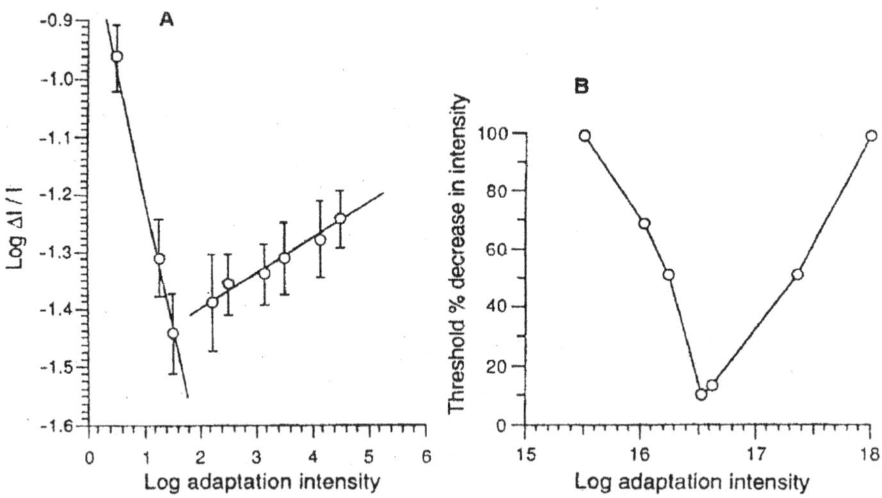

Fig. 3.4 Thresholds of instantaneous relative decreases in light intensity for positively phototactic reactions as a function of the light intensity of adaptation. (**a**) Thresholds for *D. magna* (Ringelberg et al., 1967). Adaptation light intensity is in erg cm^{-2} s^{-1}. (**b**) Thresholds for *Acartia tonsa* with adaptation intensity in photons m^{-2} s^{-1} (from Stearns and Forward, 1984a). The change in functional relation in both species is at a light intensity of approximately 0.2 and 0.07 μmol m^{-2} s^{-1}, respectively. This combination figure was originally published in Ringelberg (1995b) (Reproduced by permission of the Marine Biological Association, Plymouth)

apparent effect. Swift and Forward (1982, 1988) gave an indication of a different function of the compound eye and the pigmented ocellus in *C. punctipennis* larvae. Dark acclimated second- and fourth-instar larvae were tested for phototactic responses after a short exposure to low light intensities of different magnitude. At light intensities lower than 4.6×10^{-4} µmol m^{-2} s^{-1}, both instars moved to the lighted side. At higher intensities, the second instar, which still has an ocellus only, also moved predominantly towards the light source. In the fourth instar, however, a striking shift from positive to negative phototaxis occurred. It is tempting to suggest that the newly formed compound eye is responsible for this shift. Anyhow, these results indicate that both the visual system and behaviour at a low light intensities changes with increasing age.

It must be mentioned that a diffuse "dermal" light sense has been proposed for *Daphnia* (Schulz, 1928; Viaud, 1948; Scheffer et al., 1958; Rimet, 1960a). Physiological properties are unknown but the cuticle of *Daphnia* is transparent and nerves or the CNS might be light sensitive too.[7]

The highly comparable response curves for a planktonic copepod and a cladoceran is a hopeful sign that the extensive experimental results, obtained for *Daphnia*, can be of value for copepods too.

3.5 Reactions to Continuous Changes in Light Intensity

3.5.1 *The Stepwise Reaction Pattern*

From an ecological point of view, a one-step change in light intensity cannot be called a relevant stimulus. In nature, illumination changes gradually even when clouds suddenly cover the sun. However, instantaneous changes are simple to apply experimentally and the effects illustrate properties of photobehaviour that also hold for reactions to continuous increases and decreases. A sequence of stepwise light changes gradually converts into continuous stimulation when the intervals between steps are made shorter. This was done in experiments with *D. magna* (Ringelberg, 1964). Within a series of steps, instantaneous decreases (ΔI) and interval times (Δt) were kept constant. A single sub-threshold light decrease caused (by definition) no reaction, but several in succession led to a phototactic upward swimming. The number of steps, needed to evoke a reaction is determined by Δt and ΔI. The following conclusions were drawn:

(a) Sub-threshold decreases in light intensity have additional effects.
(b) With increased interval time, the number of decreases must also increase to initiate a reaction. This means that the effect of a single stimulus diminishes with time.
(c) The phototactic reaction does not start when a certain light intensity has been reached.

(d) When the stepwise, sub-threshold stimulation was continued long enough after a previous reaction had stopped, a second, third, and so on, swimming reaction occurred. This led to a pattern of intermittent upward swimming. This additive character of stimuli indicates that separate reactive swimmings can be linked together in a sequence.

For the marine copepod, *A. clausi,* comparable results were obtained (Johnson, 1938). Following instantaneous increases and decreases in light intensity, the copepods reacted with downward and upward swimming, respectively. If the time interval between two successive light changes was long, a moderate reaction occurred and the animals soon redistributed in the experimental tube. With shorter intervals, reactive swimming became more extensive. This indicates that effects of separate light changes are additive as in *Daphnia.*

The stimulus of a continuously changing light intensity can be defined as a rate $(dI)/(Idt)$, with reciprocal time (min^{-1}) as dimension. At low values, the evoked upward or downward swimming is stepwise or discontinuous again, as in a series of separate instantaneous light changes (Ringelberg, 1964; Daan and Ringelberg, 1969). These upward or downward reactive displacements are superimposed on the upward or downward phases of normal, oscillating swimming, discussed in Chapter 2. Illustrations are given in Fig. 3.5. Upon decreases in light intensity, the upward phases of oscillating swimming are prolonged while the downward phases become shorter. With increasing stimulus strength, the periods in between the ascending or descending phases get shorter and shorter until swimming is continuous (Ringelberg, 1964). Downward swimming reactions, caused by relative increases in light intensity, are more complicated because interrupted passive sinkings and active downward movements occur side by side. Displacement velocity of sinking is smaller than that of downward swimming. The percentage of individuals actively swimming downwards increases with increasing stimulus strength (Daan and Ringelberg, 1969). However, large latent periods in the active downward swimmers made that total overall displacement velocity is the same if a sufficiently long period is considered. Interrupted swimming and the distinction between active downward swimming and passive sinking is of importance for the interpretation of the downward migration at dawn in nature. For example, in 1990, small amplitude DVM occurred in Lake Maarsseveen (Ringelberg and Flik, 1994) and no correlation existed between this displacement and the relative increase in light intensity (see Fig. 11.5). This is, of course, to be expected if interrupted swimming with a predominance of sinking occurred. In 1992, the amplitude of migration was much larger and the downward displacement was significantly ($P = 0.001$) correlated with the rate of the relative light increase. The downward displacement velocity in the lake for these years was 4.5 cm min^{-1} and maximal 10 cm min^{-1}, respectively. Compared to sinking rates in anaesthetised animals, these overall displacements are low. For the comparable sized *D.galeata mendotae*, Brooks and Hutchinson (cited in Hutchinson, 1967) mention 6–18 cm min^{-1}, Hantschmann (cited in Hutchinson, 1967) found 0.36–27 cm min^{-1} for *D. schloderi* and Brancelj and Blejec (1994) mention 28–31 cm min^{-1} for *D. hyalina.*[8] These sinking rates suggest that the

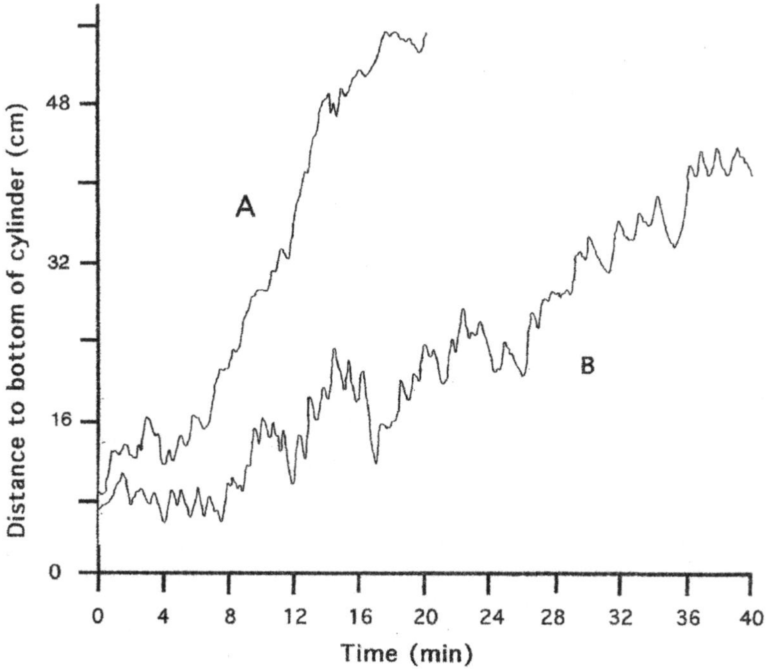

Fig. 3.5 Examples of stepwise upward swimming by *D. "longispina"* caused by different rates of relative decreases in light intensity. *A*. Stimulus $S = 0.203\ \mathrm{min}^{-1}$, displacement velocity $DV = 3.15\ \mathrm{cm\ min}^{-1}$; *B*. stimulus $S = 0.077\ \mathrm{min}^{-1}$, $DV = 1.22\ \mathrm{cm\ min}^{-1}$ (Original registrations by K. Bakker, unpublished Masters Thesis, University of Amsterdam)

downward migration in Lake Maarsseveen was not continuous. In 1990, downward displacement was within the range of velocities of 1.8–4.8 $\mathrm{cm\ min}^{-1}$, found in the experiments with *D. magna* (Daan and Ringelberg, 1969). In 1992, displacements were at a higher rate.

3.5.2 The Stimulus Strength–Duration Curve

Sub-threshold stimuli have some effect, although this is not expressed in behaviour. We can consider continuous decreases or increases in light intensity as series of sub-threshold changes. Since the (neural) effect of these changes decreases with time, it must be expected that the lower a rate of light change is, the longer the latent period must be before a reaction becomes noticeable in behaviour. This is the case, indeed. The relation between stimulus strength and the time needed to elicit a phototactic reaction is a characteristically shaped curve (Fig. 3.6).[9] In nerves, muscles and other excitable organs, the curves are all of a similar shape (Bourne, 1960).

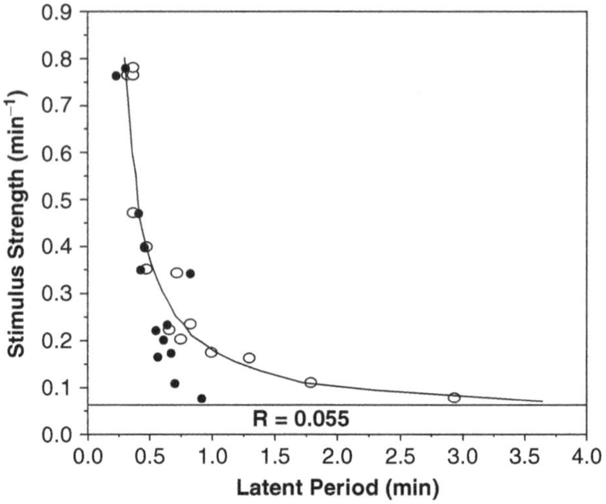

Fig. 3.6 The strength–duration curve for the positively phototactic reaction of *D. longispina*. The *open circles* indicate the latent periods of the first swimming reaction after the change in light intensity started. Each point is an average of 10–15 reactions. The curve through these points was obtained by application of the function in Fig. 3.7. The *black dots* are the average latent periods for the reactions following upon the first one. The rheobase value for the latency of the first reaction is indicated by a line at the stimulus strength of $R = 0.055$ min^{-1}

A swimming reaction can be caused by all combinations of stimulus strength and stimulus duration at the right hand side of the curve in Fig. 3.6. At the left are all sub-threshold combinations that lead to no reaction. An absolute threshold, called the rheobase (*R*) in physiology, also exists. It is defined as the stimulus that only causes a reaction if duration time is infinite. At lower rates of change, light intensity can be diminished until darkness is reached without the occurrence of a phototactic reaction. This phenomenon was already observed by Clarke (1930). Rheobase values were estimated for *D. "longispina"* (Fig. 3.7), *D. galeata x hyalina* (Van Gool and Ringelberg, 1997) and *D. magna* (Ringelberg, 1964). The values compare well: $R = 0.055$ min^{-1}, $R = 0.04$ min^{-1} and $R = 0.10$ min^{-1}, respectively. The rapid phase of morning descent and evening ascent starts at about the time the rheobase is surpassed (Ringelberg et al., 1991a; Chapter 9).

The stimulus strength–duration curve is also of theoretical interest. The form suggests a hyperbolic function and, thus, a constant product of stimulus strength and duration. However, a constant product is not conforming to the discussed experimental result of disintegrating sub-threshold effects. Especially at low rates of light change, the relatively longer time needed before a reaction starts, becomes discernable. In a theoretical model of the curve, justice must be done to this aspect of disintegration. It was assumed (Ringelberg, 1964) that a certain excitatory state (*Y*) disintegrates with time at a rate (c_2) proportional to its momentary quantity. The excitatory state is built up at a rate proportional to the stimulus strength (*S*). Therefore, the net rate of accumulation of the excitatory state is

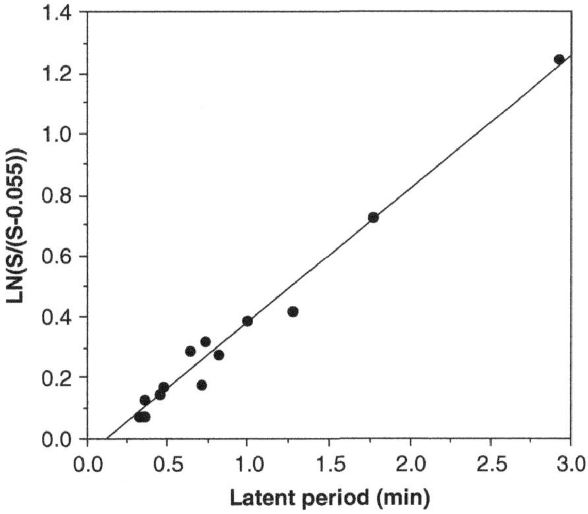

Fig. 3.7 The same latent periods for the first phototactic reactions as in Fig. 3.6 were used. The *vertical axis* was obtained by using equation (3.3) in the text and the rheobase of $R = 0.055\ \mathrm{min}^{-1}$. This value was obtained by iteration until the best fit of the line was reached. The calculated function is $y = -0.0638 + 0.43894\,x$

$$\frac{\Delta Y}{\Delta t} = c_1 S - c_2 Y \qquad (3.1)$$

The integral of this equation is

$$Y = \frac{c_1 S}{c_2}(1 - e^{c_2 t}) \qquad (3.2)$$

We can change equation (3.2) into the next form:

$$Ln\frac{S}{S - R} = c_2 t \qquad (3.3)$$

In this equation (3.3), t is a latent period that can be determined empirically for each applied stimulus S if we substitute a chosen value of R. The "best" value of R can be found by iteration until the best fitting line is obtained. The slope of the curve is the disintegration constant c_2. The result of such exercise is illustrated in Fig. 3.7.

Although the model seems to describe the empirical results rather well, the physiology of the phototactic reaction is more complicated than is apparent from the curve. The stepwise swimming at low rates of the relative change in light intensity means that after a first latent period, a second latent period is already present before a second swimming starts. This second latent period after the first reaction is followed by a third, a fourth and so on. It was found for *D. magna* that each latent period was shorter than the previous one (Ringelberg, 1964). Obviously, after a first reaction, facilitation occurs which makes the latent period shorter. In *D. longispina* facilitation occurs only once. The second latent period is shorter than the first, but the periods between the following reactions are of the same duration as the second one. For up to five successive latent periods, averages were calculated and inserted

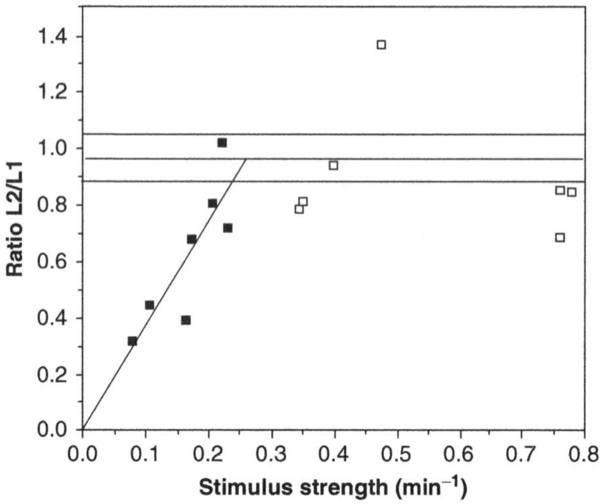

Fig. 3.8 Comparison of latent periods in the positively phototactic reaction of *D. longispina*. Squares represent the ratio of the second and the first latent period. A *line* is fitted through these ratios for rates of the relative decrease in light intensity lower than 0.25 min^{-1} (*black squares*). The function is $y = 0.0072 + 3.700\,x$. The three *horizontal lines* indicate mean (= 0.971) and upper and lower confidence limits ($P = 95\%$) for ratios of L3/L2, L4/L3 and L5/L4 (Original data from K. Bakker, unpublished Masters Thesis, University of Amsterdam)

as black dots in Fig. 3.6. The ratio of the second and the first latent period is made a function of stimulus strength in Fig. 3.8. The latent periods of the following reactions were not statistically different and the ratios between these periods L3/L2, L4/L3 and L5/L4 were averaged and the mean of 0.971 ($n = 33$) with the 95% confidence limits are inserted in the figure. For stimuli smaller than 0.25 min^{-1}, this L2/L1 is evidently smaller than one and thus, L2 is smaller than L1. Even at the higher rates of change in light intensity, the latencies of L2 are probably shorter than those of L1, as is suggested by the open squares, several of them lying below the lower confidence limit of the average of the successive ratios. The facilitation process has to be studied more extensively before details of the process can be concluded.

The phototactic reaction to continuous changes in light intensity is variable, especially at the lower rates of change. An animal might react after a very long latent period or even not at all. At another light change, the same animal might react differently. The percentage of animals reacting within the short observation time of < 15 min decreased with decreasing stimulus strength. Van Gool and Ringelberg (1997) found that for *D. galeata x hyalina* the percentage gradually increasing from 30% at a relative increase in light intensity of 0.06 min^{-1} to 100% at increases larger than 0.4 min^{-1}. These large stimulus rates exceed the natural occurring maximum of about 0.16–0.20 min^{-1}. (It was also found that in the presence of a rather high concentration of fish kairomones all animals reacted, even at the low stimulus values. We will discuss the important effect of these kairomones in Chapter 4.) The stimulus strength–duration curve is considered a physiological characteristic of a particular behaviour. As usual, before behaviour results, additional factors exert influence in the central nervous system and obscure the true nature of the physiological mechanism. The large variance in latent periods found for stimuli just exceeding the rheobase value might be considered troublesome noise from a physiological point of view. However, this variable behaviour cannot be excluded from the interpretation of migration in a lake. Probably, not all individuals migrate always, or not all genotypes migrate, and those that do migrate do not swim continuously, but possibly with interrupted bursts.[10] The large variance might be of ecological importance and a base for evolutionary selection.

3.5.3 Swimming Speed and Displacement Velocity

Since the response to changes in light intensity consists of a lengthening of the upward or downward phases of non-reactive, oscillating swimming with unreactive periods in between, a distinction must be made between swimming speed and displacement velocity. The first is defined as the velocity during these extended upward or downward reactive phases. Displacement velocity is the vertical distance covered after a long enough time to include a representative number of separate reactions including the latent periods in between these reactions. Displacement velocity is smaller or equal to the swimming speed. Swimming speed is a physiological reaction parameter, while displacement velocity is of ecological relevance. During the positively phototactic reaction of *D. magna*, swimming speed was constant at 14 cm min^{-1} up to a relative decrease in light intensity of 1.2 min^{-1} (Ringelberg, 1964). For higher stimulus values, upward swimming became continuous and the distinction between swimming speed and displacement velocity disappeared. The swimming speed increased with stimulus strength. Daan and Ringelberg (1969) found this general picture also for increases in light intensity (Fig. 3.9). For *D. "longispina"* the picture is different. From the lowest stimuli onwards, swimming speed increased with stimulus strength (Fig. 3.10). We have observed that the latent periods, with the exception of the first one, remain the same in this species and, thus, the slope of the two functions of swimming speed and displacement velocity is expected to be the same. The phototactic reaction mechanism of *D. magna* and *D. "longispina"* is evidently different. Also for the hybrid *D. galeata x hyalina*, vertical displacement velocity is a continuous function of the relative change in light intensity, at least so for increases in light intensity (Van Gool and Ringelberg, 1997). Perhaps the difference in reaction pattern has to do with adaptation to a different habitat. The large *D. magna* inhabits shallow ponds, migrates for a short time and over a short distance. On the other hand, the small pelagic species might migrate over large distances and during 1–2 h.

Reacting to relative increases or decreases in light intensity has occasionally been considered impossible for migrating animals. In most cases zooplankton lag behind a "moving" absolute light intensity during morning descend or evening ascend. In

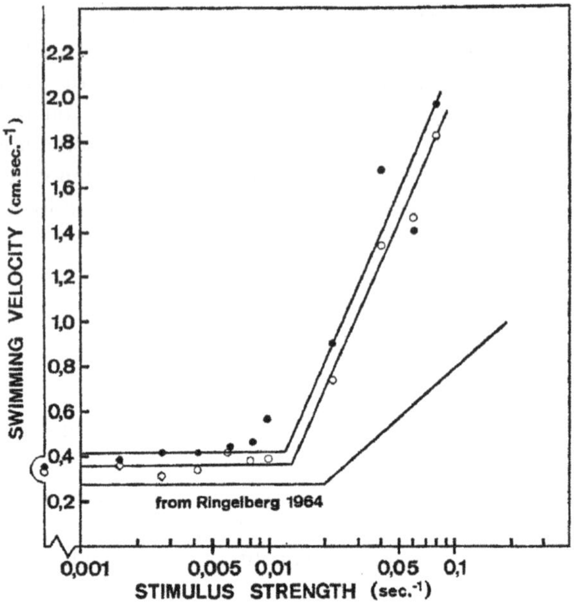

Fig. 3.9 Swimming velocity of *D. magna*, responding to relative decreases (*black dots*) or increases (*open circles*) in light intensity. The two values at a zero strength stimulus indicate velocities at the upward and downward phases of oscillatory, normal swimming if light intensity is constant (From Daan and Ringelberg, 1969)

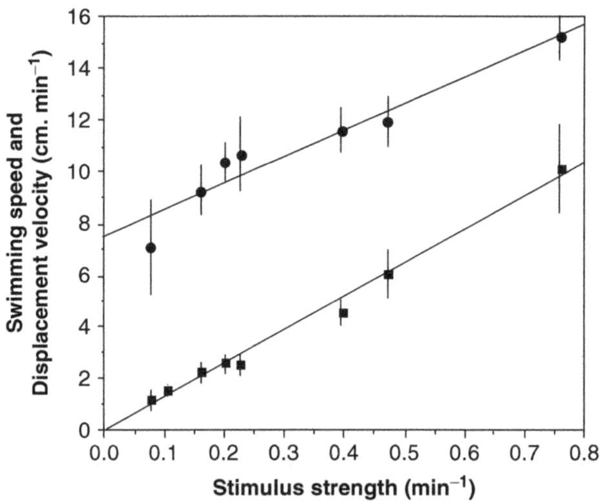

Fig. 3.10 Swimming speed (*dots*) and displacement velocity (*squares*) for the positively phototactic reaction in *D. "longispina"* as a function of the rate of relative decreases in light intensity. The function of the *lower line* is $y = -0.0763 + 12.95 \times$ ($R^2 = 0.989$, $n = 8$) and for the *upper line* $y = 7.45 + 10.30 \times$ ($R^2 = 0.924$, $n = 7$)

that case, the change in overhead light intensity can be perceived. However, when deep sound scattering layers in the oceans migrate as fast (Kampa and Boden, 1954) or even faster downwards in the morning or upwards in the evening, than a descending respectively ascending isolume (Clarke and Backus, 1956), it is difficult to conceive that a light intensity increase or decrease is the responsible stimulus. For example, if during evening ascend an animal swims faster than the upward moving isolume, it experiences a light intensity increase instead of a decrease. A stepwise, interrupted swimming, as in *Daphnia*, might offer the solution. The stimulus can be perceived again during the short interludes of no swimming. The problem will be discussed again in Chapter 11. It will be shown in Chapter 5 that under certain circumstances, a migration more rapidly than isolumes move is possible and can be understood along this line of reasoning.

3.6 Photobehaviour in Other Planktonic Species

Systematic analysis of behaviour in other zooplankton species has rarely been done and no consistent body of knowledge is available. A nice example is, however, the extensive research done with the marine copepod *A. tonsa*. An account can be found in the doctoral thesis of Donald E. Stearns (Stearns, 1983; see also Stearns and Forward, 1984a, b). His determination of the threshold instantaneous decreases in light intensity at different acclimation intensities showed strikingly similar results with that of *D. magna* (Fig. 3.4). Schallek (1942) had already described that *A. tonsa* swam upwards after sudden decreases in light intensity and moved downwards when light intensity was increased. Stearns studied how the stimulus can best be defined: as an absolute intensity, an absolute change or a relative change in light intensity. His conclusion was in favour of the last one, although absolute intensity did play a role, at least as a determinant of the threshold value as is apparent from Fig. 3.4. Also other aspects of behaviour are comparable to that of *Daphnia*. In constant light or darkness, the hop-and-sink mode of swimming prevails, for example. Reactive upward swimming is interrupted by small periods of sinking and looks like a prolongation of the "normal" swimming, discussed in the previous chapter. The downward movement is by passive sinking only. As Stearns mentioned (p. 290), this was the first experimental demonstration of the relative change in light intensity as a light cue in a marine organism. Obviously, the *Daphnia* model can be used more generally. Stearns (1983) had no opportunity to study the effect of continuous changes in light intensity. One problem was the orientation of this copepod. Abnormal light fields often lead to aberrant behaviour in many animal species (Verheijen, 1958). Schallek (1942) already observed that a lamp, placed directly above an experimental cylinder, made *Acartia* persistently positively phototactic. Verheijen (1958) would have called this an example of the "trapping effect" of an abnormal angular light distribution (ALD). With oblique illumination and the resulting more diffused light field, animals showed the expected negatively phototactic downward swimming when light intensity increased.[11] Stearns' experimental set up realised the natural ALD as best as possible, however, at the expense of a large vertical swimming distance. Nevertheless, a 3-day analysis of the timing of upward migration of *A. tonsa* in the field made it plausible "that most of the migrating *A. tonsa* begin their ascent under background light conditions that are optimal for eliciting upward swimming with a minimal decrease in light intensity" (p. 282). The minimal percentage decrease, determined from the experiments, was 8.9% at an acclimation intensity of 4.309×10^{16} photons m^{-2} s^{-1}. In the field, 50% of the animals had left the bottom at 4.0×10^{16} photons m^{-2} s^{-1}. Relative decreases at that time were 33.2, 15.4 and 14.6% for the 3 days of study, and thus certainly above the experimentally determined threshold.

In a series of papers, Forward and co-workers (Forward, 1974; Cronin and Forward, 1980; Forward and Cronin, 1980; Forward et al., 1984; Forward, 1985, 1986) studied photobehaviour in larval stages of the estuarine crab *Rhithropanopeus harrisii.* At metamorphosis, the larvae have to be in the estuary sufficiently close to the habitat of the adults. They have an endogenous, tidal vertical migration pattern with a maximum depth at the time of low tide and a minimum depth at high tide. In addition, a diel vertical migration of the normal pattern exists, which is to a large extent independent from tidal currents, although the details of the pattern change in relation to tidally induced cycles of currents and salinity (Cronin, 1982, cited in Forward et al., 1984).

Phototaxis is superimposed on the tidal rhythms. Light acclimated animals are positively phototactic at high and moderate light intensities but become negatively phototactic at low intensities. These surprising results made Forward (1974) hypothesise a reversed diel vertical migration pattern with an ascent at sunrise and a descent at sunset. However, the larvae were found near the surface at night and deeper down during the day. The unexpected behaviour in the laboratory proved to be due to a light field, which differed strongly from the angular light distribution (ALD) present in the natural habitat. When the originally used, horizontally placed cuvette with light from one side was replaced by an apparatus designed to produce a light field comparable to that occurring under water, swimming changed (Forward et al., 1984). In an extensive series of experiments, Forward (1985) observed that animals, acclimated in darkness or at a light intensity near the lower threshold of 8.65×10^{-7} W m^{-2}, moved slowly upwards at low light intensities. At higher intensities, they moved slowly down (< 0.3 mm s^{-1}).[12] Experiments were done, among others, to establish the threshold lowest light intensity for vision (see Section 3.3). With the preferendum explanation of DVM in mind it was thought that the upward movement occurred at the threshold intensity and a downward movement at light intensities exceeding this threshold. In this way, larvae would stay at about the threshold isolume during the day (Forward et al., 1984). This same idea was at the background of comparable experiments with *S. hispida* (Sweat and Forward, 1985) and *Pleuromamma sspec* (Buskey et al., 1989). The latter authors supposed that differences in absolute light intensity thresholds could explain the difference in timing of the morning descent of two *Pleuromamma* species in the Sargasso Sea. However, the underwater light intensities at the daytime depths of these species were much higher than the absolute threshold. Both *Pleuromamma* species certainly could have gone down sufficiently deep in the ocean. The hypothesis of migrating at a visual threshold intensity was refuted and experiments with changing light intensities started (Forward, 1985). In *Rhithropanopeus* larvae, neither absolute intensity nor the absolute change in light intensity was found to be the appropriate stimulus, but relative decreases were. The optimal acclimation intensity for the positively phototactic reaction was just above the lower absolute threshold for vision (8.65×10^{-7} W m^{-2}), which suggests that relative changes operate as soon as light is perceived. This result seems to be comparable to the results of threshold studies in *Acartia* and *Daphnia* as presented in Fig. 3.4. Another similarity with the *Daphnia* model was that the time before a reaction started, decreased with increasing stimulus strength. Observations at other acclimation light intensities have to be done before it can be concluded that a similar curve is present also in this larval crab. A difference with *Daphnia* was found in the orientation component. For the upward swimming reactions to relative decreases in light intensity orientation is not to light, but to gravity because movements of *Acartia* were also upwards when light was coming from below.

Increases in light intensity elicited downward swimming in *Rhithropanopeus* larvae but the descents occurred at a specific threshold light intensity (on average 1.87×10^{-7} W m^{-2}), independently of the relative rate of intensity increase. On the other hand, orientation was now to incident light, not to gravity, since illumination from below made the larvae swim upwards, thus away from light, indeed. Summarising, the stimulus for ascent seems to be a relative decrease in light intensity with gravity as cue for orientation. The stimulus for descent movements seems to be a specific absolute light intensity and not a change in light intensity; on the other hand, orientation is optically.

Not all decreases in light intensity cause a positively phototactic reaction in a natural ALD. Forward (1986) demonstrated that upon strong decreases, the crab larvae reacted with sinking and downward swimming. This is the so-called "shadow response", a reaction present in many animals to prevent predation (Von Buddenbrock, 1931, 1952). This phenomenon indicates that the mechanism underlying photobehaviour serves several purposes and must be quite complicated.

Species like *Rhithropanopeus* are interesting and will reveal reaction patterns that can be applied to the behaviour of pelagic species but they are not the real representatives for these species (neither is *D. magna*, a pond dweller).

The same holds for brine shrimp larvae (*Artemia*) studied, among others, by Forward (1993) and Forward and Hettler (1992). More will be said about theses species in the next chapter. Obviously, phototaxis or, more generally, photobehaviour is not confined to planktonic crustaceans. Extensive field and experimental research was done with phantom larvae (*Chaoborus* species), and the results of Swift and Forward (1988), Haney et al. (1990) and Wagner-Döbler (1988) will be discussed in Chapter 11. Experimental research with marine planktonic organism remains rare, perhaps due to technical constrains. An example is the research of Arkett (1985) with the hydromedusan *Polyorchis penillatus*. These animals appear in the water column only after sunset. Arkett found that upon relative decreases in light intensity, as after sunset, the frequency of swimming pulses increased, leading to upward displacements. Predawn increases inhibited the pulses and resulted in sinking. For hydromedusa and most other animals, it holds that the role of light in the initiation and control of vertical migration is not well understood. To repair that, laboratory experiments and field observations must be combined. Due to the limitations of experimental cylinders, only reactions of short duration and over short distances can be studied in the laboratory. For this reason, some ecologists have doubts about the relevance of such studies for phenomena that take place in the vast expanse of oceans and lakes. However, the large-amplitude migrations in nature can be considered as successions of the short reactions observed in the experiments. At dawn and dusk, these responses are initiated over and over again by relative light intensity changes, continuing for nearly 2 hours. Reconstructions of migrations can be made with mechanistic models. These models can, with parameters derived from small-scale experiments, simulate ecologically relevant migration amplitudes. An example of a simulated photobehaviour response is given in Fig. 3.11. For the same two stimulus values that evoked the swimming tracks of real animals presented in Fig. 3.5, the characteristics of the phototactic reaction were applied. Disintegration of the excitation is evident from the stepwise swimming. This is not surprising, of course, because the theory was applied in the model. In Chapter 5, the models are treated more in detail but an example of the power of the model to simulate migration is presented in Fig. 3.12. Relative increases of a natural dawn were used as input stimuli. Morning descent is reproduced although the distance covered is small. This can be repaired by introduction of the effect fish chemicals have on photobehaviour. This will be done in the next chapter.

I will close this chapter with a discussion of terminology. So far we have used predominantly the traditional terms positive and negative phototaxis to indicate movements towards and away from a light source as elicited by an aspect of light. Per definition, the orientation component of phototactic movements had to be

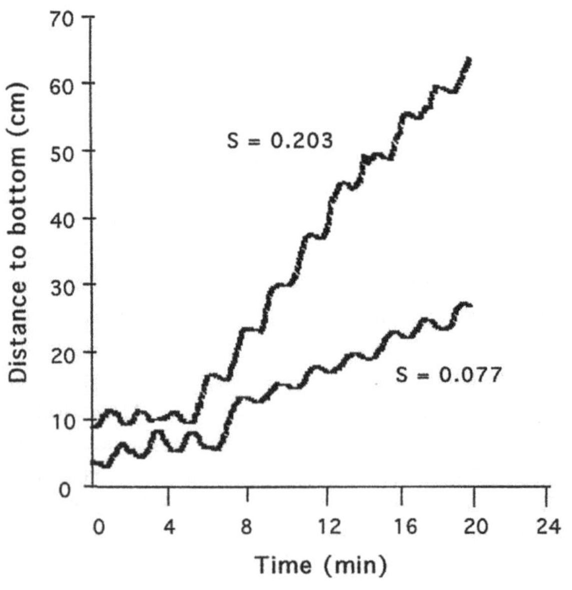

Fig. 3.11 Two examples of simulated positively phototactic reactions to relative decreases in light intensity with Stimulus strength $S = 0.203\ \text{min}^{-1}$ and $S = 0.077\ \text{min}^{-1}$. Compare with the examples of experimentally obtained swimming tracks in Fig. 3.5. Displacement velocities are 3.11 and 1.16 cm min^{-1}, respectively

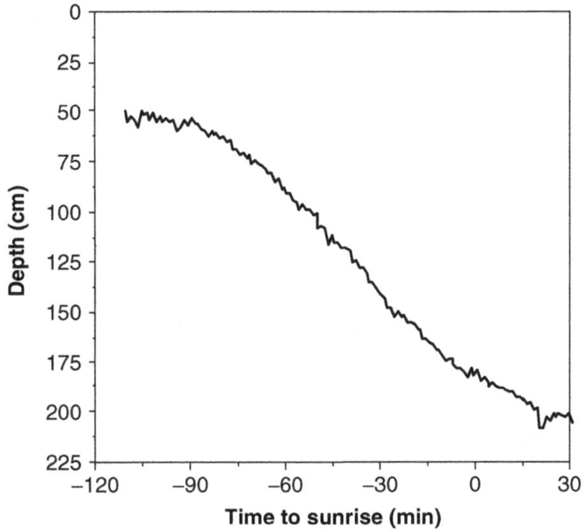

Fig. 3.12 Simulation of a diel vertical migration with the natural relative increase in light intensity of dawn as input. Light intensity was measured 0.3 m below the water surface of Lake Maarsseveen on 30 May 1990

directed by incident light. However, we also described a swimming reaction caused by light changes but directed by gravity. Strictly speaking, the term phototaxis cannot be applied in this case. The more neutral term "photobehaviour" can be used to indicate any response caused by whatever aspect of the factor light, although this term is less precise than phototaxis. I will dub photobehaviour mechanism 1 (PBM 1) as the mechanism behind phototactic swimming discussed in this chapter. This implies that also PBM 2 exists, which is the topic of the next chapter.

Notes

1. In electrophysiology, spike frequency in the axons of tonic receptor neurons is proportional to the strength of a constant environmental circumstance and this factor is also called a stimulus. For example, if the eye of the horseshoe crab (*Limulus*) is illuminated by a one-step increase in light intensity, a high frequency of action potentials in the optic nerve fibres results. This initial burst of spikes declines rapidly to a steady-state frequency, which is proportional to the *logarithm* of the intensity of the stream of light (Hartline and Ratliff, 1972). In sensory physiology, the logarithm of the light intensity might be called a stimulus, responsible for the nerve activity during steady state. It does not characterise the effect of the change in light intensity. Moreover, behaviour often returns to the prestimulus state, and, thus, at the border of sensory ecology and behaviour, I prefer to ignore this habit of calling a constant environmental factor a stimulus and speak of a stimulus only if a change in the factor occurs.
2. The time of upward migration after sunset in the field can be determined only roughly and, thus, the light intensity at the mud surface must be an inaccurate value. Moreover, it is questionable whether absolute light intensity is the right factor for the initiation of DVM. Chaston (1969) found that in January–February the threshold was lower than in April and explained the difference (and reasoned the difference away) by posing that the time between coming

out of the mud and the start of upward swimming was longer in winter. Since temperature was considered of no influence, reaction time must have been different. Indeed, the result of (preliminary) experiments with animals kept for a week under different daylength regimes showed that reaction time under a short day was longer. The experiment has to be repeated.

3. A conversion factor of 1 klux = 19.5 μmol cm^{-2} s^{-1} (daylight and the 400–700 nm spectral range) was used (LI-COR manual 1979).
4. In fact, the animal was within a tube that was suspended in a larger tube with water. As soon as the daphnid started to swim upwards after a one-step decrease in light intensity, this inner tube was lowered in the water column at the same speed as the animal was swimming upwards. Thus the distance to the overhead light source was kept constant, and, consequently, the light intensity at the position of the animal was kept constant.
5. Secchi depth is the depth in metres at which the Secchi disk, a kind of soup plate attached to a piece of string, disappears when lowered into the water.
6. At that time, experiments to determine a discrimination intensity, with criterion the path taken by an animal between two light sources (tropotaxis, see Fraenkel and Gunn, 1961), the so-called "zwei Lichter Versuch" was often done.
7. Schulz (1928) observed that the so-called "Lichtrückenreflex" by which *Daphnia* turns the dorsal side towards a horizontally placed light source is absent if the compound eye is removed. On the other hand, a reaction to a large decrease in light intensity ("Schattenreflex") is present in eyeless daphnids. Blinded animals also gather in yellow-green light and swim down if illuminated with UV-radiation. Also the frequency of the hart changes when illuminated and metabolism might be influenced by light intensity. It is obvious that the results of this early research are promising. Repetition will certainly lead to a better understanding of the role of light in *Daphnia*. The excellent paper by Schulz (1928) provides a starting point.

   The "sensibilité dermatoptique" plays an important role in the papers of the French authors. A rather deviant terminology was used to interpret photobehaviour.
8. For the larger sized *D. pulex*, sinking rates of 39.63 cm min^{-1} or 18 cm min^{-1} are mentioned by Gorski and Dodson (1996) and Dodson (1996), respectively.
9. In nerve and muscle excitation this characteristic is also called the rheobase–chronaxie curve, named after the two parameters that determine the curve. The rheobase is the smallest intensity of the stimulus that is able to cause a response when applied for an infinite time. The chronaxie is the time after which a stimulus, twice the rheobase, causes a reaction.
10. To get at the underlying physiology as best as possible, filtering of the experimental data must take place. A *t*-test was used to determine the probability that very small or large values belonged to the estimated statistical population. If belonging to the estimated statistical population, a new mean and standard deviation was calculated and the next larger or smaller one was evaluated.
11. More will be said about orientation in plankton animals and the importance of the angular light distribution in Chapter 7 on orientation.
12. This result seems at variance with the one presented in Forward et al. (1984). In this paper, stages I and IV zoeae show a slight negative phototaxis at the lower intensities and at higher light intensities ($>3.43 \times 10^{-7}$ W m^{-2}) a pronounced positive phototaxis, thus upward swimming. The result must be due, however, to the unnatural light field in the experiments, as was discussed. If tested in a natural ALD, as Forward and Rittschof (1999a) did, descent photoresponses were obtained for light intensities higher than 2.9×10^{11} photons m^{-2} s^{-1} ($= 3.3 \times 10^{-7}$ W m^{-2}).

Chapter 4
A Decision-Making Mechanism

Contents

4.1 Introduction

As described in the previous chapter, changes in light intensity suffice to cause phototactic swimming reactions in zooplankton organisms. No predators or kairomones are needed, only changes in light intensity. We deal with a fixed behavioural response and I called this photobehaviour mechanism 1 (PBM 1) to free it from the per definition necessary optical orientation component inherent to the term phototaxis. Some photoreactive swimming is directed by gravity, for example, in *Rhithropanopeus* larvae or *Acartia*. Also passive sinking, as in *Daphnia*, can be indicated by the new term as long as it is caused by light. The old term, phototaxis, is covered and can be used if it is known that the orientation component is directed by the angular light distribution. It was possible to simulate a DVM, using a train of photoresponses, although the resulting vertical amplitude of 1.5–2 m was small. Such small migrations certainly occur in nature but the "real" ones cover much larger distances. This failure of PBM 1 to describe large amplitude DVM made me realise that the mechanism was insufficiently described and was reason to look for new approaches. It is now generally accepted that the adaptive significance of DVM is avoidance of visually hunting fish (Lampert, 1989). Field work in Lake Maarsseveen revealed that the seasonal period of DVM lasted as long as large shoals of 0^+ perch were present in the lake (Ringelberg et al., 1991a). Dodson (1988) had shown that several *Daphnia* species were lower in a test cylinder if water from an aquarium with fish was added. The difference of a few centimetres was small but the experiments demonstrated for the first time that predator-mediated chemical substances are perceived by daphnids.

J. Ringelberg, *Diel Vertical Migration of Zooplankton in Lakes and Oceans*,
DOI 10.1007/978-90-481-3093-1_4,

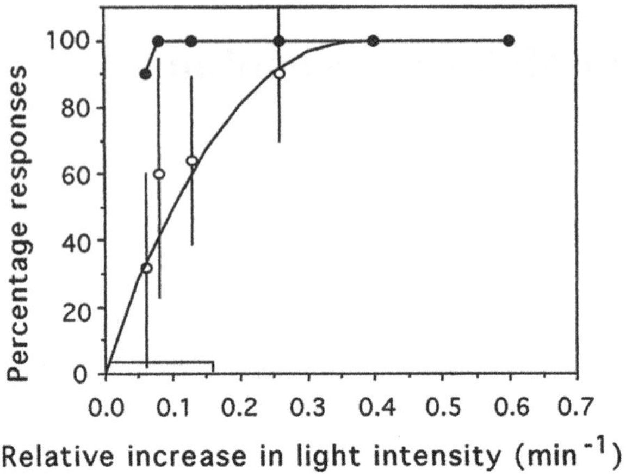

Fig. 4.1 Mean percentage of *D. galeata x hyalina* responding to relative increases in light intensity. Error bars represent 95% confidence limits of the mean. *Open dots* represent responses in the absence of fish kairomone; *black dots* are responses in the presence of fish kairomones. The function of the curve $y = 2.93 + 541.83\ S - 761.7\ S^2$ holds for $S = 0$–0.4. The *horizontal bar* indicates the range of relative increases in light intensity occurring at dawn (modified after Van Gool and Ringelberg, 1997)

Experiments with *D. hyalina* and juvenile perch (*Perca fluviatilis*) from Lake Maarsseveen confirmed the perception of these kairomones and showed its effect on photobehaviour (Ringelberg, 1991a). The effect of the substance on the response to relative increases in light intensity was tested. *Daphnia* reacted very strongly, even creeping in between small pieces of charcoal that covered the bottom of the cylinder. It was also observed that an increased percentage of *Daphnia* reacted to a given light change. Without fish kairomone, this percentage gradually increased with increasing stimulus strength but in the presence of fish kairomones, 100% reacted at all stimulus values (Fig. 4.1). Van Gool devoted his thesis research to a further analysis of photobehaviour in the presence of, among others, fish kairomones (Van Gool, 1998). One of the important results was that these infochemicals did not affect the physiology of the reactions: the stimulus strength – duration curve remained the same in the presence of perch kairomones, also the linear function between stimulus strength and displacement velocity was not altered. We concluded that PBM 1 operated independently of fish predators or fish chemicals and, thus, that PBM 1 was not primarily a strategy to prevent predation. It was hypothesised that an extension of the photobehaviour mechanism had to be present, probably elicited by another releasing stimulus. The nature of the light stimulus had to be studied again.

The chemical structure of the fish kairomones was (and to some extent still is) unknown (see Chapter 13). This hampered the quantification of the kairomone effect on light-induced swimming and migration behaviour. On the other hand, we felt the need to close in on the natural situation and to test lake water, with and without fish chemicals, and to quantify the 0^+ perch population during the crucial migration period in Lake Maarsseveen. Step by step, research moved in the direction of ecology.

4.2 The Interdependence of Two Successive Stimuli

In Section 3.5.1, I discussed that a succession of sub-threshold instantaneous, stepwise decreases in light intensity leads to a normal phototactic reaction if the time intervals between the steps are sufficiently short. Obviously, the effect of individual stimuli can be added. At low rates of the relative changes in light intensity, the reactive displacement is continuous but the separate reactions are still visible. I supposed that the rest excitation of a previous reaction influenced the next one.

Of a completely different mutual influence are accelerations and decelerations in rates of the relative changes in light intensity. Accelerating and decelerating stimuli are familiar to us. An example is the effect of a starting elevator. The perception of movement disappears as soon as velocity becomes constant, but reappears when deceleration sets in.

A first step to study the effect an acceleration in stimulus can be made if a suprathreshold relative increase in light intensity (s_1) is followed by a larger relative increase (S_2) (I use the notation with lower and upper case to indicate the difference in rate: $s_1 < S_2$, for example.). If the displacement velocity DV(S_2) is identical, whether or not it was preceded by a stimulus s_1, this indicates that s_1 was without effect on DV(S_2). Let us denote the displacement velocity of a reaction that follows upon a previous one by DV(s_i, S_{i+1}). A change in displacement velocity, related to the previous one, can be denoted by DV(s_i, S_{i+1})/DV(s_i). In Fig. 4.2, this change is

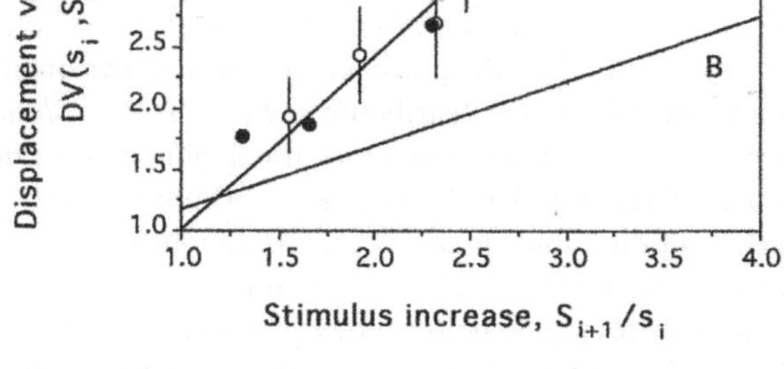

Fig. 4.2 *A*. The relation between the ratio of successive continuous relative increases in light intensity S_{i+1}/s_i (capital *S* indicates that stimulus is larger than lower case *s*) and the increase in displacement velocity, DV. Each point represents the mean of 17–30 measurements performed with different individuals of *D. galeata x hyalina.* Error bars represent 95% confidence limits of the mean. For the *open circles* hold that $i = 1$. The function through these points is $y = -0.395 + 1.408\,x$; $R^2 = 0.972$, $n = 8$, $P < 0.001$. For the *black dots* hold that $i = 2$. *B*. Line *B* ($y = 0.663 + 0.516\,x$) represents an effect in the absence of a stimulus acceleration and is based upon the empirical function DV = 1.63 + 31.34 *S* (see Note 1)

made a function of the ratio of both stimuli S_{i+1}/s_i. If no special interaction would be present, the ratio of DV's would consist of the DV's at S_{i+1} and s_i if determined separately. In that case, the lower line in the figure holds. This lower line was calculated from the results of experiments with single stimuli.[1] Evidently, the swimming velocity after a combination of two stimuli is enhanced. The same enhancement occurs if a third stimulus follows upon a second one, as is apparent from the fact that the black dots in Fig. 4.2 lie on the line of the effect of the second stimulus. Thus no additional enhancement occurs after the first (Van Gool, 1997). A tentative conclusion was that accelerating light increases induce a certain fixed enhancement of the phototactic swimming reactions.

During dawn, the initial acceleration in relative increases in light intensity changes into a deceleration about 50 min before sunrise (Fig. 6.8). It proved that at decelerations the displacement velocity also enhanced photo-induced displacement velocities, although less so than accelerating light stimuli do (Van Gool, 1997). This enhancement of displacement velocities (caused by decelerations) is a factor 0.73 lower than the enhancement caused by accelerations.

4.3 The Influence of Kairomone, Food and Temperature

Haney (1993) presented some tests designed to examine the interaction of temperature, food and fish (kairomone) on the day and night distribution of *D. carinata* in a 2.30m water column. The daily light intensity changes were considered the primary factor responsible for the changes in vertical position and the mentioned factors were thought to modify the photoresponses leading to the particular depth distributions. After 12 h of starvation and without fish, the animals remained day and night above the temperature gradient (Fig. 4.3). After an additional 12 h of starvation, the distribution changed and most individuals went below the temperature gradient during the day and especially during the night. After 2 days of starvation *Daphnia* moved upwards at night, spreading out over the epilimnion. On day 4, food was added to the epilimnion and after a few days a distribution as found at the beginning was resumed. Then the presence of fish was investigated. To exclude direct fish consumption effects on distributions, *Lebistus reticulatus* too small to predate on *Daphnia* was used. On the first day, the daphnids went down below the temperature gradient where the fish did not come. Although they were starved no nightly move into the epilimnion occurred. This behaviour was enhanced on days 2 and 3, although the upward move at night, as seen in the column without fish, also occurred but the animals remained below the temperature gradient at all times. The day and night distributions on day 8 were the same as on day 3, although food was added to the column. The conclusion was that the presence of food was overruled by the presence of fish.

The results make clear that both food and fish kairomone are able to modify vertical distributions. Light intensity changes induce diel vertical displacements and the temperature gradient acts as a barrier. Probably, the thermocline makes the distinction of different behaviour more visible because preventing vertical

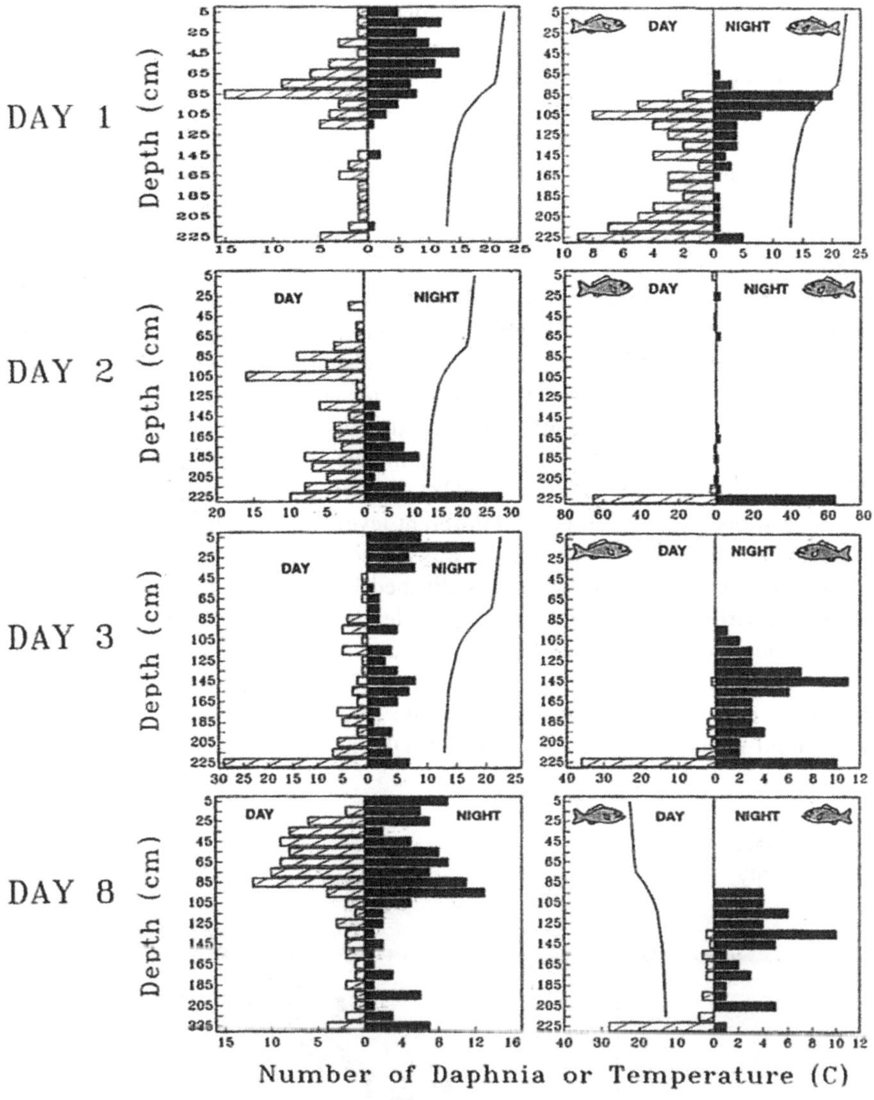

Fig. 4.3 The effect of starvation on the vertical distribution of *D. carinata* with and without fish. Food algae were added to the epilimnion of the cylinder on day 4. See text for further explanations (from Haney, 1993; Reproduced by permission of E. Scheizerbart'sche Verlagsbuchhandlung)

dispersion as soon as the light intensity changes are over. The distributions are the result of behaviour that in itself is difficult to discern and experiments of another type had to be done in which responsive swimming can directly be measured. Therefore, we continued with photobehaviour experiments at different concentrations of food, kairomones and temperature.

The extent of enhancement of the displacement velocity caused by accelerations and decelerations in rate of relative changes in light intensity was found to depend on the presence of fish kairomones and food (Van Gool and Ringelberg, 1998). This was an important result because now we have an idea how a proximate mechanism realises the adaptive goal of DVM! Food availability and fish predation have been suggested to be the two important factors that determine whether DVM is profitable or not. All mathematical models about the adaptive significance of migration depend on these factors (Gabriel, 1993). The general opinion is that DVM is largely a flexible, environmentally modified phenomenon. Consequently, *Daphnia* and other migrating zooplankton must somehow assess food availability and predation pressure. The obtained information must then be used to "decide" when and how deep to migrate. The amplitude of migration is very variable indeed; it varies between individuals of the same species and between species. It changes with season and depends on the characteristics of the lake. A first approach to a behavioural mechanism that might be responsible for "assessing" profitability of DVM has been made by Van Gool (1998). *D. galeata x hyalina* were acclimated at different food concentrations and different levels of perch kairomones. Because the chemical nature of fish kairomones was unknown, specific concentrations could not be made. Different levels were established by dilution of water from a standardised aquarium with juvenile perch. Accelerations[2] in rates of relative increases in light intensity were realised by rate increases of 10% every 20 s. For each S_i, displacement was measured over the pertinent 20 s. An example, at a particular food concentration and kairomone level of the displacement velocity of an individual daphnid is given in Fig. 4.4. As usual, DV increases with increasing stimulus

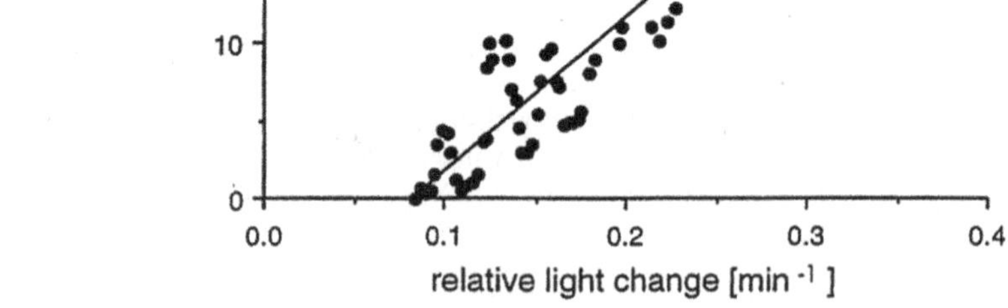

Fig. 4.4 Displacement velocity (DV) of *D. galeata x hyalina* as a function of relative increases in light intensity (*S*), accelerating by 10% every 20 s. The linear function is DV = –7.98 + 98.67 S, thus PC = 98.97. The experiment was performed at a food concentration of 0.5 mg Cl^{-1} (*Scenedesmus acutus*) and a 10% kairomones level (see Fig. 4.6) (after Van Gool and Ringelberg, 1998; reproduced by permission of Oxford University Press)

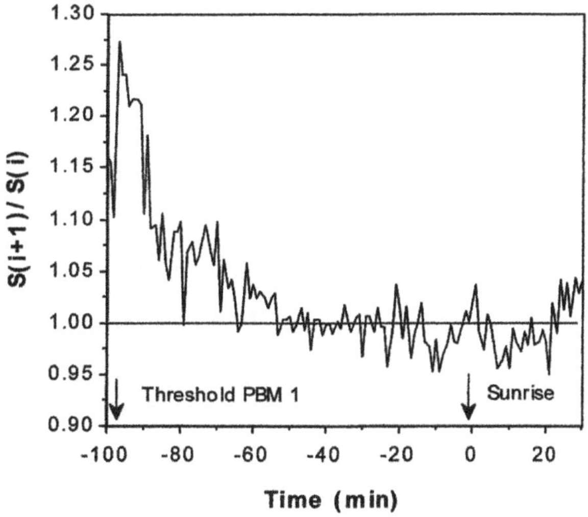

Fig. 4.5 The ratio of two successive rates of the relative increase in light intensity around sunrise on 30 May 1990, as measured in Lake Maarsseveen. The maximum value of accelerating rates occurs when the threshold for phototaxis (PBM 1) is surpassed. See also Fig. 6.9. The change from acceleration into deceleration is of course at the maximum of relative increases at –50 min before sunrise. About 30 min after sunrise, the relative increases in light intensity become sub-threshold again

value. The extent to which this occurs when rates of the relative increase accelerate depends on food concentration and kairomone level. The slope of the regression line in Fig. 4.4 is a measure of this extent and was called the Phototactic Coefficient (PC) (Van Gool and Ringelberg, 1998). Accelerations[3] during a natural dawn are presented in Fig. 4.5. Noticeable is, that maximum acceleration occurs very early and coincides with the moment the rheobase threshold of PBM 1 (see also Fig. 6.8) is surpassed. Acceleration values gradually decrease until 50 min before sunrise. At that time the rate of the relative increase in light intensity is at a maximum. Thereafter deceleration sets in and is present for the rest of dawn. At about 30 min after sunrise, rates of the relative light increase fall below the rheobase of PBM 1 again.

The experimentally applied $S_{i+1}/s_i = 1.33$ over 1 min is high, but not far from the maximum during dawn. Van Gool (1998) tentatively explored also accelerations of 5% (ratio = 1.15) and 20% (ratio = 1.6). The PC values obtained with the 5% increases did not differ from those with accelerations of 10% per 20 s. This indicates that we possibly deal with an "all-or-nothing" effect of accelerations. Experiments with the highest percentage failed, however, because behaviour became aberrant after a few accelerations. Animals made somersaults and seemed unable to maintain a normal body posture. This is a sure sign of optical disorientation (Ringelberg, 1964). Somersaulting can easily be induced by sudden large light intensity increases. During sunrise, a ratio of 1.6 only exceptionally occurs when a cloud cover suddenly opens.

PC values at different kairomone levels and two food concentrations are presented in Fig. 4.6. The increase in PC at the two food concentrations is linear and the two lines are nearly parallel. This means that the interaction between kairomone level and food concentration is weak. When fish kairomone level is zero, PC ranges between 26.57 and 52.88 for food concentrations between

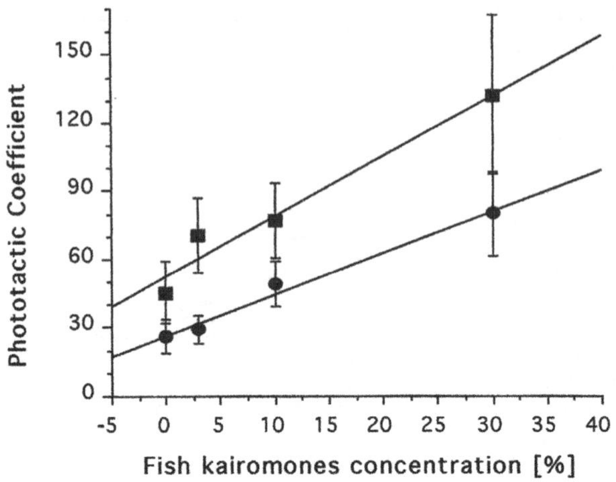

Fig. 4.6 Enhancement of the phototactic reaction, as expressed by the phototactic coefficient, of *D. galeata x hyalina* to accelerating relative increases in light intensity. Experiments were done at different levels of fish kairomone and the two algal (*Scenedesmus acutus*) concentrations of 0.05 mg C l^{-1} (*black dots*) and 0.5 mg C l^{-1} (*black squares*). Values are means ($n = 23$–26) with 95% confidence limits. The functions of the regression lines are as follows: PC = 26.57 + 1.82 × kairomones concentration (*dots*) and PC = 52.88 + 2.65 × kairomones concentration (*squares*) (modified after Van Gool and Ringelberg, 1998)

0.05 mg C l^{-1} and 0.5 mg C l^{-1}. A PC = 30.74 was found in experiments with different *S* values but no accelerations.

Migrating zooplankton have to come up in the evening again. Starting from the hypolimnion, internal and external conditions are different from those in the epilimnion at the end of the night. Light intensity is higher and the animals are not dark acclimated, which means a different threshold for the phototactic reaction. Temperature of the hypolimnion is lower and these poikilothermal animals are less active. Since food is scarce at the day depth, the gut might be more or less empty and satiation low. But above all fish kairomone concentration is lower or even absent because juvenile perch do not venture as deep. Thus, reactive swimming to relative decreases in light intensity is affected. It was found, however, that at low food concentrations the phototactic coefficient (PC) was significantly higher for decreases than for increases in light intensity (Fig. 4.7) (Van Gool and Ringelberg, 1998). The differential effect might compensate for the absence of kairomones in the hypolimnion and facilitates a return to the epilimnion. The enhancing effect of fish kairomone was precisely the same for accelerating increases and decreases in light intensity. In addition, kairomones, once perceived, exert the enhanced reactivity for longer periods (more than 1 day), which facilitates swimming between the fish-free hypolimnion and the fish-containing epilimnion. Whether the food effect is caused by a lower algal concentration in the water or due to the physiological condition of the animals like an empty gut or a low sugar concentration in the blood is not clear. At low food concentration, the effect of accelerating rates of the relative increases

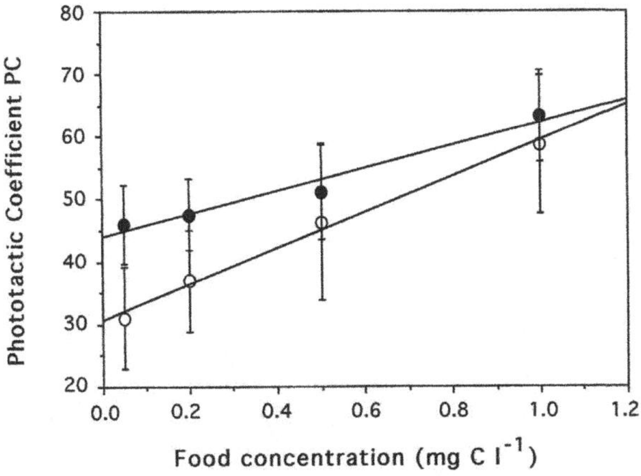

Fig. 4.7 The average value of the phototactic coefficient (PC), with 95% confidence limits, at different food (*Scenedesmus acutus*) concentrations. *Open dots* represent effects of relative increases in light intensity (PC = 30.45 + 29.26 × food concentration); *black points* represent relative decreases in light intensity (PC = 44.01 + 18.12 × food concentration). The regression coefficients of both functions are significantly different ($P < 0.025$). Convergence is at 1.22 mg C l^{-1} (from Van Gool and Ringelberg, 1998)

in light is nil (Figs. 4.6 and 4.7), which prevents an extensive downward swimming to increasing light in the morning.

The adaptive significance of this is clear: starved animals should not migrate deep into the hypolimnion where food concentration is even worse. Forward and Hettler (1992) found a comparable food effect for brine shrimp larvae (*Artemia*). In this species, starvation also activated the positively phototactic response to decreases in light intensity but did not affect the negatively phototactic reaction to increasing light intensity.

As is apparent from Fig. 4.8, temperature also influences the phototactic coefficient (PC). Contrarily to predator kairomones, temperature is not considered a factor in the predation-prevention strategy as far as I can see. Sufficiently high fish populations and thus kairomone concentrations are not always present. On the other hand, the physical factor temperature is always "present". Temperature influences metabolism and activity. In the present case, temperature influences the reaction of *Daphnia* to light intensity changes and must be considered a modifier of the PC-food function. Experiments were done in the absence of fish kairomones but at the high food concentration of 0.5 mg C l^{-1} (Van Gool and Ringelberg, 2002). The relation between PC and temperature is the same for light intensity increases and decreases, and, when averages are used, a quadratic function fits the data perfectly (Fig. 4.8). At 17°C, two other food concentrations were tested. The PC values are inserted in the figure and parallel quadratic functions were drawn, suggesting that the temperature effect depends on food concentration. This is a strange phenomenon and, if true, would indicate that temperature plays a role in determining the migration amplitude.

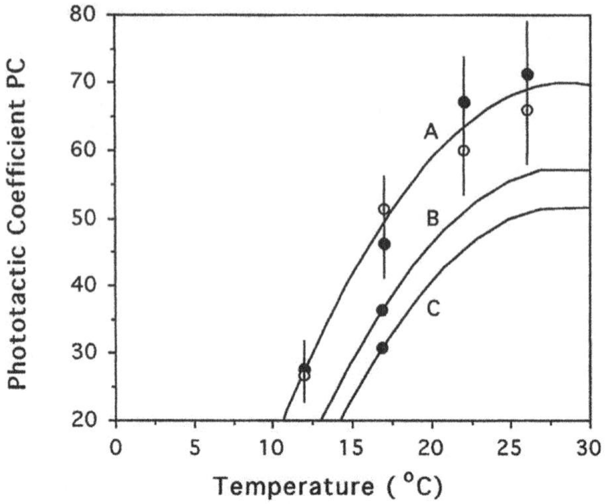

Fig. 4.8 *A*. Average values of the phototactic coefficient (PC), with 95% confidence limits, at different temperatures and a food concentration of 0.5 mg C l^{-1}. Black points represent effects of relative increases in light intensity; *open dots* represent effects of relative decreases in light intensity. The quadratic regression function was calculated for averages of increases and decreases in light intensity ($PC = -57.39 + 8.92\,T - 0.156\,T^2$; $R^2 = 1$). (From Van Gool and Ringelberg, 2002); *B* and *C* are lines drawn parallel to line *A* through empirical values (*black dots*) of 0.2 and 0.05 mg C l^{-1}, respectively. PC values for *B* and *C* at $T = 0$ are –70.20 and –75.76; otherwise, the functions are the same (Reproduced by permission of Oxford University Press)

Extensive experiments with nauplia of the brine shrimp (*Artemia*), a species quite different from *Daphnia,* revealed that fish kairomone induces variable photoresponses also in this species (McKelvey and Forward, 1995). Odour of larvae of the marine fish Menhaden (*Brevoortia tyrannus*) enhanced the reaction to relative increases in light intensity (Fig. 4.9). Also for this species, the concentration of the kairomone is important. The smallest rate at which the nauplia reacted significantly decreased with increasing concentration of the kairomone. One can think off two possible ways to explain the results: (1) different individuals have a different sensitivity for fish odours, while the threshold of the phototactic mechanism itself is not affected; (2) kairomones affect the photobehaviour mechanism by lowering thresholds at which nauplii react. Anyhow, a chemical substance produced by Menhaden is conditional for a negatively phototactic reaction caused by step-wise increases. Light intensity decreases cause upward swimming only if the animals are starved (Forward and Hettler, 1992) and kairomones are of no influence (Fig. 4.10). On the other hand, downward movements are enhanced as in *Daphnia*. Feeding seems to be of no influence (Fig. 4.11). *Artemia* larvae might not be considered true zooplankton, although Lenz (1980) described DVM in adults and small planktivorous fish readily predate on the larvae. Thus a comparable mechanism seems to be available to prevent predation. Similar results were obtained with infochemicals from other visual planktivorous fish such as Mummichog (*Fundulus heteroclitus*) and Pinfish (*Lagadon rhomboides*), but megalopa crab larvae (*Callinectes sapidus*) or the chaetognath *Sagitta elegans* were of no

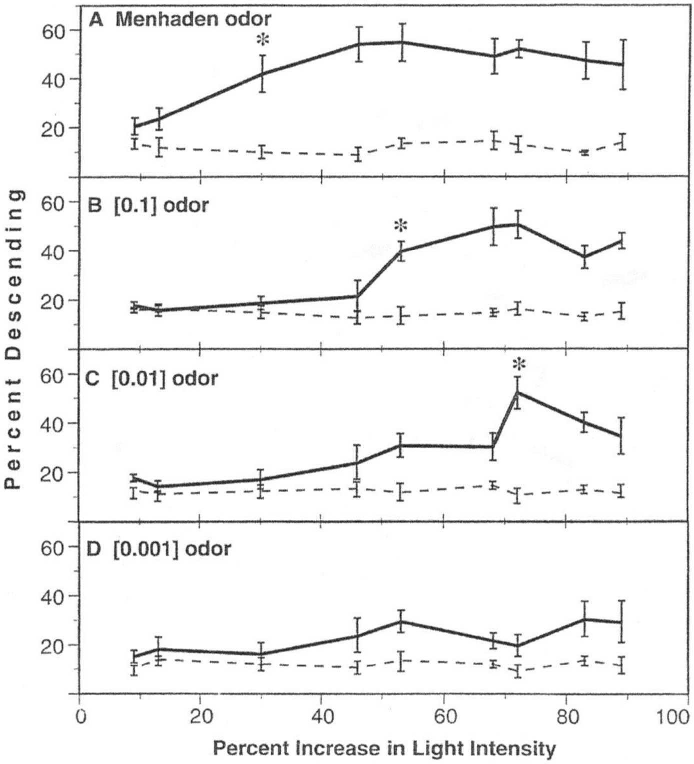

Fig. 4.9 Percentage of *Artemia* nauplii showing a descent response upon step relative increases in light intensity after 1 h exposure of undiluted Menhaden odour (*A*) and to dilutions with clean seawater of 10% (*B*), 1% (*C*) and 0.1% (*D*). The *solid line* shows descent responses upon light stimulation, while the *dashed line* shows the percentage descending prior to stimulation (control). An asterisk indicates the smallest percent increase in intensity to evoke a response significantly ($P < 0.05$) different from the control level. Means ($n = 5$) and standard errors are plotted. After McKelvey and Forward (1995) (Reproduced by permission of Oxford University Press)

influence. Surprisingly, also a chemical substance of the ctenophore *Mnemeopsis leidyi* enhanced the response, which is unexpected because *Mnemeopsis* does not hunt by sight. McKelvey and Forward (1995) considered the phenomenon a contradiction to the hypothesis that the adaptive significance of DVM is avoidance of visual planktivores. Lass and Spaak (2002) wrote that the reaction of *Artemia* is inconsistent with the idea that anti-predator responses are very specific. They probably had the specific difference in response to fish and invertebrates, like *Chaoborus* and *Notonecta*, in mind. Specificity of a response is not determined by taxonomy but by the kind of danger. According to Mann and Swift (1988), *Mnemeopsis* predate near the water surface. If a diel predation pattern is also present, the hypothesis that DVM is a strategy to avoid *visually* hunting predators has to be extended into a strategy to avoid predators that operate in daytime in well-lit surface water. We must know more about the activity pattern of the Ctenophore.

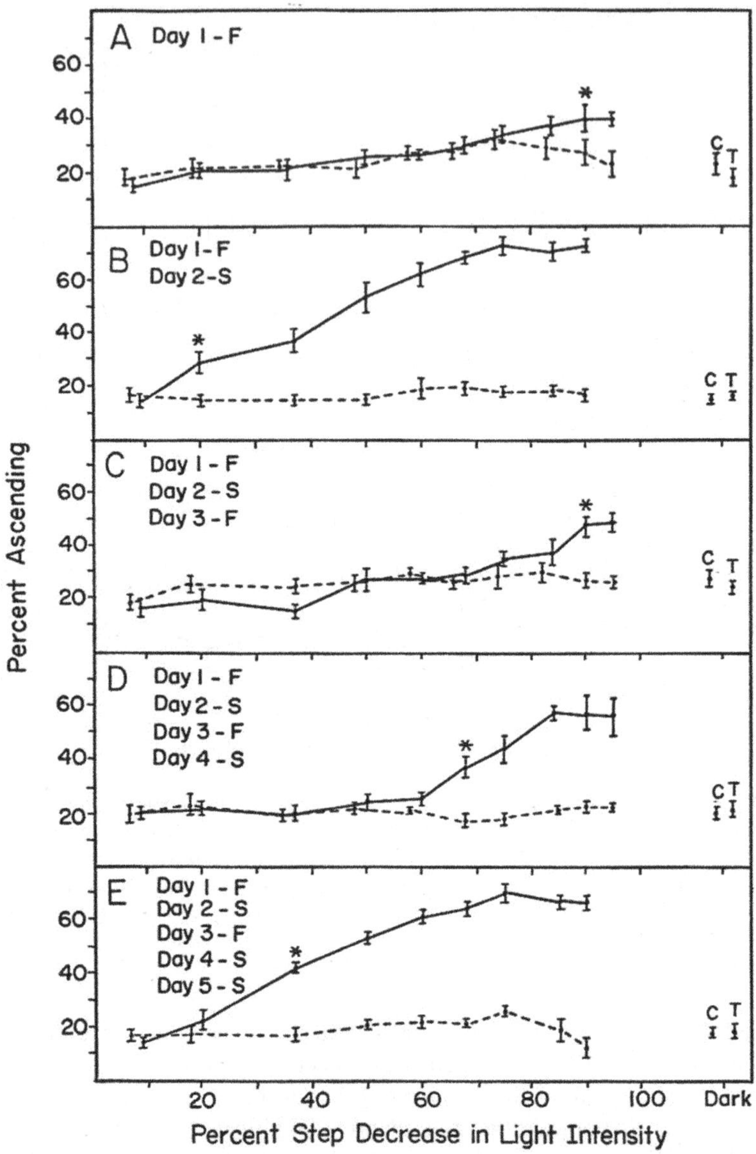

Fig. 4.10 Percentage of *Artemia* larvae showing an ascent response upon step relative decreases in light intensity. Conditions prior to testing (F = fed and S = starved) are shown in each panel. Dark: Control(C) and test (T) responses with light intersity decreasing to total darkness. Further information as in Fig. 4.9. After Forward and Hettler (1992) (Reproduced by permission of the American Society of Limnology and Oceanography, Inc.)

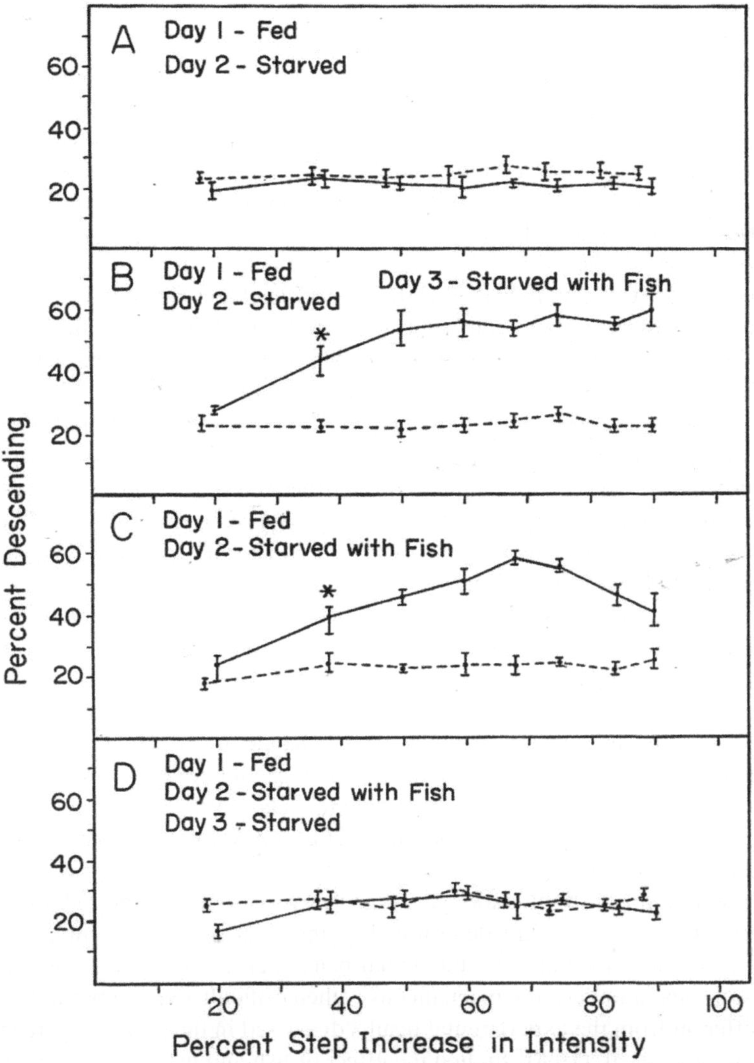

Fig. 4.11 Percentage of *Artemia* nauplii showing a descent response upon instantaneous relative increases in light intensity and various pretreatment conditions. After Forward and Hettler (1992) (Reproduced by permission of the American Society of Limnology and Oceanography, Inc.)

4.4 Other Kinds of Experiments to Study the Role of Kairomones and Food

The experiments with *Daphnia* and *Artemia* (and also larvae of the crab *Rhithropanopeus harrisii*, Forward et al., 1984; Cronin and Forward, 1980; Forward and Rittschof, 1999b) were done with the preconception that light changes are the proximate cause of swimming responses and that fish kairomone is possibly a modifying factor. Other factors were kept constant and thus a physiologically oriented research approach is recognised. I have long wondered why this approach is rare in migration research. A reason might be that these types of experiments are considered too far away from a natural situation. Reactions are of short duration and considered irrelevant to migrations that last for a few hours. Moreover, the physical circumstances in a short cylinder are not comparable to those of a water column in a lake or ocean. On the other hand, making experimental conditions as "natural" as possible is not feasible and the applied tricks may lead to a high degree of artificiality going unrecognised. For example, Harris and Wolfe (1956) used Indian ink in a short cylinder to create light intensity differences between surface and bottom comparable to many metres in a lake. The resulting turbidity was very high, of course, with attenuation coefficients of 1–1.8 m^{-1} or 17–34 times those found, for example, in mesotroph Lake Maarsseveen. A very high change in light intensity over a short vertical distance is present and must be perceived by a daphnid when it swims spontaneously or reactively upwards or downwards. Consequently, the feedback generated by the swimming response is very high and inhibits swimming (consult the information flow chart in Fig. 3.2).

Dawidowicz and Loose (1992) compressed a temperature difference of 10°C within 15 cm, whereas in a lake this would extent over a metalimnion of approximately 5 m. The temperature curve certainly looks like the curve of a stratified lake and divides the cylinder into an "epi-, meta- and hypolimnion". This is the appearance, however, because the naturally occurring change in temperature of 0.02°C per cm is increased by 33 times. An individual animal cannot be compressed, and swimming through such dense temperature gradient leads to very unnatural changes per unit time realising strong stimuli that may act as a physiological barrier inhibiting vertical displacements.

If the experimental light–dark cycle extents over a long time period, often 24 h, and vertical distributions are determined at intervals of a few hours, results of responses are described and not the actual behaviour. As with field data, drawing conclusions about underlying mechanisms is then difficult, even impossible. This is quite different from the experimental results discussed in the previous paragraph.

Dawidowicz et al. (1990) studied the effect of fish and fish kairomone on fourth instars of *Chaoborus flavicans* in a cylinder with a halfway temperature gradient of 8°C over an unmentioned but probably short distance. The light period lasted for 9 h and the dark period for 15 h and the transition was in steps over a period of 30 min. In the absence and presence of fish or fish kairomone, the average individual was higher in the cylinder during the dark period (Fig. 4.12). In the presence of fish chemicals the day–night differences in depth became larger. The authors concluded: "It appears that the stimulus for vertical migration is chemical, not visual or mechanical" (p. 163).

In my opinion, the light changes were responsible for the upward and downward movements, while the kairomones enhanced these photoreactions. I also think that the strong temperature gradient prevented a re-distribution over the column as soon

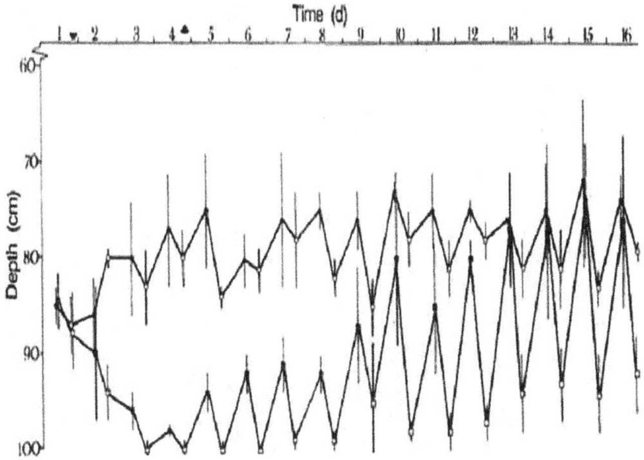

Fig. 4.12 Vertical distribution (mean depth ± 1 SD) of *Chaoborus* larvae in fish-free (*dots*) and caged-fish (*squares*) aquaria during the day (*open symbols*) and at night (*black symbols*) over a period of 15 days. *Arrowheads* indicate times fish were introduced and removed (from Dawidowicz et al., 1990; reproduced by permission of the American Society of Limnology and Oceanography, Inc.)

as the changes in light intensity were over, thus establishing a dual migration-like distribution pattern over 24 h.

Similar experiments, also with *C. flavicans,* were performed by Tjossem (1990) and the results were comparable. In the light period and without fish cues, more larvae were up in the water column than in the presence of kairomones. Although small, a migration occurred also in absence of predators or predator kairomones.

In these experiments with *Chaoborus* larvae we probably do not deal with simple, causal responses to changes in light intensity because an endogenous rhythm is also involved. In daytime, the fourth instar is buried in the mud and first has to come out in the evening in order to migrate upwards. Within the mud, light intensity might be too low to perceive an evening decrease and an endogenous rhythm seems to be responsible for emerging (LaRow, 1968). If coming out of the mud occurs when light intensity is still too high or, changes in light intensity are still sub-threshold, the larvae sometimes bury again. Parma (1971) found that larvae, coming out of the mud in the dark, do not swim upwards in the tubes. They float without motion in the water, rest on the mud surface or even bury again. It seems that a stimulus for upward migration is absent.

Interesting experiments with different levels of fish kairomones, temperature and food concentration were performed by Loose and Dawidowicz (1994). The upper half and the lower half of a 100-cm-long tube was separated by a temperature gradient of about 6°C over 17 cm. "Sunrise" and "sunset" were imitated by changing the overhead light intensity gradually over a period of an hour. A flow of fresh medium through the tube realised a more or less constant food concentration and kairomone level in the upper and lower half. At low kairomones levels, *D. magna*

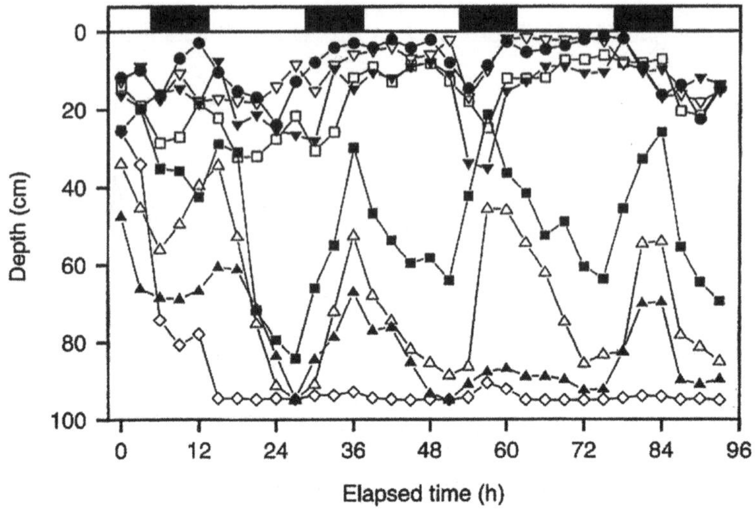

Fig. 4.13 The effect of eight different fish kairomone concentrations on the position of *D. magna* in a cylinder with a strong mid-way temperature gradient. Bars at the top reflect the day/night rhythm. The symbols represent the control with no-fishwater (*black dots*); kairomone level 5000 l per fish (*inverted open triangle*); 1000 l per fish (*inverted black triangle*); 500 l per fish (*open square*); 100 l per fish (*black square*); 50 l per fish (*open triangle*); 10 l per fish (*black triangle*); and 5 l per fish (*open diamond*). After Loose and Dawidowicz (1994) (Reproduced by permission of the Ecological Society of America)

remained in the upper half of the cylinder but at higher concentrations moved down with "sunrise" into the lower half to rise again after the light intensity decreased (Fig. 4.13).

Although the authors did not make this conclusion, the diel change in average depth must have been caused by the changes in light intensity. Without the temperature barrier, dispersion over the whole tube would probably have occurred after light intensity was constant again. Now the strong difference in temperature halfway the tube made that the animals stayed in the compartment where they had been driven to by the light stimulus. Therefore, the experiment is also about the opposing effect of two stimuli, i.e. changes in light intensity and changes in temperature. The latter are experienced while swimming downwards or upwards through the gradient. The results showed that fish kairomone enhanced the phototactic swimming response because with increasing concentration more animals overcame the counteracting temperature stimulus and moved through the barrier. Loose and Dawidowicz (1994) did these experiments to assess the costs of predator avoidance, in particular the effect of an extended time at a lower temperature. They were not concerned with causal factors as light or temperature changes that must have been responsible for the displacements and the maintenance of the diel distributions, respectively. They either took causation for granted or believed that only the fish kairomone was responsible for the "migrations", as Dawidowicz et al. (1990) did. To a certain extent this is true because the differences in distributions depended on the concentration of the fish chemical. However, with fish kairomone, a constant light intensity and no temperature gradient, the results would not have been obtained.

There seems to be a difference in opinion about the role of light stimuli and predator kairomones in causing DVM. The line of division is drawn by the preference for proximate or ultimate aspects and, thus, between physiologically oriented or ecologically oriented researchers. Physiologically oriented researchers analyse behavioural responses of individuals to controlled changes of the environment and relate causes with effects. They thus try to describe how physiological–behavioural mechanisms realise DVM. From the result of their analysis was concluded that light changes causes upward or downward swimming, which can be modified by environmental factors such as fish kairomones and food concentrations. Ecologically oriented researches study the *results* of behaviour and try to relate these outcomes to the extent of adaptation in different environmental settings. Their interest is in why a particular behavioural result was evolved. From the result of their studies was concluded that predator evasion is the principle adaptive reason to migrate. Since DVM diminishes the availability of food, the outcome of migration behaviour is studied for optimality.

In the next paragraph is discussed how the behavioural mechanisms realise possible optimal migrations.

4.5 A Decision-Making Mechanism

In the previous paragraphs, it was shown that fish kairomones enhance swimming reactions to changes in light intensity. Experiments with lake water strongly suggest that a chemical is sometimes present in the epilimnion, enhancing phototactic reactions in the same way as in the photobehaviour experiments (see Chapter 9). Therefore, evidence is sufficiently strong to say that kairomones, mediated by visually hunting fish, have to be present to make large amplitude diel vertical migrations in lakes possible. It is known that migration amplitudes differ from lake to lake and even vary with time within one lake. Migration is costly and the variable amplitudes can be considered adaptive. However, no systematic set of quantitative data about costs and benefits coupled to amplitude is available. Predator avoidance is traded-off with less food and to make the best of both, flexible behaviour is necessary. A logic deduction from the presence of environmentally induced plastic behaviour and the resulting differential migrations is a supposed ability of an individual to "choose" from different alternatives. This leads to the hypothesis of a neural, decision-making mechanism fed by relevant information about the environment.

Predator–prey interactions have got much attention and many papers have been written about the subject reviewed, among others, by Lima and Dill (1990), Dill (1987), Lima (1998) and Kats and Dill (1998). The discussions happen predominantly on an ecological level and it is not always clear that anti-predator decision making must be based on a physiological–behavioural mechanism. For example,

Lima (1998) writes "that animals might actually use some *m*/*f*-like rule in their decision making" (p. 222). This rule states that the rate of mortality *m* per unit feeding rate *f* has to be minimised in order to maximise fitness. The formulation can be used as a guiding principle for ecological research but has to be reformulated for an analysis of decision-making behaviour. Dill (1987) gives another example of decision making in the ecological context as he writes that animals ought to behave as if they compare costs and benefits of behavioural alternatives and trade them off in reaching their decisions (p. 803). Also Blumstein and Bouskila (1996) wrote that "decision making includes weighing costs and benefits of alternative behaviors..." (p. 570). This is all very well but goal-directed reasoning. Costs and benefits become apparent only after some time, for example, a higher or smaller fecundity in the course of *ecological time*. Costs and benefits represent ultimate aspects of behaviour and cannot be part of a proximate mechanism. Our present concern is to explore decision making as a purely physiological–behavioural activity in *real-time*. In Fig. 4.14 a flow scheme for such decision-making process is presented. It starts with sensory perception of relevant environmental factors, is followed by assessment of obtained information in the central nervous system and the end is behaviour. We deal with the sensor-actor chain of Fig. 1.1 again. Physiological–behavioural mechanisms like that were shaped by selection in *evolutionary time* and the result can be studied now

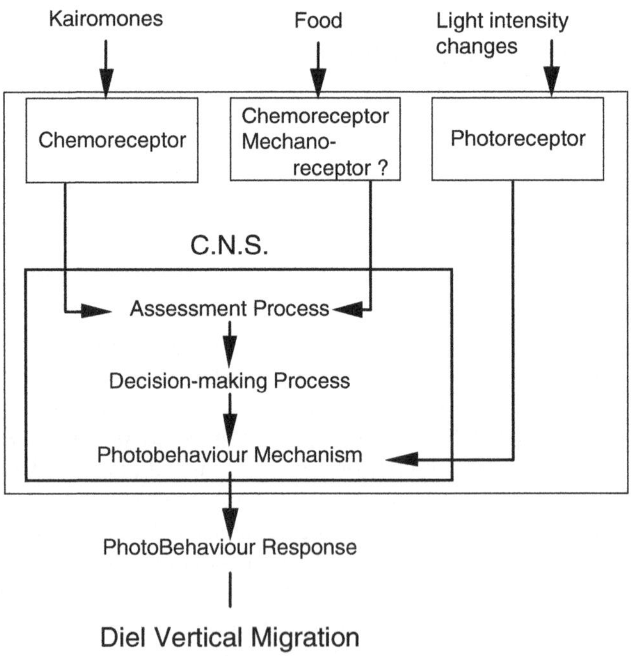

Fig. 4.14 Simplified flow scheme of the decision-making process in *Daphnia* leading to diel vertical migration

in *real-time*. How are the relevant, information conveying factors from the environment to be formulated qualitatively and quantitatively? The "risk of predation" is not a perceptible factor but chemicals, mediated by fish or other predators, are. Differential behaviour makes plausible that *Daphnia* and other species can estimate the concentration of fish kairomones. Since this concentration is correlated with total fish biomass (Ringelberg et al., 1997; Van Gool and Ringelberg, 2002; see Chapter 9), and fish abundance determines predation risk, kairomone concentration can be thought to represent a "risk of predation". Also food availability can be perceived, for example, as blood sugar concentration (satiation/hunger) or as pressure perception in gut filling. Both information conveying environmental factors modify the response to light intensity changes (Fig. 4.6). Finally, a reaction results, which probably is a response that leads to an adaptive migration that is a migration tuned to the actual situation in the lake. Of the processing in the central nervous system, nothing is known: the model is based on first physiological principles and experimentally derived responses and the animal, represented by the outer square in Fig. 4.14, is a black box. Since one particular behaviour out of a set of possible alternatives is performed, it can be said that the individual has made a decision. Quantification is essential to "decision-making" and to make a mechanistic model, it is also necessary to formulate mathematically the quantitative relations. Although some results are preliminary, I thought it necessary to have a try at formalisation and to indicate the route of future DVM research, as I see it.

In the experiments by Van Gool, described in Section 4.3, different kairomone levels and food concentrations were used and accelerated light intensity changes were applied to evoke phototactic swimming reaction (Van Gool and Ringelberg, 1998). The results are laid down in the two regression lines of Fig. 4.6. The functions describing these lines (see legend) were rearranged[4] and used to formulate part of a decision-making algorithm:

$$\mathrm{PC} = 23.65 + 1.73k + 58.46f + 1.84k \times f \tag{4.1}$$

where k = kairomone level and f = food concentration. The value of the ordinate when $f = k = 0$ is the value of the regression coefficient in the relation between displacement velocity and the relative change in light intensity, found when relative changes in light intensity are given without accelerations. The contribution of the factor food and kairomone depends, of course, on the concentration of these factors. The contribution of the interaction factor is small and also an analysis of variance (ANOVA) of the PC values in Fig. 4.6 revealed that interaction was not significant and food and kairomones are thus independent albeit additive factors (Van Gool and Ringelberg, 1998).

Predator avoidance in *Daphnia* is relatively simple. In fact, DVM is predation *prevention*, although an enhanced escape reaction to the sucking force of a pipette has been described in animals that had been for some time in "fish water" (De Meester and Pyanowska, 1997). Nevertheless, small (starting at about 15 mm length) 0^+ perch eat adult daphnids readily and migration is a more reliable way to avoid predation.[5] Food and kairomone concentration determine decision-making but temperature also influences the Phototactic Coefficient that determines displacement velocity. At a food concentration of 0.5 mg C l^{-1} but without fish kairomone, PC was determined at various temperatures (Van Gool and Ringelberg, 2002). The result is presented in Fig. 4.8. The following equation describes line A:

$$\mathrm{PC} = -57.39 + 8.92\,T - 0.16\,T^2 \tag{4.2}$$

According to this temperature function, PC = 49.11 at 17°C. This compares reasonable well with a PC = 52.88, which can be calculated from the decision-making algorithm at a concentration of 0.5 mg C l^{-1}. The experiments at the base of this algorithm were performed at 17°C. The phototactic coefficient PC increases rapidly from zero at 7.8°C to 70 at 25°C and then levels off. The curve is the same for relative light intensity increases and decreases. The curve is probably different for other food concentrations. At the lower food concentrations of 0.2 and 0.05 mg Cl^{-1}, PC was determined also at 17°C. The values[6] are given in Fig. 4.8 and two curves, parallel to the original one, were inserted. The quadratic curves start with PC = 0 at 7.4, 9.4 and 10.4°C, respectively. This suggests that no DVM is possible at lower temperatures. With increasing temperature in spring, PC rapidly increases. A seasonal increase in amplitude of migration is often observed in lakes, although the increased abundance of juvenile fish is a factor not to be dismissed. Some authors have stressed the role of absolute light intensity and temperature in determining vertical distribution, in freshwater as well as the marine environment (Moore, 1955). The PC-temperature curves show of what nature the functional relationship of temperature might be. However, both factors have no obvious information conveying properties (the role of water clarity is discussed in Chapter 5) and, thus, are not part of the predation-prevention strategy. However, Geller (1986) brought DVM and temperature in a causative relation. He found in Lake Constance that several species, migrating or not, occurred within rather limited, species-specific temperature ranges during the summer season of changing thermal stratification. He thought that temperature (and food) caused and controlled DVM in this lake. The adaptive significance of DVM should be a maximisation of metabolism (see also McLaren, 1963, 1974, in this respect) and the predator avoidance hypothesis was rejected. Geller's opinion is not shared by most authors, for example, Stich and Lampert (1981) and Lampert (1989, 1993) who based their opinion also on a study of DVM in Lake Constance. Nevertheless, Geller's careful study of vertical distributions deserves attention. With the mechanistic model, presented in Chapter 5, the temperature influence can be explored further.

Values of PC calculated from the decision-making algorithm but holding for temperatures different from 17°C should be corrected for temperature. The following correction is suggested: the PC = 49.11 at 17°C is set to 1. The temperature correction factor becomes

$$T(\mathrm{corr}) = \frac{-57.39 + 8.92T - 0.16T^2}{49.11} \tag{4.3}$$

If the temperature function depends on food concentration, the correction factor also depends on food concentration. The equation that includes food is an extension[7] of (4.3)

$$T(\text{corr}) = \frac{-78.06 + 41.08F + 8.92T - 0.16T^2}{28.45 + 41.08F} \tag{4.4}$$

The correction at a food concentration of $F = 0.5$ mg C l^{-1} and a temperature of $T =$ 17°C is $T(\text{corr}) = 1$, as it ought to be. The high algal concentration of 0.5 mg C l^{-1} is not often come across in unpolluted lakes. Therefore, the (4.4) is of importance to correct PC. It will be used in modelling DVM in Chapter 5.

Notes

1. The relation from which the lower line was calculated is DV = 1.63 + 31.34 S. (Van Gool and Ringelberg, 1997). Example: let $S_1 = 0.15$ and $S_2 = 0.3$, then $S_2/S_1 = 2.0$; $DV(S_1) = 6.33$ and $DV(S_2) = 11.03$, so $DV(S_2)/DV(S_1) = 1.74$. Compare with $DV(S_1, S_2)/DV(S_1) = 2.06$.
2. An acceleration is defined as a difference in velocity per unit time, with dimension $l\ t^{-2}$. The result of moving at a particular speed is a distance covered. Relative changes in light intensity are velocities with dimension $(\Delta I\ I^{-1})\ t^{-1}$ and accelerations are likewise second derivatives with dimension $(\Delta I\ I^{-1})\ t^{-2}$.

   However, the result is a physiological reaction and not comparable to a distance. Preliminary results indicate that the result of an acceleration, displacement velocity, is not independent of the value of the first stimulus. Therefore quotients of stimuli and not differences were used to study the reactions and the latter were likewise represented as a ratio.
3. "Real" accelerations, as defined in physics as increasing differences per time of velocities, are presented in Figs. 6.8 and 6.9. Dimensions are s^{-2}. The increases in rate, used in the experiments, are stepwise percentages and without dimension.
4. PC at different experimental kairomone concentrations (Fig. 4.6) was made functions of food concentration. Two problems arise due to the lack of data: 1. only two food concentrations were studied, 2. the highest concentration of 0.5 mg C l^{-1} is above the insipient limiting concentration of ILC = 0.26 mg C l^{-1} (Mück and Lampert, 1984). From this ILC onwards, food uptake is constant and thus, the food condition of the *Daphnia* does not alter. If information about the environmental food condition is derived from, for instance, the amount of food in the gut, the phototactic coefficients at different food concentrations would not be different above the ILL. The value of B in PC = $A + B \times$ food concentration would become larger in that case. If, however, information about food in the environment would be obtained from the number of food particles arriving in the food groove of the animal, a continuous increase with the environmental algal concentration would hold and the straight line (?) between the lowest and highest experimental concentrations would be valid. This was assumed, which led to the two functions for A and B of $A = 23.65 + 1.73 \times \text{kc}$ and $B = 58.46 + 1.84 \times \text{kc}$ and, finally, to the presented decision-making algorithm. PC and thus the enhancement of the phototactic swimming reaction are underestimated compared to an algorithm based on a constant PC for food concentrations higher than the ILL. Therefore, also the amplitude of a DVM, based on the used algorithm, will be smaller.
5. This even holds for calanoid copepods, although some marine calanoids, provided with myelinated sensory and motor axons, and thus having shorter reaction times to brief hydrodynamic disturbances, do not migrate. On the other hand, species without myeline sheath and longer reaction times, for example, *Pleuromamma xiphias*, are extensive migrators (Hays et al., 1997; Lenz et al., 2000).

6. At 17°C, a PC = 36.30 is available for a food concentration of $F = 0.2$ mg l^{-1}, and at the same temperature but for $F = 0.05$ mg l^{-1}, two values from different experiments viz. PC = 31.91 and 29.57 were averaged and inserted in Fig. 4.7.
   Temperature for which PC = 0 depends on food concentration. The three values $T = 7.4$, 9.4 and 10.4 (at $F = 0.5$, 0.2 and 0.05 mg l^{-1}) lie on a line $T = 10.733–6.667\ F$ with a perfect fit ($R^2 = 1.00$).
7. The value of the ordinate if $T = 0$ depends on food concentration. The three values available for PC (at $T = 0$) are –57.393, –70.20 and –75.76 (at $F = 0.5$, 0.2 and 0.05 mg l^{-1}). A line with function $X = 78.055–41.083F$ ($R^2 = 0.999$) can be drawn through these points. X can be substituted in the general form of the temperature correction function:

$$T(\mathrm{corr}) = (-X + 8.9229\,T - 0.15636\,T^2)/(-X + 106.50)$$

This leads to the correction function with food included:

$$T(\mathrm{corr}) = \frac{-78.055 + 41.083\,F + 8.9229\,T - 0.15636\,T^2}{28.45 + 41.083\,F}$$

as presented in the text.

Chapter 5
Mechanistic Models

Contents

5.1 Introduction

"What makes models so fashionable?" is the first sentence of an excellent essay by Gabriel (1993) on the use of models in studying DVM. As is usual with fashionable things, it is difficult to give a single answer. The incorporation of mathematics in one's research has the scent of sophistication and it is a "must" in grant proposals to announce that a model will be made of the results yet to obtain. And DVM is an attractive topic because the phenomenon seems to be simple. Most modellers divide the water column into two parts that supposedly represent clearly distinct habitats, the epilimnion and the hypolimnion. On a daily basis, zooplankton are supposed to migrate from one part to the other. If the factors of importance for the life history of the migrating animals, such as food, predation and temperature, are differentiated between the parts, the adaptive advantage of either staying in a layer or migrating between them is easily formulated, especially so because quantitative data for conformation are scarce. Thus, the ultimate aspects are fashionable research topics and most models deal with those. However, I like to present a model addressing the proximate aspects, based on physiological–behavioural mechanisms. The model describes *how* zooplankton migrate, not *why*. Proximate mechanisms operate during the relatively short periods of transition from one depth zone into the other. The scarce mechanistic models available are mostly not based on empirically obtained data that elucidate the mechanisms involved. For example, as long as the vertical displacement velocity of zooplankton is made a function of the changing

DOI 10.1007/978-90-481-3093-1_5,

light intensity at dawn and dusk, a picture emerges that mimics DVM qualitatively. The three models of light-mediated migrations presented by Richards et al. (1996) may illustrate this. The authors write that "...the exact ways in which individuals react to it (i.e. *a change in light*) are yet to be completely determined". I hope that the previous chapters have been sufficient for the reader to frown at the absolutism of this statement, but to make up for their lack of knowledge, Richards et al. (1996) go on to make various assumptions. The first one was that animals stayed at a preferred light intensity throughout the day. Here the old paradigmatic explanation of the first half of the last century is pulled out of the cupboard again. This results of course in a perfect vertical migration because the animals are pinned to one particular isolume that moves up and down with physical precision during the day.

Since cybernetics invaded technology as well as biology some 50 years ago, we know that keeping to a certain factor value means responding adequately to a deviation from that value. Such mechanism was in fact at the base of the second model proposed by Richards et al. (1996): individuals were assumed to minimise deviations from a current light intensity. The principle of a " just noticeable difference", discussed in Section 3.3, is applied here: "for an individual to notice that the level of light has changed means that it has moved off the isolume" (p. 2208). In this model, the current light intensity is not a preferred one, as in the first model, but an arbitrary one depending on where an animal happened to be at some point in time. The difference between the two models is not large: in both the animals try to stay at an absolute light intensity. However, the pattern obtained with the second model is less clear-cut and the migration pattern becomes fuzzy.

In the third model, zooplankton is assumed to respond to the rate of relative changes in light intensity. This sounds familiar: vertical swimming speed is assumed to be proportional to the difference between the observed relative rate of light change and a rheobase or threshold, as proposed by Ringelberg (1964). Since the rheobase is a constant, this corresponds to the stimulus concept that has been worked out in Section 3.3. With these elements, the simulated migration is of maximum fuzziness (Fig. 5.1). Concentrations of zooplankton only occur for short periods after dawn, when the day depth is reached and also after dusk when surface layers are reached.

The authors are reluctant to make a choice. They reason that insufficient descriptions of real migrations are available: "little data that are of sufficient resolution in both space and time have been recorded to compare fully with the results of the model" (p. 2216). There have been several continuous registrations of migrating sound scattering layers (see Chapters 10 and 11). However, I think that even high-frequency observations may never make a definite choice possible. If a migrating SSL in nature remains closely associated with an isolume all the time, the assumptions underlying both first and second model would seem justified. However, also the third model might still be right because the response mechanism to relative changes in light intensity is sometimes adjusted in such way that the animals remain within a small range of light intensities. When a migration pattern cannot be associated with a particular light intensity, the conclusion can only be that models one and

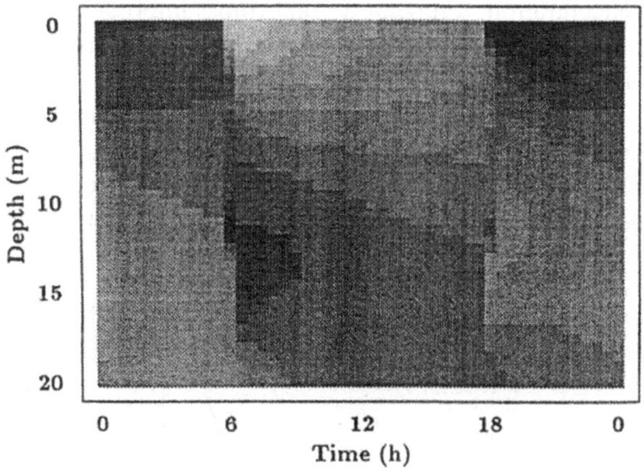

Fig. 5.1 Simulation by a model of Richards et al. (1996) of the daily movement of a zooplankton population when animals respond to relative changes in light intensity. Outside the periods of maximal displacement, a dispersion factor is responsible for the diffuse pattern. Concentrations occur at the end of the morning descend and after return to the surface in the evening (Reproduced by permission of Oxford University Press)

two are inappropriate. The literature on migrations of SSL shows that moving more slowly than an isolume occurs often but even moving faster than an isolume has been observed and this poses a problem, of course. Han and Straskraba (1998) remarked that none of these models of Richards et al. (1996) make it possible to understand the general control mechanism underlying widely varying DVM because predation pressure has not been taken into consideration. Richards et al. (1996) were aware of this limitation and listed several environmental factors of possible importance: "in conclusion, the patterns of vertical distribution of zooplankton observed in the field may be generated by many mechanisms" (p. 2218). We seem to be back to square one, but I doubt whether it is as bad as that.

The distinction between ultimate and proximate aspects is not always clear-cut and several published models have a hybrid character. The one developed by Han and Straskraba (1998) may serve as an example. They consider light intensity and predation pressure essential ingredients for the realisation of DVM and the central property of their model is the product of both, which they call the "realised predation pressure". There is logic in this approach because if predators are abound but light intensity is below the visual threshold, the "realised predation pressure" is nil. To make the model a mechanistic one the next assumption is that an animal senses only part of the "realised predation pressure" and they call this part the "sensed predation pressure". We recognise an echo of Von Uexküll's (1921) ideas about the "Umwelt" and the sensorially perceived part of it, the "Merkwelt" (see Section 1.2). As abstractions these two notions may be valuable but their meaning in physiological terms is difficult to grasp. In reality, the authors appear to have sensory physiology in mind because the perceived predation depends on the sensitivity of the prey species to the presence of predators. To identify a physiological base, chemoreception of fish kairomones presents a good candidate. Also food conditions and temperature were considered factors of influence: "...when food is limited, zooplankton tend to pay more attention to food than to fish predation" (p. 1464). Thus the calculated migration velocity depends on changes in light intensity, food concentration, temperature and, most importantly, on the "sensed predation pressure". With this mix the model was used to simulate different amplitudes of DVM under changes in light intensity during 24 h.

The effect of food concentration on migration amplitude is illustrated in Fig. 5.2. With increasing food concentration, the animals migrate deeper, as is to be expected if swimming velocity is made a function of food concentration. The model is supposed to be a mechanistic one but is in fact not based on established physiological

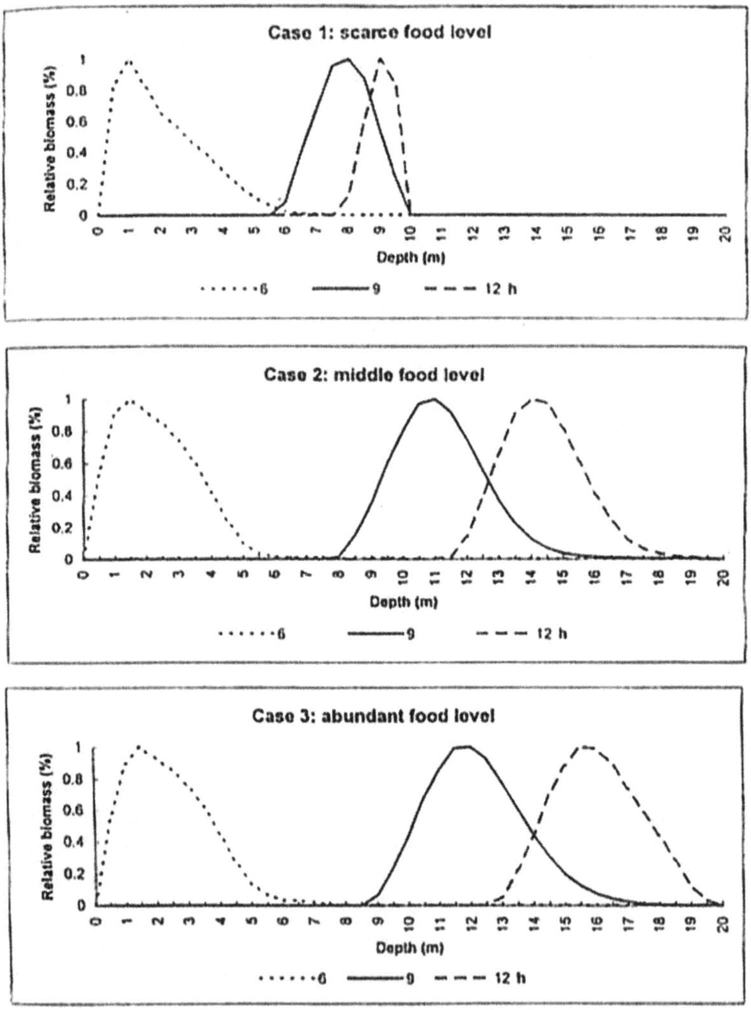

Fig. 5.2 Simulation by a model of Han and Straskraba (1998) of the vertical distribution of relative biomass of migrating animals at different times of the day and for three different food levels. Food is decreasing linearly with depth and food concentrations at the surface are 0.2, 2.0 and 3.0 mg C l^{-1}, respectively (reproduced by permission of Oxford University Press)

and behavioural mechanisms. The way in which water clarity is used may illustrate this. Water clarity is (indirectly) part of the visual system of the predator and thereby predation success, but is not used as input for the visual system and thereby swimming velocity of the migrating animal, as one would expect from the effect of light attenuation on photobehaviour of *Daphnia* (as discussed in Chapter 3).

A formal description of the physiological–behavioural mechanisms is an important precondition of a good mechanistic model, but requires sufficient knowledge before it can be used in models. If too many assumptions have to be made, the mark is easily overshot. As long as displacement velocity is made a function of light-intensity changes during sunrise and sunset, changing vertical positions of the animals will always mimic DVM qualitatively but the model will have no added value and nothing new will be revealed. In the models developed in the next paragraphs, the empirical knowledge of the physiology and behaviour of *Daphnia* determining its swimming reactions in response to light changes represents the core element. Model development is split in three successive parts, the first dealing with "normal", non-reactive swimming as discussed in Chapter 2. This model is then extended by the incorporation of phototactic reactions to changes in light intensity because of the link between photobehaviour and migration (Chapter 3). Finally, the decision-making mechanism (Chapter 4) is added to model DVM under a broader set of environmental conditions. Although a few assumptions have to be made, these have been kept to a minimum.

I used Stella Software version 4 (High Performance Systems, Inc., Hanover, NH, USA) to merge the available experimental data into one functional model system. Lists of the equations and constants used together with comments are presented in the Appendix for the phototactic behaviour and the DVM model, respectively.

5.2 A Model of Oscillating Swimming

The alternating upward and downward displacements of *Daphnia* often observed under constant conditions have been the argument for choosing a coupled oscillator as the basic mechanism underlying swimming (Ringelberg and Lingeman, 2003). Imagine two "neural groups" in the central nervous system that either inhibit or facilitate the action of the other group, thus generating oscillations in the output of both. Such networks of coupled neurons are supposed to exist in physiology and are called "central pattern generators" (Strogatz and Stewart, 1993). The extent of interaction between the neural groups, called the coupling strength, determines the frequency of the rhythm. Our model has to describe not only "normal" swimming but also has to allow for the variety of frequencies and amplitudes of positive and negative phototaxis that has been described in the literature. This can be realised by changing the value of one parameter only, viz. the strength of the coupling of the two "neural groups". Figure 5.3 provides a flow chart of the most important model features.

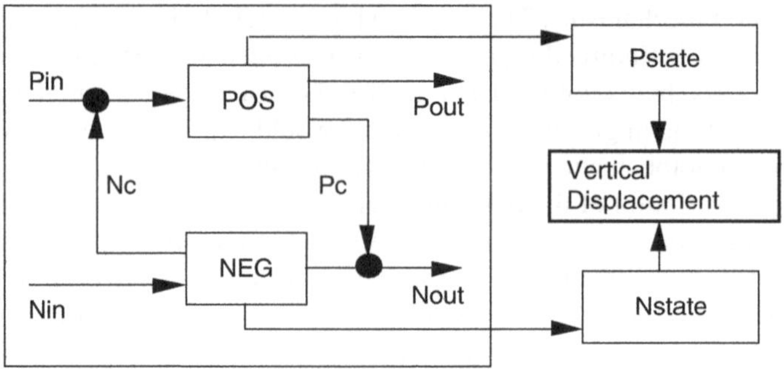

Fig. 5.3 Flowchart of a mechanistic model of small-scale, oscillating swimming in *Daphnia*. The central part, within the large square, consists of two blocks that represent the origins of the upward phase (POS) and the downward phase (NEG) of this swimming. Both have a non-rhythmic input and output. The blocks are connected such that NEG facilitates POS and POS inhibits the action of NEG, which leads to an oscillating signal in both blocks. Capacitor-like units (Pstate and Nstate) are loaded by the oscillator blocks and determine the direction and velocity of the vertical displacement (after Ringelberg and Lingeman, 2003)

The large square encases what represents the "central pattern generator. POS and NEG are the fictitious "neural groups" with non-rhythmic input and output and if they work independently, no oscillating swimming occurs. Oscillations arise if both are coupled. In Fig. 5.3, NEG facilitates POS through the connection Nc with the input of POS: if the value of NEG increases, the value of POS increases. However, if the value of POS increases, connection Pc enhances the output of NEG whose value decreases.[1] Of importance is that by changing the intensity of these couplings (by a factor f_c), a continuous range of frequencies of the emerging oscillations can be realised. The couplings are also responsible for a time lag, and this makes that POS and NEG oscillate with a phase difference. POS and NEG are short for positive and negative or upwards and downwards and are closely linked to positive and negative phototactic reactions caused by changes in light intensity in the photobehaviour model. These upward or downward photoreactive swimmings were described as superimposed on the sinus-like, non-reactive "normal" swimming (see Chapter 2). The neural oscillators are connected to a Pstate and a Nstate, representing the excitatory states determining both the extent and the direction of vertical swimming. The existence of excitatory states can also be inferred from the numerous experiments with changes in light intensity. They function like capacitors, powered by POS and NEG, and emptying gradually at an empirically found rate,[2] as calculated from phototaxis experiments.

The default configurations of the model are based on the analysis of normal swimming of *D. "longispina"* and *D. magna* (Ringelberg, 1995). With a period length of $T = 120$ s (see Chapter 2), the coupling intensity factor f_c was given the arbitrary mean value of 1 changed every 2 min using a standard deviation of 0.2. This led to a phase difference of 30 s. The average vertical displacements were calibrated to correspond to the empirically found amplitude of 2.9 cm. Figure 5.4

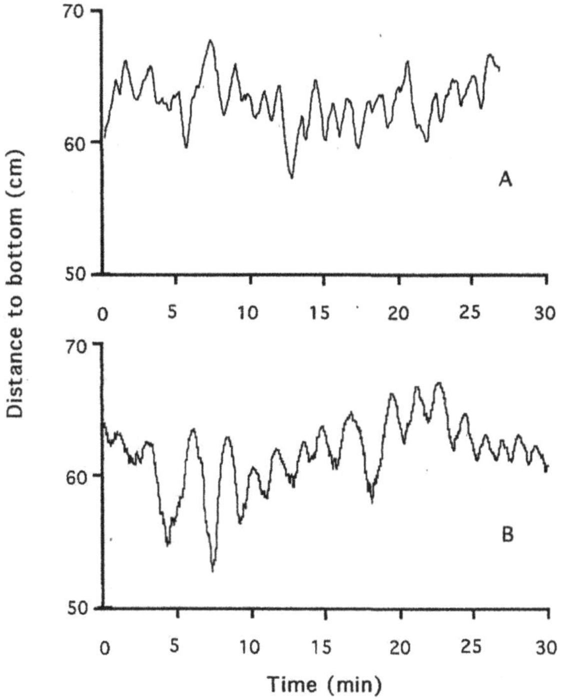

Fig. 5.4 A comparison of a normal swimming pattern of *D. "longispina"* (**a**) and a computer simulation (**b**) with the default value of $f_c = 1$ of the coupling intensity

provides a comparison of observed and simulated oscillating swimming. The model appears to be also capable of simulating aberrant swimming patterns reported in the literature. After roughly estimating period length from the published registrations of *Daphnia* swimming the coupling factor was calculated assuming a fixed relationship between the two parameters.[3] Then, the model was run to simulate the particular swimming pattern observed. Two examples are given in Figs. 5.5 and 5.6. In the first one, a registration of swimming by *D. pulex*, as given by Viaud (1938), is well reproduced with the two coupling factors made $f_c = 0.35$, instead of the default $f_c = 1$. In Fig. 5.6, a peculiar but often observed behaviour of *D. magna* (Clarke, 1932) is reconstructed. In a horizontal tube *D. magna* stays at the side away from the light source during the first minutes of the registration but then suddenly swims to the lighted side where it may remain for more than half an hour and then shifts again to the side away from the light source. A low coupling value of $f_c = 0.03$ was needed to realise this behaviour. Clarke (1932) called this swimming pattern of *D. magna* "primary phototaxis" because it occurred under constant light intensity. The duration of the positive and negative phases varies in different animals and depends on – mostly unknown – circumstances. Daphnids showing this type of behaviour do not respond to environmental stimuli such as changes in light intensity. Therefore, this behaviour may be considered abnormal and I hypothesise that it is due to a temporary neural disorder with the extent of the malfunction reflected

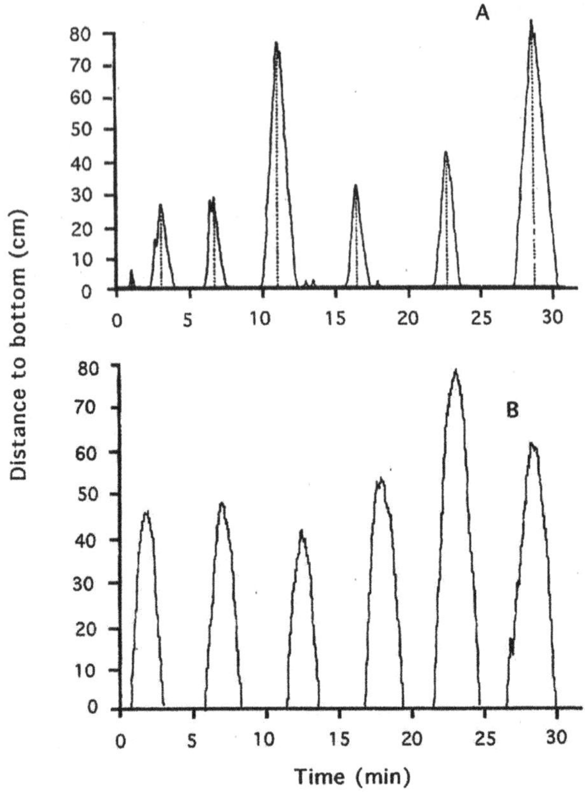

Fig. 5.5 A comparison of a swimming pattern of *D. pulex* (**a**), illustrated in Viaud (1938), and a simulation (**b**) with a coupling factor of $f_c = 0.35$ (from Ringelberg and Lingeman, 2003)

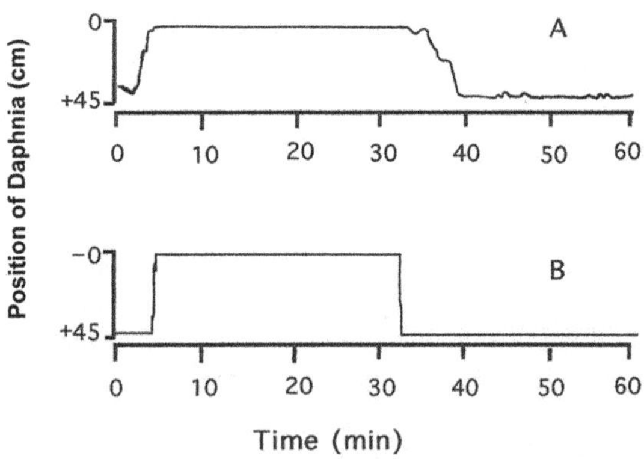

Fig. 5.6 A swimming pattern of *D. magna* (**a**), as illustrated in Clarke (1932) is simulated by the model (**b**) with the coupling factor fixed on $f_c = 0.03$ (from Ringelberg and Lingeman, 2003)

in the frequency of the oscillation. Although it does not look like an oscillation, the swimming in Fig. 5.6 can be considered as such with a long period length of $T =$ 33 min = 1980 s. With a longer observation time and a longer experimental tube, the sinus-like character might have been recognisable. If indeed two neural groups represent the physiological counterpart of the coupled oscillator in the model, the interpretation is that for some reason the mutual influence or coupling of the "positive" and "negative" nuclei would have become temporarily weak. Dumont et al. (1985) isolated individuals of *D. magna* that were either permanently positively or negatively phototactic and these aberrant animals have been studied at length by De Meester (1991, 1993a, b). An important outcome was that the traits seemed to be genetically fixed. However, these observations still need confirmation. Interbreeding of permanently positively phototactic and permanently negatively phototactic animals produced offspring in which the normal behaviour was restored. My suggestion is that these "forced movements" in one direction[4] are the result of defects in the genome affecting either one of the couplings and not an evolutionary adaptation to specific environmental circumstances as supposed by De Meester.

Also other zooplankton species display oscillating swimming (see Chapter 2 for examples). Figure 5.7 compares a swimming track of the freshwater calanoid *Eudiatomus gracilis* with a model simulation. In this case, the default relationship between period time and coupling factor, as based on the analysis of *D. "longispina"* swimming, have to be changed, which is generally true for other species with a different physiology. It also holds for the hybrid *D. galeata x hyalina*, where the coupling factor had to be larger than one (Ringelberg and Lingeman, 2003).

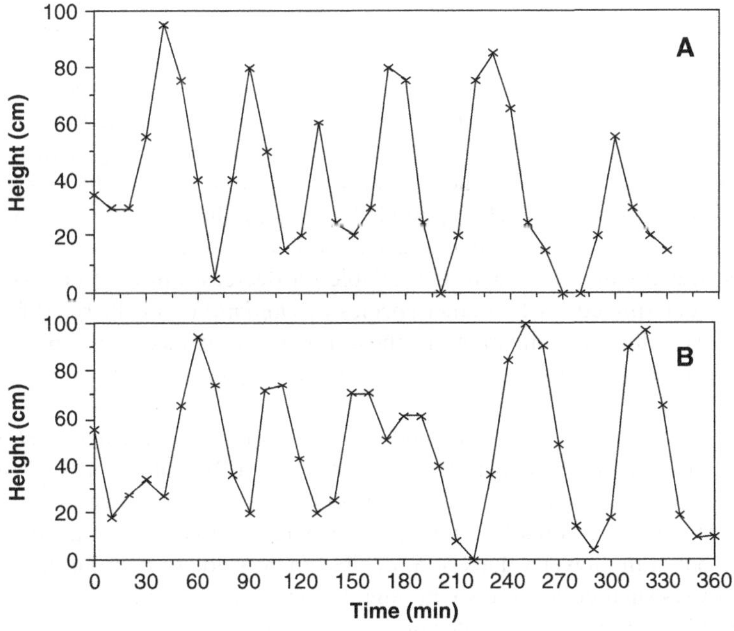

Fig. 5.7 Comparison of a vertical swimming track (Van Gool, not published) of *Eudiaptomus gracilis* (**a**) and a simulation (**b**) with a coupling factor $f_c = 0.037$

5.3 The Photobehaviour Model

Modelling of phototactic swimming under changing light conditions was first done without incorporating the oscillations of normal swimming under constant light because calibration was easier without the variation caused by the latter. The stimulus strength–duration relation, discussed in Section 3.5.2, is the nucleus of the model. Because we deal with two opposite reactions, two excitatory states were assumed, one for relative increases in light intensity and the other for relative decreases. These excitatory states, called POSitive and NEGative, are functionally comparable with those states in the oscillation model. The POS and NEG states are "loaded" at a rate proportional to the value of the relative decrease and increase in light intensity, respectively. Excitatory states disintegrate with time at a rate proportional to the existing quantity and a rate constant. The difference equation for the change in NEG with time is given as example:

$$\mathrm{NEG}(t) = \mathrm{NEG}(t - \Delta t) + [c_1 S - c_2 \mathrm{NEG}(t - \Delta t)]\Delta(t) \quad 5.1$$

The value of the disintegration constant $c_2 = 0.123\ \mathrm{min}^{-1}$ was based on the stimulus strength–duration curve for *D. galeata x hyalina* (Van Gool and Ringelberg, 1997). The proportionality constant c_1 was obtained through comparison of the latent periods of model runs with those of the empirical stimulus strength–duration curve. According to the theory behind this curve, a particular threshold value of the excitatory states must be surpassed before a reaction occurs. Each combination of stimulus strength and latent period leads to this threshold. Because the excitatory states are of a hypothetical nature, we can set the threshold arbitrarily at 1 and run the model to determine the value of c_1 that has the time needed to reach the threshold at different stimulus strengths corresponding with the latent periods observed. Figure 5.8 shows that with $c_1 = 4$, a good fit was obtained between the thus estimated latent periods for six stimulus strengths and the empirical curve as based on experimental data.[5] Although for *D. galeata x hyalina* no stimulus strength–duration curve is available for decreases in light intensity, these have been determined for two other species of *Daphnia* (Ringelberg, 1964, 1993) and they were highly comparable to those for increases (see Fig. 3.6, for example). Therefore, the constants c_1 and c_2 have also been assumed to apply to the POS state.

The next step was to attune the model to the swimming velocities of *D. galeata x hyalina* observed in the experiments (Van Gool and Ringelberg, 1997). Actually, not the swimming speed but the displacement velocities were determined for this species. At the higher stimulus values, when swimming is continuous, the two are the same but at the lower values of *S*, displacement velocity is, from the nature of things, always smaller because swimming is discontinuous. The model would have become rather complicated if interrupted swimming was incorporated. Moreover, because the final objective was to simulate DVM, modelling vertical displacement was preferred.

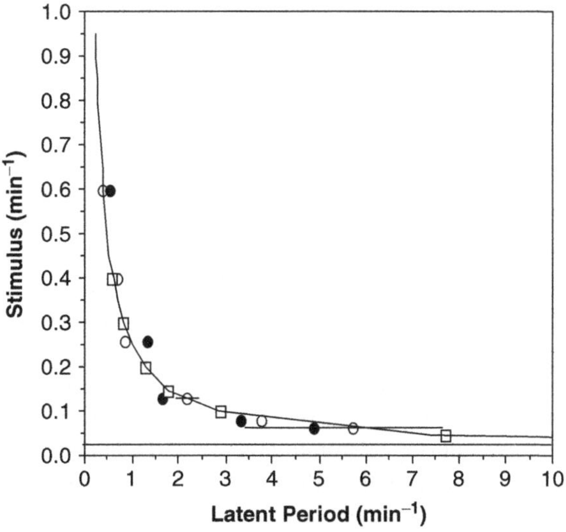

Fig. 5.8 The stimulus strength–duration curve. The model values (*squares*) are compared with the results found in experiments with the hybrid *D. galeata x hyalina* by Van Gool and Ringelberg (1997). *Open circles* represent values obtained in the absence of fish kairomone; *black dots* were obtained in experiments with fish kairomone. The line is drawn according to the equation $\text{LN}(S/(S–R)) = 0.002 + 0.123$ LP, in which S is the rate of the relative increase in light intensity, R is the rheobase $= 0.03 \text{ min}^{-1}$ and LP is the latent period or the needed duration time of S to elicit a swimming reaction. *Horizontal lines* at two points represent 95% confidence limits

The empirical relation between S and displacement velocity DV was found (Chapter 4) to be

$$\text{DV} = \text{constant } 1 + \text{PC} \times S = 1.63 + 31.34S \qquad 5.2$$

where PC is the phototactic coefficient in the absence of specific kairomone enhancement effects.

After the start of a relative change in light intensity, the excitatory state rapidly builds up and as soon as the value 1 is reached, the vertical displacement starts. In the model, DV is not made a function of S but of the difference of the excitatory states of POS and NEG. This is necessary for the incorporation of the previous model of oscillatory swimming because then, both states are activated.[6] Therefore, the relation between the difference of POS and NEG and S was determined first:

$$\text{ExcitatorState} = \text{constant } 2 + A \times S = 0.00058 + 31.67S \qquad 5.3$$

Neglecting constant 2 because of its small value compared to constant 1, the relation between DV and the ExcitatoryState in the model is obtained by substitution of (5.3) in (5.2):

$$\mathrm{DV} = 1.63 + \frac{\mathrm{PC}}{A}\,\mathrm{ExcitatoryState} \quad 5.4$$

In the model, PC/$A = c_3$, while without enhancement of the photoresponse, PC is A. The displacement velocity reaches an asymptote for each value of S and these DV values were compared with the empirically found values. Although the modelled values are a factor 0.91 lower than the empirical values (Fig. 5.9), no further corrections have been made.

In Section 3.4, I argued that the response of an upward or downward swimming of *D. magna* in response to an instantaneous change in light intensity is influenced by the perceived change in light intensity while swimming. If the reacting animal was kept at a constant depth and, thus, no change in light intensity could be perceived during swimming, the reaction time gets longer and the swimming distance larger (Ringelberg, 1964). A simple explanation of this effect is that the change in light intensity perceived during swimming owing to the change in depth leads to an excitatory state that counteracts the effect of the original stimulus. For example, a single, sudden decrease in light intensity causes an increased value of POS. If the assumed threshold of 1 is surpassed, upward swimming follows and consequently the animal perceives an increase in light intensity, which causes an increase in NEG state. Since the difference between the two states determines the swimming reaction, reaction time must be longer without an increased NEG state. The effect depends on water clarity and the product of swimming velocity and attenuation coefficient determines the relative change in light intensity due to reactive swimming. This product is included in the model and therefore the model was run with and without this effect of swimming. The results are largely comparable with the experimental data (Table 5.1). The difference, expressed as the ratio between the responses with

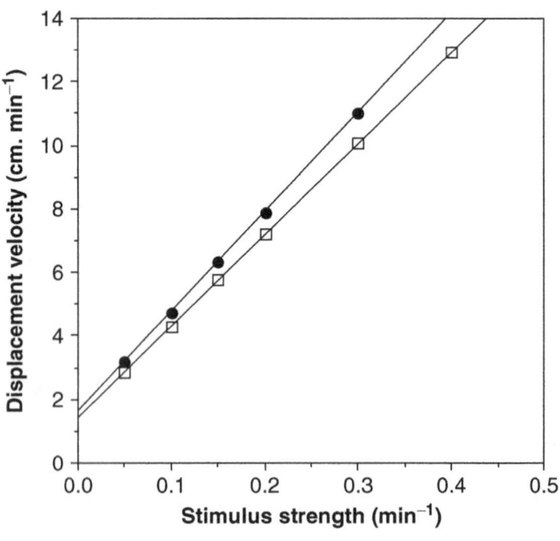

Fig. 5.9 A comparison of the displacement velocity obtained with the model (*squares*) with those from experiments with *D. galeata x hyaline* (*black dots*) (Van Gool and Ringelberg, 1997). The equations, describing the lines are DV(model) = 1.44 + 28.77 S and DV(*Daphnia*) = 1.63 + 31.34 S

Table 5.1 Ratio between reaction parameters in absence and presence of a change in light intensity while swimming. The first row contains experimental data (with an unknown attenuation coefficient) from *D. magna* (Ringelberg, 1964)

Attenuation coefficient	Ratio swimming velocity	Ratio response time	Ratio swimming distance
? (1)	1.04	1.58	1.63
0.004	1.01	1.25	1.25
0.01	0.97	1.67	1.61

and without the feedback, depends on the attenuation coefficient. With increasing turbidity, the ratio gets larger for the response time and for the swimming distance but swimming velocity is not affected when instantaneous light intensity changes are applied.

Because it has been thought for many years that migrating zooplankton followed a certain optimal light intensity,[7] it is of interest to compare the displacement velocity (V_D) of an animal that reacts to changes in light intensity with the displacement of an isolume (V_L). Ringelberg (1964) compared the two by making $(V_L - V_D)/V_L$ a function of the relative decrease in light intensity. A peculiar figure emerged: at a certain value of *S*, the difference seemed to be zero and thus, *D. magna* swam as fast as an isolume moved upwards. This was the first experimental evidence ever for the potential of zooplankton to stay in an isolume. However, for lower and higher values of *S*, the animals lagged behind and thus were found at continuously lower light intensities. Moreover, it seemed that the critical value of *S* changed with the turbidity of the water. To investigate the possible effect of turbidity, the model was run at different attenuation coefficients.[8] In Fig. 5.10, the results are compared with the figure obtained originally with *D. magna* (Ringelberg, 1964). Although the critical point is at a different relative decrease in light intensity and a zero difference between V_D and V_L is not reached, there is some similarity between both figures. Ratios are a bit misleading because values change owing to changes in either the nominator or the denominator. Figure 5.11 gives a better insight into how displacement velocities change with stimulus strength and in the role of the attenuation coefficient. At low relative changes in light intensity and up to a certain value, DV of the model daphnid increases at the same rate as the DV of an isolume. The difference between the two rates remains nearly constant, which means that the nominator in Fig. 5.10 remains approximately the same. Because the denominator increases, the ratio gets smaller up to the critical point. Above this critical value of *S*, the rate of increase of DV becomes considerably less, and thus, the ratio $(V_L - V_D)/V_L$ increases again. Displacement velocity of the model daphnid (and of a real *Daphnia*) is not high enough for the animal to stay at a certain isolume, especially not so for relative changes in light intensity above the critical stimulus value. This critical value depends on the attenuation coefficient: in more turbid water it occurs at a higher relative change in light intensity.[9]

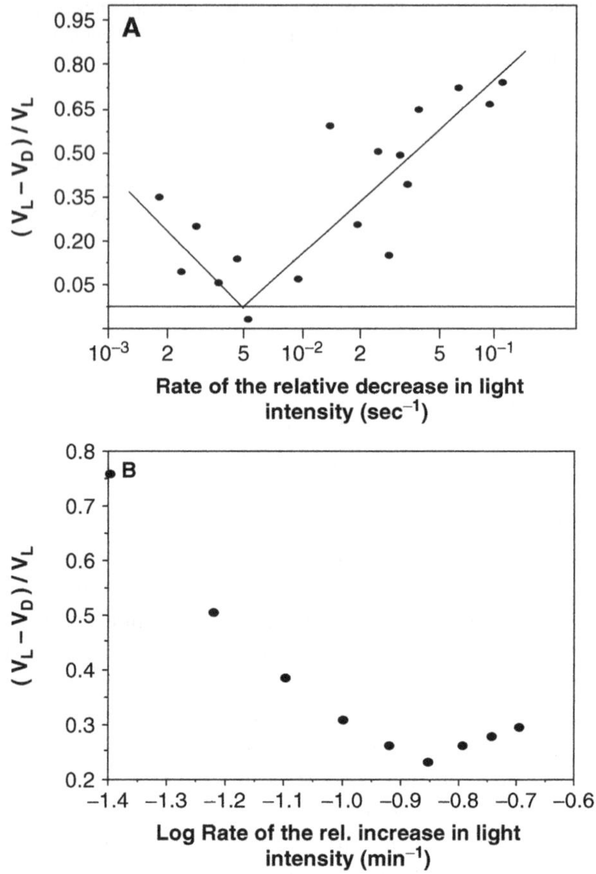

Fig. 5.10 A comparison of the relative difference in velocity (V_D) at which *D. magna* swims upwards during a certain relative decrease in light intensity and the displacement velocity (V_L) of the absolute light intensity in the water column as (**a**) observed in experiments (from Ringelberg, 1964) and (**b**) simulated for a model daphnid

Combining the photobehaviour submodel with the submodel for normal, oscillating swimming is realised by adding the difference between the oscillator units POS1 and NEG1 to (5.4). The effect of this is visible in two swimming tracks illustrated in Fig. 5.12. At the lowest stimulus of $S = 0.07\ \text{min}^{-1}$, upward swimming proceeds in steps as described in Section 3.5.1 (see also Fig. 3.5). At higher stimulus values, swimming becomes more continuous. The pattern is irregular, as it is in actual swimming patterns observed in *Daphnia*. In the model, these irregularities are realised by varying the value of the coupling factor of the oscillator. After each cycle, a new value[10] is randomly chosen from a normal distribution with average 0.8 (s.d. = 0.4). Since the coupling factor determines both amplitude and period time of the oscillation, irregularities in the swimming tracks occur (see Chapter 2, especially Section 5.2). Frequencies observed in natural systems may be distributed according to a probability function (Winfree, 1987) and the behaviour of the system depends on the width of such a distribution. The photobehaviour model allows for sudden jumps in depth, as real *Daphnia* often exhibit in tubes. Also the inhibition

Fig. 5.11 Displacement velocities V_D and V_L obtained from model runs are compared for two attenuation coefficients: $K = 0.04$ cm^{-1} (*top panel*) and $K = 0.01$ cm^{-1} (*bottom panel*)

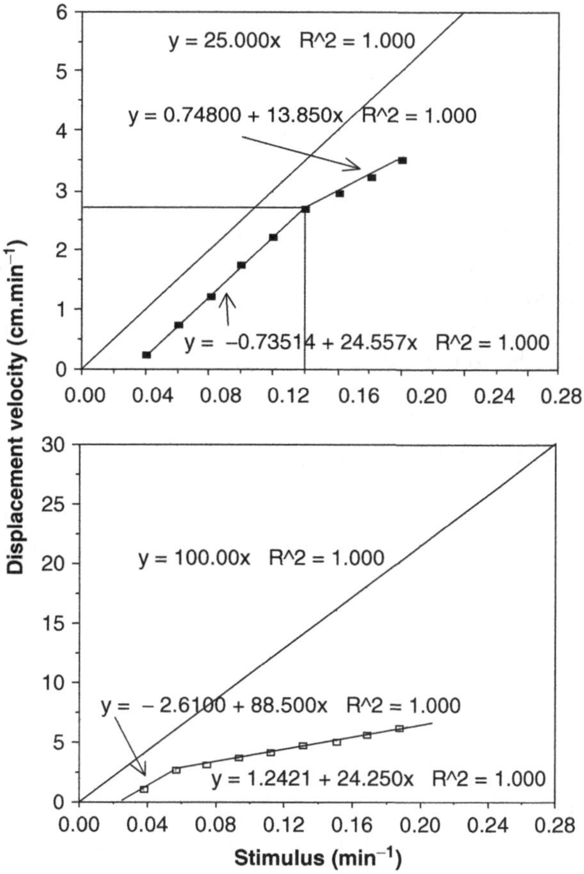

Fig. 5.12 Two runs with the phototaxis model at different relative decreases in light intensity. Compare with the two real swimming tracks of *D. "longispina"* presented in Fig. 3.5

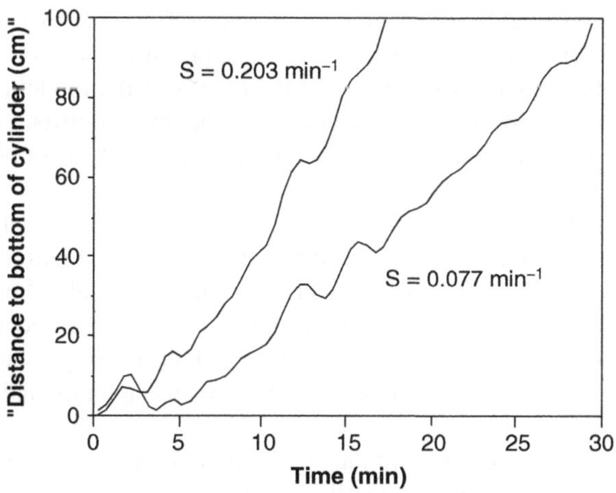

of reactions to changes in light intensity can easily be simulated. With a coupling factor of $f_c = 0.03$ (as used to simulate the swimming pattern in Fig. 5.6) and a stimulus strength of $S = 0.07$ min^{-1} (as in Fig. 5.12), no reaction is discernible. The model system is completely dominated by the oscillatory part. If we were dealing with real daphnids, we would say that primary phototaxis inhibits reactivity to changes in light intensity.

Up till now relative changes in light intensity have been assumed to continuously determine swimming as soon as the threshold is surpassed. The effect of the counterstimulus also is thought to be effective immediately. However, we do not know the reaction well enough to exclude the possibility of an initial short " all-or-nothing" response before the counterstimulus becomes effective. Details like that may differ for other zooplankton like euphausiids, for example.

5.4 The DVM Model

To change the photobehaviour model into a DVM model, the input of a constant relative change in light intensity was replaced by the change in light intensity observed during dawn or dusk, incorporating the algorithm for the "decision-making" mechanism (equation 4.1) to calculate the phototactic coefficient (PC, (equation 4.4)) as well as the temperature correction for this coefficient. This DVM model differs from most others because its construction is based almost entirely on experimentally derived units of photobehaviour (Chapters 3 and 4). Ascent and descent migrations may be viewed as a sequence of beads on a string: the separate responses to relative changes in light intensity are wired together by the light changes of dawn and dusk. The stepwise character of the responses is maintained and this appears to be essential for understanding the variety in diel migration patterns. Stepwise swimming originates in the oscillatory pattern, described for *Daphnia*, but oscillating swimming has also been observed in other planktonic animals (Chapter 2). Because the values of the response parameters have been provided by experiments, assumptions could be kept to a minimum. Of course, experimental data are not without errors and uncertainties, but the approximations derived from these are preferable above mere assumptions. If the results of the model are able to describe the actual DVM patterns observed and the model thus contributes to elucidating hitherto badly understood patterns, I think that with the analysis of photobehaviour DVM research is on the right track.

To simulate a migration observed in a lake, values of the relevant environmental factors such as temperature, light attenuation and food concentration have to be available. These were available for Lake Maarsseveen, only the concentration of fish kairomones could not be measured and had to be guessed. Van Gool (1998) used relative values of kairomone presence to investigate its influence on the decision-making mechanism in daphnids and his values were used in the model.

With the values of the environmental factors introduced into the model, the value of the phototactic coefficient (PC) can be calculated. To include the enhancing effects of accelerations and decelerations in relative light intensity changes, every minute a new rate is calculated by the model based on the light-change signal and compared with the previous one. The resulting acceleration or deceleration is used in the calculation of PC. Over the nearly 3-h-long change in light intensity, the corresponding swimming velocity is calculated and the vertical distance covered is integrated. The process can be visualised as a developing dark-room picture of migration or simply as a list of depth values attained by an individual. Finally, the total swimming distance can be compared with the one observed in the real lake-world, and, if necessary, the relative kairomone concentration can be changed until the calculated and observed migration amplitudes are the same. This procedure may be unsatisfactory but is inevitable as long as kairomone concentrations cannot be determined chemically.

Examples of actual and modelled migrations are presented in Figs. 5.13 and 5.14. The descent and ascent migration were observed on 10 and 22 June 1992, respectively, both in Lake Maarsseveen. The total distances correspond precisely, which

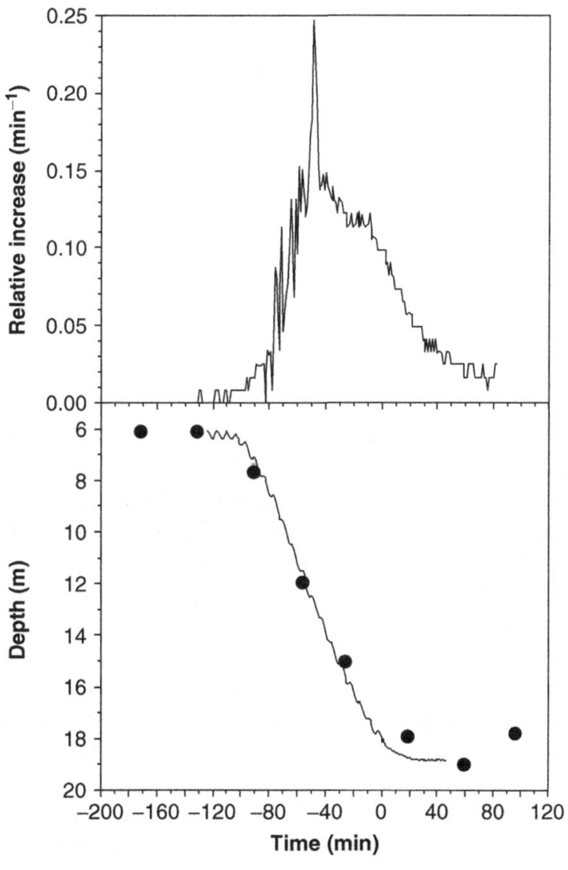

Fig. 5.13 Simulation (drawn line *lower panel*) of the descent migration of *D. galeata x hyalina* (*black dots*) on 10 June 1992 in Lake Maarsseveen, evoked by the observed relative changes in light intensity (*upper panel*), relative to sunrise (time 0) at 5 h 17 min local time. Kairomone level used in the model was 44% and food concentration was 0.15 mg C l^{-1}

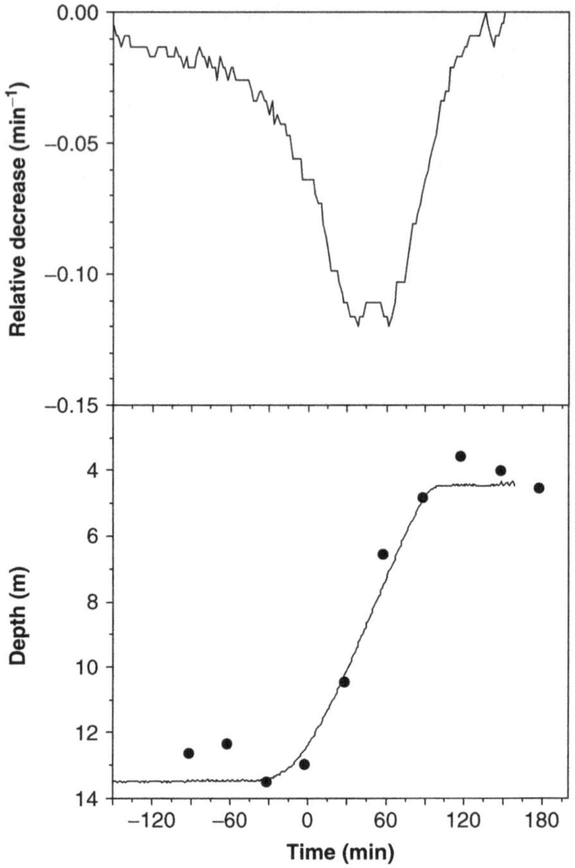

Fig. 5.14 Simulation (drawn line *lower picture*) of ascent migration of *D. galeata x hyalina* (*black dots*) in Lake Maarsseveen on 22 June 1992. Sunset (time 0) was on 22 h 03 min local time. Kairomone level used in the model was 32% (see text) and used food concentration was 0.05 mg C l^{-1}

is not surprising because the relative kairomone concentration was chosen to realise this perfect correspondence. However, the influence of the kairomone concentration is still based on experiments and the route taken in the lake and the model from night depth to day depth and day depth to night depth is similar. The interrupted way of downward swimming is also visible in Fig. 5.13. After each downward movement, swimming stops for a short time and is then resumed as long as the change in light intensity lasts. This interrupted directional swimming is the result of the disintegration of the "excitatory state" and of the negative feedback caused by the "observed" decrease or increase in light intensity while swimming downwards or upwards, respectively. In the observed migrations, such details are of course lost because we can only present population averages.

The similarity of modelled migration and lake observations provide strong support for the conclusion that phototactic responses to relative changes in light intensity are at the base of descent and ascent migration of *Daphnia*. Nevertheless, the story is far from complete. The effect of accelerations surely needs further analysis and the influence of environmental factors on the positively phototactic reaction also has to be studied further. At the day depth in the hypolimnion, the kairomone concentration is supposed to be lower than in the epilimnion, but this might be of little influence because of a "memory" effect (Ringelberg and Van Gool, 1995). And, of course, a population of *Daphnia* in a lake is not genetically homogenous but consist of genotypes with different migration amplitudes, different life histories (Reede and Ringelberg, 1998) and variations otherwise (Spaak and Hoekstra, 1993; Spaak and Ringelberg, 1997; Ringelberg et al., 2004; see also Chapter 12). Photobehaviour responses seem to differ among genotypes (Van Gool and Ringelberg, 1998), but further studies are needed.

Modelling results depend, of course, on the values of the reaction parameters. The velocity constants of the synthesis and disintegration of the "excitatory state" were deduced from reactions in *Daphnia*, but will be different in other planktonic animals. A 20% difference in calibration factor 1 or disintegration constant leads to a 10% change in the calculated amplitude. The effect of a change in the denominator of calibration factor 3 depends on the kairomone level. At large values, the effect of a 20% change is very small, but at a relative kairomone level of only 25% amplitude changes by about 10%. Generally speaking, altering one of the three constants in the model by 20% results in a 10% change in amplitude. This is about the range of confidence limits at $P = 0.05$ as estimated for amplitudes in the field. In the model, amplitude depends on food concentration and kairomone level, but the former has far less influence. At kairomone levels smaller than 2.5%, amplitude rapidly drops and the effect of the food concentration would be difficult to recognise under natural circumstances. At high kairomone levels, amplitudes approach a maximum that depends on the food concentration. Within the level range of 0–100% kairomone, the asymptote is reached at a food concentration of 0.5 mg C l^{-1}. Thus, the model predicts that at kairomone levels higher than those used in Van Gool's experiments, no further increase in amplitude would be possible. Additional experiments have to be performed to verify this prediction.

To give an idea of the effect of temperature on migration amplitude, the model was run at two temperatures at a food concentration of 0.1 mg C l^{-1} but in the absence of kairomones. At 10°C, no downward swimming occurred, while the displacement at 20°C was over 2.8 m. The difference between a late spring and full summer migration in lakes is of course much larger because food and kairomone concentration will also differ. For instance, with a food concentration twice as high and a kairomone level of 15%, the amplitude predicted by the model increases to nearly 9 m.

The hypothesis that DVM helps to avoid visually hunting predators implies that migration amplitude has to increase with increasing water clarity. Indeed, Dodson (1990) found a linear relation between Secchi disk depth and amplitude reported for 21 lakes in the literature. According to the author, Secchi disk depth explains most of the migration amplitude for a wide range of lakes and thus only the simplest of data would be required for predicting amplitude. Neither kairomone nor food concentration nor temperature appears to play a major role. So why should we do it the hard way of carrying out extensive experiments and constructing a complex model if that old invention of an Italian admiral, dropping a soup plate overboard (so the legend of the origin of the Secchi Disk has it), suffices to account for 84% of the variation in migration amplitude? For three oligotrophic lakes Dodson's relation breaks down, probably because of low food concentration and the absence of fish, as Dodson (1990) suggested. This possibility has been verified with the DVM model by converting Secchi disk depth into attenuation coefficients.[11] If food concentration for Lake Muskellunge was set at 0.01 mg C l^{-1} (lowest value of chlorophyll for oligotrophic lakes in Wetzel (2001), converted into carbon) and assuming a July temperature of 20°C, the DVM model would predict an amplitude of 2.39 m in the absence of fish kairomone (fish is minimal in this lake, according to Dodson). This amplitude is close to the value of 2.35 m, reported by Dodson (1990). Although the presented statistics of the correlation may be right, from a biological point of view the explanatory power of Dodson's relation is poor.

The current model does not take care of the slow drifts occurring in the afternoon, nor is the role of food deprivation ("hunger") during the ascent migration properly included. The experimental results with regard to these aspects are as yet insufficient for sophisticated modelling. The lack of food at the day depth and the resulting hunger could well be a strong motivation to react to decreasing light intensity (Pearre, 2003). A decreasing light intensity below the threshold for phototaxis might be the origin of slow drifts.

There is an old controversy about the relevance of relative changes in light intensity as the most important stimulus for inducing DVM (see Section 11.2.3 for an additional discussion). Although George L Clarke (1932) first thought that phototactic swimming in response to relative changes in light intensity could be at the base of vertical migration, he later rejected this idea because sound scattering layers (SSL) sometimes moved faster than isolumes displaced (Clarke and Backus, 1956, 1964). Thus, he reasoned that swimming down at sunrise into lower light intensities could not be governed by relative increases in light intensity, nor could swimming upwards in the evening into higher intensities be caused by decreases in light intensity. This seems a logical conclusion if one assumes that the animals swim continuously but is not necessarily true if swimming is interrupted as the phototaxis experiments indicate. The DVM model was used to test if the three potential patterns – lagging behind the isolume descent, matching the descent and surpassing the descent – could be simulated by varying the importance of the feedback of the counterstimulus perceived while swimming and the influence of the kairomone concentration as an external factor. All three patterns could be easily obtained (Fig. 5.15). Lagging behind (1) occurs when the influence of the counterstimulus is strong, which leads to a shorter reaction time. Decreasing this inhibition effect leads, of course, to a faster descent and may result in a displacement that approximately matches the descent of an isolume. Exceeding the descent of an isolume may be achieved, for example, by a higher kairomone level. At the end of the migration period, when stimulus values decrease, the isolumes are crossed in the opposite direction. Similar patterns can be observed in the migration of SSL presented by Clarke and Backus (1964), see Fig. 10.9.

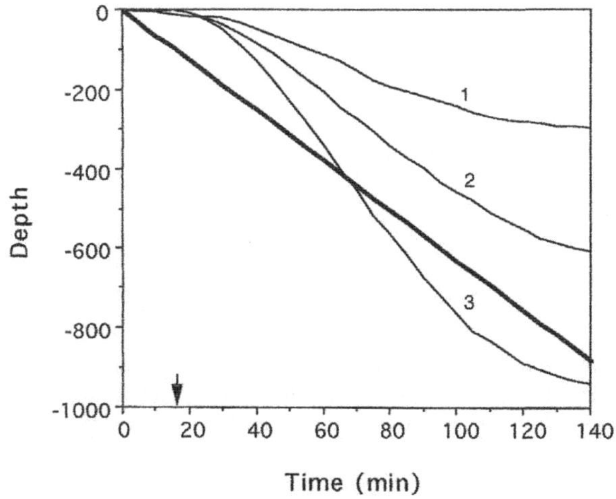

Fig. 5.15 By varying the parameters of the photobehaviour and DVM model, simulations of descent migrations at different velocities are possible that have nothing to do with descending isolumes of "optimal" light intensity. Migrations 1 and 2 are at the same kairomone level but the influence of the feedback of the counterstimulus is higher in 2. With the same feedback as in 2 but a higher kairomone level descent migration 3 exceeds the descent of the isolumes. The *thick line* is a reference isolume. The *arrow* indicates when the threshold of the behaviour response is surpassed on 10 June 1992 at Lake Maarsseveen (= 94 min before sunrise)

As a side product, the simulations also show the presence of the threshold of the photobehaviour model, which has to be surpassed before the descent movement starts, viz. at 94 min (17 min in Fig. 5.15) before sunrise as measured on 10 June 1992 in Lake Maarsseveen.

Of course, all patterns depend on the displacement velocities of isolumes and the migrating zooplankton and could have been obtained by simply varying their respective velocities. However, this would be of no interest because no part of the reaction mechanism would be involved, nor the influence of an environmental factor that is known to profoundly affect the behaviour. It is evident too that isolumes or absolute light intensity as such is of no direct influence and its inclusion in Fig. 5.15 only serves as a reference to facilitate the comparison among the speeds of the model *Daphnia* obtained. Several other factors (e.g. the attenuation coefficient and food availability) determine whether an SSL will move faster than isolumes, but first of all the potential swimming velocity of the species involved must be sufficiently high.

The models presented do not pretend to provide the final answers to all intricacies of DVM. However, they appear to explain a wealth of information collected over many years by many specialists about this intriguing phenomenon. More importantly, they do provide a frameworks for further research, which hopefully will allow further refinements in due course.

Notes

1. The essential equations of the coupled oscillator are as follows:

$$\mathrm{dPOS/dt = P_{in} \cdot N_c \cdot NEG - P_{out}}$$

and

$$\mathrm{dNEG/dt = N_{in} - N_{out} \cdot P_{\mathit{c}} \cdot POS}$$

Additional equations can be found in the original article of Ringelberg and Lingeman (2003).

2. Continuous changes in light intensity evoke, after some latent period, a swimming reaction in *Daphnia*. The latent period increases with decreasing rate of the light change and the relation between both is reflected in the stimulus strength–duration curve. From the equation of this curve, a disintegration rate can be calculated. See Chapter 3 for more details.
3. In the default situation, the different parameters of the model are mathematically fixed. For example,

$$T^{-1} = 1/120 \cdot f_{\mathrm{c}}$$

4. Loeb (1893) found that zooplankton often performed persistent swimming, mostly towards a source of light, which he called "forced movements". He thought that this behaviour was normal and therefore represented part of diel vertical migration. Later, several authors (Bohn, 1906; Franz, 1912; Russell, 1934) showed that this behaviour was caused by adverse laboratory conditions. The persistent positive or negative phototaxis of *D. magna* in the studies of De Meester (1991, 1993a, b) is of another nature.
5. In the model, time is expressed in minutes and distance in centimetres. The model was run at steps of dt = 0.1 min.
6. There is also an empirical experience at the background. The latent period depends to a small extent on the moment a relative change in light intensity is started. If, for example, an increase is started during the upward phase of normal swimming, it takes more time before the reaction becomes visible, than when the animal was already going down. During the experiments in which the stimulus strength–stimulus duration curve was determined, the rule was that the stimulus had to be started when the animal was swimming in the opposite direction from the one expected. Nevertheless, this normal swimming contributes to the variation in latent periods and is larger at lower stimulus strengths as is illustrated in Fig. 5.8 by the different confidence intervals of two latencies.
7. In fact, the idea was considered as the (proximate) explanation of DVM.
8. When the displacement of the reacting animal results in a maintenance of a certain perceived light intensity, $\mathrm{d}I/\mathrm{d}t = 0$. The speed of displacement is then $V = S/K$, where K is the attenuation coefficient. For an elaborate mathematical derivation, see Ringelberg, 1964, p. 358.
9. The relation between Scritical and the attenuation coefficient is *Scrit.* = 0.034 + 2.68 K.
10. For *D. galeata x hyalina*, the default coupling factor of 1, as based on *D. "longispina"*, was made smaller to meet the higher frequency usually observed in this species. Also the displacement velocity is higher for this species as compared to that of tracks of *D. "longispina"* in Fig. 3.5. Whether this is a species-specific effect or due to a possibly higher turbidity in the cylinders with *D. "longispina"* is unknown.
11. In the literature, several conversion factors can be found. For example, $K = 1.44/\mathrm{SD}$ (Holmes, 1970); $K = 4.38/\mathrm{SD}$; $K = 0.0665 + 2.95/\mathrm{SD}$ (George, 1983).

Appendix: The Used Model (Stella Software, Research 1996, High Performance Systems, Inc.)

Environment
Attenuation_Coeff = 0.004
DOCUMENT: The calculation of the attenuation coefficient as a function of the Food concentration (POC in mg per l of algal cells < 35 μm) is based on Fig. 2 in Lampert et al. (1986). To change Secchi disk values into attenuation coefficient the relation K = 2.67/SD was used. The value 2.67; s(x) = 0.881 is an average of 1.44 (Holmes, 1970), 4.38 and 2.2 (IBP handbook Primary Production). K in cm. The original function derived from the figure in Lampert is SD (m) = 1.9006 + 1.1408/POC, R^2 = 0.772.

PC = 31.34
DOCUMENT: The value PC = 31.34 is the default based on the slope in Fig. 2 in Van Gool and Ringelberg (1997). With this basic value phototactic reactions occur in the absence of kairomone. Experiments were done at a food conc. of F = 0.5 mg C l^−1. It is possible that at lower values also PC is somewhat less. The possibility was ignored, however.

Excitatory State
NEG(t) = NEG(t – dt) + (NEG_flow – Disintegration_NEG) * dt
INIT NEG = 0

DOCUMENT: NEG represents an excitatory state resulting from stimulation by a relative increase in light intensity.

INFLOWS:
NEG_flow = Calibr1*(Relincrease + 0.4*AttenuatxSvelo)

DOCUMENT: The external rel. increase in light intensity is diminished by the (secondary) light change resulting from swimming. Upon a rel. increase in light intensity (positive) a downward swimming with a negative sign occurs and thus the product atten. x swimming velo is subtracted from the external stimulus. The factor 1 is of no consequence but indicate that the full secondary light change is used in the feedback. Usual, the feedback enters the excitatory state through POS when the stimulus is an increase.

OUTFLOWS:
Disintegration_NEG = Disintegration_constant*NEG
POS(t) = POS(t – dt) + (POS_flow – Disintegration_POS) * dt
INIT POS = 0

DOCUMENT: POS represent an excitatory state resulting from stimulation by a relative decrease in light intensity.

INFLOWS:
POS_flow = Calibr1*0
DOCUMENT: In the present model, POS is switched off

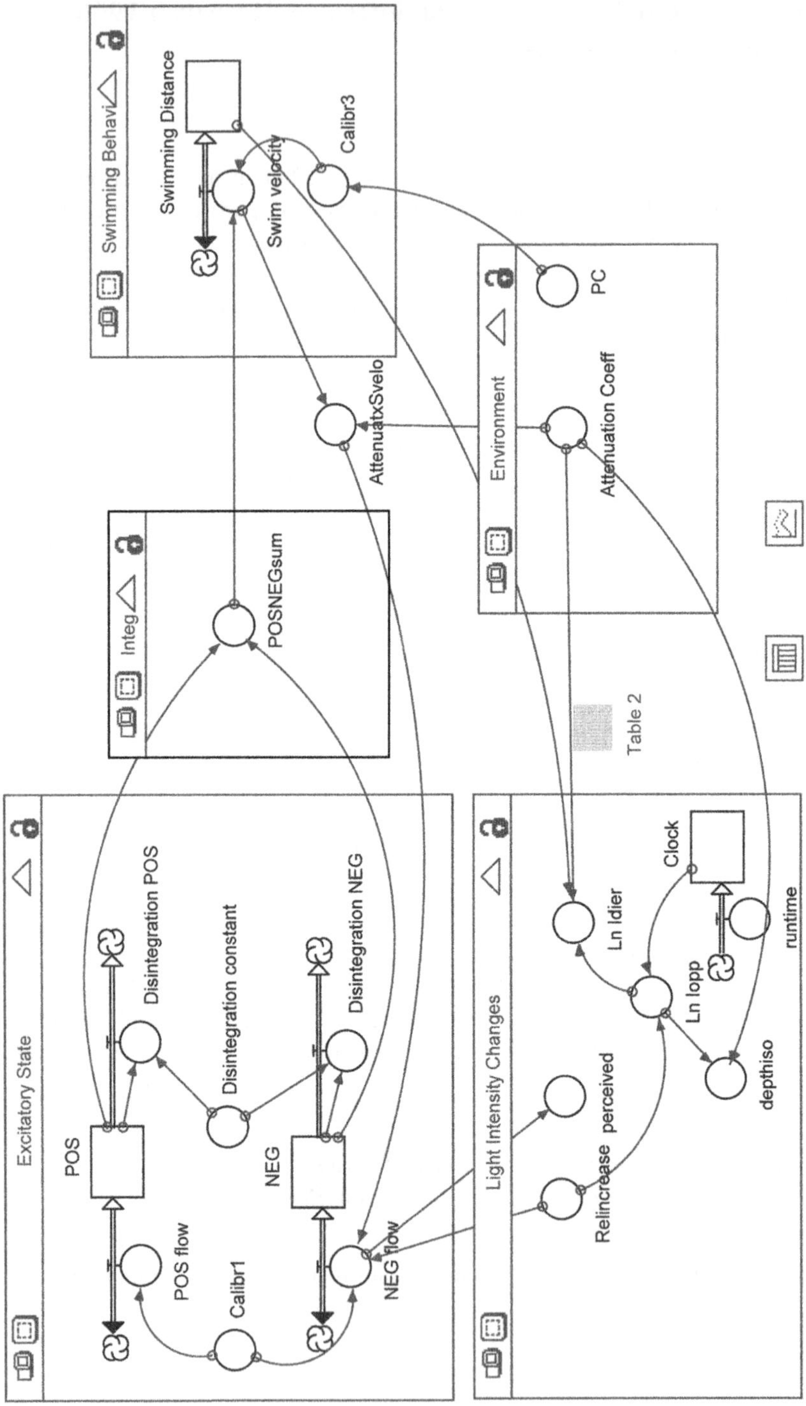
Swimming Behavi
Swimming Distance
Swim velocity
Calibr3
Integ
POSNEGsum
AttenuatxSvelo
Environment
Attenuation Coeff
PC
Graph 1
Table 1
Table 2
Excitatory State
POS
Disintegration POS
Disintegration constant
Disintegration NEG
NEG
POS flow
Calibr1
NEG flow
Light Intensity Changes
Ln Idier
Clock
runtime
Ln Iopp
perceived
RelIncrease
depthiso

OUTFLOWS:
Disintegration_POS = Disintegration_constant*POS
Calibr1 = 4

DOCUMENT: The calibration factor = 4 is based on the emprically derived formula LN(S/(S–R)) = 0.002 + 0.123 LP. Relative light change S in min^–1 and LP in min. Rheobase R = 0.03 min^–1. It was assumed that the threshold for a phototactic reaction was reached at a value of the excitatory state of 1, to be reached within 30 min. This is an arbitrarily chosen value. The calibration factor was determined by trial and error to get the LP, calculated with the empirical formula, as given above. A good fit with the experimentally obtained data is achieved. See Fig. 5.8

Disintegration_constant = 0.123
DOCUMENT: This value is obtained from the experiments with *D. galeata x hyalina* by Van Gool and Ringelberg (1997). The dimension is min^{-1}.

Integration
POSNEGsum = POS–NEG
DOCUMENT: The difference in excitatory states is supposed to govern the swimming response. The sign of the difference determines swimming direction: – is downwards and + is upwards; the amount determines the swimming velocity.

Light Intensity Changes
Clock(t) = Clock(t – dt) + (runtime) * dt
INIT Clock = 0

INFLOWS:
depthiso = –1*(Ln_Iopp/Attenuation_Coeff)

DOCUMENT: I(d,t) = I(0,t)*exp(–kd). Make I(d,t) = 1, then Ln i(d,t) = 0 = LnI(0,t)–kd. Thus d = Ln(I(0,t) /k. 0r Ln Iopp/attenuation coeff). Since depth is negative in the model, the quotient is multiplied by –1.

Ln_Idier = Ln_Iopp–(Attenuation_Coeff*–1*Swimming_Distance)
DOCUMENT: Ln I(dier) = Ln opp–K*swim.distance. Swimming distance is negative and thus has to be multiplied by –1 in the equation.

Ln_Iopp = Clock*Relincrease
DOCUMENT: I(0,t) = I(0,0)* exp (rel.increase * t). I(0.0) = 1. Thus Ln (0,t) = Relincr * clock

perceived = 0.25*NEG_flow
DOCUMENT: The calibration factor (=4) does not apply to the light intensity change perceived bij de animal. Therefore NEG flow multiplied by 0.25.

Relincrease = 0.13
DOCUMENT: The stimulus in this model is a constant relative increase in light intensity. The objective of this model was to explore the possibility to migrate (go down in this case) faster than an isolume moves down and thus to swim into lower light intensities and, nevertheless, to react to increases in light intensity. Essential is that swimming down is interrupted or stepwise with the result that the stimulus

can be percieved again. The three values of the stimulus were choosen from a dawn increase as presented in Fig. 6.9.

Swimming Behaviour
Swimming_Distance(t) = Swimming_Distance(t – dt) + (Swim_velocity) * dt
INIT Swimming_Distance = 0
INFLOWS:
Swim_velocity = IF POSNEGsum> = 1 then (Calibr3*(POSNEGsum + 1.63)) else IF POSNEGsum<–1 then (Calibr3*(POSNEGsum–1.63)) ELSE 0
DOCUMENT: The empirical relation between Displacement Velocity DV and the rel. increase in light intensity S is DV = 1.63 + 31.34 S Van Gool and Ringelberg (1997). The relation of the excitatory state and S as found in the model, proved to be Exc.State = 0.00058 + 31.72 S (at 30 min). Therefore, DV = 1.63 + 0.99 Excit. State. or DV = 1.63 + Excit.State.

Calibr3 = 8*PC/31.67
DOCUMENT: Calibr. 3 is a measure for the enhancement by fish kairomone, food concentration and temperature as found in the experimental study of Van Gool and Ringelberg (1998).
The phototactic coeficient PC as a function of above mentioned factor values is divided by the standard PC = 31.34.
In the present model, swimming velocity had to be larger than that of *Daphnia* and thus was multiplied by 8.

Not in a sector
AttenuatxSvelo = Attenuation_Coeff*Swim_velocity

DOCUMENT: This product represents the rel. light change as a consequence of vertical swimming. The feedback brings the turbidity of the water as a factor in phototaxis.

Photobehaviour Model with Oscillating Swimming

Environment
Attenuation_Coeff = 0.003
DOCUMENT: The calculation of the attenuation coefficient as a function of the Food concentration (POC in mg per l of algal cells < 35 μm) is based on Fig. 2 in Lampert et al., 1986, To change Secchi disk values into attenuation coefficient the relation K = 2.67/SD was used. The value 2.67 (s(x) = 0.881) is an average of 1.44 (Holmes, 1970), 4.38 and 2.2 (IBP handbook Primary Production). K in cm. The original function derived from the figure in Lampert is SD (m) = 1.9006 + 1.1408/POC R^2 = 0.772.
food_conc = 0
kairomone_conc = 0
PC = 31.34

Excitatory State
NEG(t) = NEG(t – dt) + (NEG_flow – Disintegration_NEG) * dt
INIT NEG = 0

INFLOWS:
NEG_flow = Calibr1*(relincrease + 1*AttenuatxSvelo)
OUTFLOWS:
Disintegration_NEG = Disintegration_constant*NEG
POS(t) = POS(t – dt) + (POS_flow – Disintegration_POS) * dt
INIT POS = 0

INFLOWS:
POS_flow = Calibr1*feedback
OUTFLOWS:
Disintegration_POS = Disintegration_constant*POS
Calibr1 = 4
DOCUMENT: The calibration factor = 4 is based on the empirically derived formula LN(S/(S–R)) = 0.002 + 0.123 LP. Relative light change S in min^{-1} and latent period LP in min. Rheobase R = 0.03 min^{-1}. It was assumed that the threshold for a phototactic reaction was reached at a value of the excitatory state of 1, to be reached within 30 min. This is an arbitrarily chosen value. The calibration factor was determined by trial and error to get the LP calculated with the empirical formula, as given above. A good fit with the experimentally obtained data is achieved. See Fig. 5.8

Disintegration_constant = 0.123
DOCUMENT: This value is obtained from the experiments with D. galeata x hyalina by Van Gool and Ringelberg (1997). The dimension is min^{-1}.

Integration
POS1NEG1sum = 0.34*(POS1–NEG1)*0
DOCUMENT: POS1NEG1sum determines normal, oscillatory swimming. The scale factor 0.34 attunes the oscillation to an amplitude of 2.9 cm. This is the amplitude of 2.9 cm (s.d. = 1.39, n = 98) as found for *D. "longispina"*. Period length is 2 min at a coupling factor of 1 (Ringelberg, 1995).
Calibrated at an attenuation coefficient of k = 0.003 cm^{-1}
For the DVM model oscillation was switched off.

POSNEGsum = (POS–NEG)
Light Intensity Changes
Clock(t) = Clock(t – dt) + (runtime) * dt
INIT Clock = 0

INFLOWS:
runtime = 1
d_isoI = (Ln_Iopp–1)/Attenuation_Coeff
Ln_Idepth = Ln_Iopp–(Attenuation_Coeff*(–1*Swimming_Distance))
Ln_Iopp = EXP(Clock*(–1*relincrease))
relincrease = –0.02

oscillator
NEG1(t) = NEG1(t – dt) + (inputneg – outputneg) * dt
INIT NEG1 = 0

DOCUMENT: The value of NEG is used to drive the downward swimming drive

INFLOWS:
inputneg = neginputfactor
OUTFLOWS:
outputneg = negoutputfactor*Poscoupling*POS1
POS1(t) = POS1(t – dt) + (inputpos – outputpos) * dt
INIT POS1 = 0

DOCUMENT: The value of POS is used to drive the upward swimming drive.

INFLOWS:
inputpos = posinputfactor*NEG1*Negcoupling
OUTFLOWS:
outputpos = posoutputfactor
Fishkairomoon = 1.0
Negcoupling = 1*couplingfactor*10
DOCUMENT: Every 2 min a new coupling factor value is entered.

neginputfactor = 5.2*6
DOCUMENT: The basic input for the negative part NEG of the oscillator was chosen to realise an oscillation time of 2 min (with a coupling factor of 1) as was empirical found for *Daphnia* (Ringelberg, 1995). See also negoutputfactor.

negoutputfactor = 0.52*6
DOCUMENT: The basic output for the negative part NEG of the oscillator was chosen to realise a oscillation time of 2 min (with a coupling factor of 1) as was empirical found for *Daphnia* (Ringelberg, 1995). Oscillation time is independent of DT, however, the time values differ with the value of DT. Amplitude of the oscillation also alters and the calibration factor has to be changed if DT is changed.

Poscoupling = 1*Fishkairomoon*couplingfactor*10
DOCUMENT: Each 2 min a new coupling factor is used randomly choosen from a normal distribution with mean = 1 and standard deviation 0.2.

posinputfactor = 0.52*6
DOCUMENT: The basic input for the positive part POS of the oscillator was chosen to realise an oscillation time of 2 min (with a coupling factor of 1) as was empirical found for *Daphnia* (Ringelberg, 1995). See also negoutputfactor.

posoutputfactor = 5.2*6
DOCUMENT: The basic output for the positive part POS of the oscillator was chosen to realise a oscillation time of 2 min (with a coupling factor of 1) as was empirical found for Daphnia (Ringelberg, 1995).See also negoutputfactor.

Swimming Behaviour
Swimming_Distance(t) = Swimming_Distance(t – dt) + (Swim_velocity) * dt
INIT Swimming_Distance = –1

INFLOWS:
Swim_velocity = IF POSNEGsum>= 1 then –1*(Calibr3*(POSNEGsum + 1.63) + POS1NEG1sum) else if POSNEGsum<–1 then –1*(Calibr3*(POSNEGsum–1.63) + POS1NEG1sum) ELSE –1*POS1NEG1sum
DOCUMENT: The empirical relation between Displacement Velocity DV and the rel. increase in light intensity S is DV = 1.63 + 31.34 S (Van Gool and Ringelberg, 1997). The relation of the excitatory state and S as found in the model, proved to be Exc.State = 0.00058 + 31.72 S (at 30 min). Therefore, DV = 1.63 + PC/A. Excit.State. See text Chapter 5.

Calibr3 = PC/31.34
DOCUMENT: Calibr. 3 is a measure for the enhancement by fish kairomone, food concentration and temperature as found in the experimental study of Van Gool and Ringelberg (1998).
The phototactic coeficient PC as a function of above mentioned factor values is divided by the standard PC (=A) = 31.

Not in a sector
couplingfactor(t) = couplingfactor(t – dt) + (input1 – output1) * dt
INIT couplingfactor = 0

DOCUMENT: Each 2 min a new value is obtained and as a coupling factor used. With Kutta-Runge2 a rather strange (unknown method) value is calculated. This is not so with Euler. However, with Euler a tendency is introduced in the oscillation which is more grave an error. The "strange" values of the couplingfactor seem to have a mean (equal? 1), however. From n = 47 values a mean = 0.9578 and a standard deviation = 0.15275 was found, instead of 1 and 0.2, respectively.

INFLOWS:
input1 = IF teller = 0 then distribution else 0
DOCUMENT: Each time teller is 0 a new value from "distribution" is used.

OUTFLOWS:
output1 = If teller = 0 then couplingfactor/dt else 0
DOCUMENT: Couplingfactor is emptied with intervals of 2 min.

AttenuatxSvelo = –1*Attenuation_Coeff*Swim_velocity
distribution = normal(1.0,0.2)

DOCUMENT: "Distribution" realises values from a normal distribution with a mean of 1 and a standard deviation of 0.2.

feedback = 1*AttenuatxSvelo
DOCUMENT: The model was developed for light intensity increases. Reactive downward swimming makes that the external rel. increase in light intensity is probably perceived at a lower value. The consequences for behaviour are not well known but swimming could become less. The extent depends on the turbidity of the water. In this way can the role of the attenuation on the swimming distance be explained. For intensity decreases the same holds but in the model must the feedback made negative. The zero must be removed in that case.

perceived_S = 0.25*NEG_flow
teller = counter (0,2)
DOCUMENT: The teller covers 0–120 s or 2 min, which is about the oscillation time. After 120 s the counter is set to 0 and starts again.

DVM Model

Environment
Attenuation_Coeff = 0.00558
food_conc = 0.01
kairomone_conc = 0
PC = (23.65 + 1.73*kairomone_conc + 58.46*food_conc + 1.84*kairomone_conc*food_conc)*Temp_correction
Temp = 20
Temp_correction = (–78.06 + 41.08*food_conc + 8.92*Temp–0.16*Temp^2)/(28.45 + 41.08*food_conc)

Excitatory State
NEG(t) = NEG(t – dt) + (NEG_flow – Disintegration_NEG) * dt
INIT NEG = 0

INFLOWS:
NEG_flow = Calibr1*(Rel_increase2 + 1*AttenuatxSvelo)
OUTFLOWS:
Disintegration_NEG = Disintegration_constant*NEG
POS(t) = POS(t – dt) + (POS_flow – Disintegration_POS) * dt
INIT POS = 0

INFLOWS:
POS_flow = Calibr1*feedback
OUTFLOWS:
Disintegration_POS = Disintegration_constant*POS
Calibr1 = 4
Disintegration_constant = 0.123

Integration
POS1NEG1sum = 0.34*(POS1–NEG1)

POSNEGsum = POS–NEG

Light Intensity Changes
Local_time_2(t) = Local_time_2(t – dt) + (runtime1) * dt
INIT Local_time_2 = –111

INFLOWS:
runtime1 = 1
acdeceleration? = IF Rel_increase_3> = Rel_increase2 THEN 1 ELSE 0.73
runtime = DT
LN_I = GRAPH(runtime)
(0.00, –11.3), (1.00, –11.2), (2.00, –11.1), (3.00, –11.0), (4.00, –10.9), (5.00, –10.8), (6.00, –10.7), (7.00, –10.6), (8.00, –10.5), (9.00, –10.4), (10.0, –10.3), (11.0, –10.2), (12.0, –10.1), (13.0, –10.0), (14.0, –9.93), (15.0, –9.83), (16.0, –9.73), (17.0, –9.63), (18.0, –9.53), (19.0, –9.43), (20.0, –9.33), (21.0, –9.23), (22.0, –9.13), (23.0, –9.03), (24.0, –8.93), (25.0, –8.83), (26.0, –8.73), (27.0, –8.63), (28.0, –8.53), (29.0, –8.43), (30.0, –8.33), (31.0, –8.23), (32.0, –8.13), (33.0, –8.03), (34.0, –7.93), (35.0, –7.83), (36.0, –7.73), (37.0, –7.63), (38.0, –7.53), (39.0, –7.43), (40.0, –7.33), (41.0, –7.23), (42.0, –7.13), (43.0, –7.03), (44.0, –6.93), (45.0, –6.83), (46.0, –6.73), (47.0, –6.63), (48.0, –6.53), (49.0, –6.43), (50.0, –6.33), (51.0, –6.23), (52.0, –6.13), (53.0, –6.03), (54.0, –5.93), (55.0, –5.83), (56.0, –5.73), (57.0, –5.63), (58.0, –5.53), (59.0, –5.43), (60.0, –5.33)
Rel_increase2 = GRAPH(Local_time_2)
(–111, 0.0144), (–110, 0.00863), (–109, 0.003), (–108, 0.00863), (–107, 0.00288), (–106, 0.0173), (–105, 0.00863), (–104, 0.0144), (–103, 0.0288), (–102, 0.023), (–101, 0.0201), (–100, 0.0259), (–99.0, 0.0259), (–98.0, 0.0317), (–97.0, 0.0403), (–96.0, 0.0489), (–95.0, 0.0374), (–94.0, 0.0719), (–93.0, 0.023), (–92.0, 0.0547), (–91.0, 0.0547), (–90.0, 0.0547), (–89.0, 0.0518), (–88.0, 0.0691), (–87.0, 0.0834), (86.0, 0.0604), (85.0, 0.0604), (84.0, 0.0604), (–83.0, 0.095), (–82.0, 0.0978), (–81.0, 0.0691), (–80.0, 0.0921), (–79.0, 0.0834), (–78.0, 0.0892), (–77.0, 0.101), (–76.0, 0.101), (–75.0, 0.109), (–74.0, 0.0777), (–73.0, 0.127), (–72.0, 0.101), (–71.0, 0.115), (–70.0, 0.112), (–69.0, 0.135), (–68.0, 0.167), (–67.0, 0.132), (–66.0, 0.129), (–65.0, 0.138), (–64.0, 0.124), (–63.0, 0.15), (–62.0, 0.138), (–61.0, 0.144), (–60.0, 0.147), (–59.0, 0.147), (–58.0, 0.127), (–57.0, 0.184), (–56.0, 0.144), (–55.0, 0.144), (–54.0, 0.158), (–53.0, 0.155), (–52.0, 0.152), (–51.0, 0.144), (–50.0, 0.155), (–49.0, 0.141), (–48.0, 0.158), (–47.0, 0.144), (–46.0, 0.144), (–45.0, 0.161), (–44.0, 0.138), (–43.0, 0.144), (–42.0, 0.155), (–41.0, 0.141), (–40.0, 0.147), (–39.0, 0.121), (–38.0, 0.138), (–37.0, 0.138), (–36.0, 0.138), (–35.0, 0.129), (–34.0, 0.138), (–33.0, 0.138), (–32.0, 0.138), (–31.0, 0.135), (–30.0, 0.138), (–29.0, 0.141), (–28.0, 0.127), (–27.0, 0.141), (–26.0, 0.135), (–25.0, 0.0978), (–24.0, 0.129), (–23.0, 0.129), (–22.0, 0.112), (–21.0, 0.135), (–20.0, 0.118), (–19.0, 0.106), (–18.0, 0.0834), (–17.0, 0.104), (–16.0, 0.115), (–15.0, 0.132), (–14.0, 0.0978), (–13.0, 0.109), (–12.0, 0.0863), (–11.0, 0.0921), (–10.0, 0.0921), (–9.00, 0.0863), (–8.00, 0.0806), (–7.00, 0.0863),

(–6.00, 0.0806), (–5.00, 0.0806), (–4.00, 0.0691), (–3.00, 0.0633), (–2.00, 0.0748), (–1.00, 0.0863), (0.00, 0.0691), (1.00, 0.0633), (2.00, 0.0748), (3.00, 0.0806), (4.00, 0.0748), (5.00, 0.0748), (6.00, 0.0806), (7.00, 0.0633), (8.00, 0.0633), (9.00, 0.0691), (10.0, 0.0575), (11.0, 0.0575), (12.0, 0.0518), (13.0, 0.0518), (14.0, 0.0575), (15.0, 0.0518), (16.0, 0.0575), (17.0, 0.0518), (18.0, 0.0518), (19.0, 0.0518), (20.0, 0.046), (21.0, 0.0518), (22.0, 0.0403), (23.0, 0.046), (24.0, 0.046), (25.0, 0.046), (26.0, 0.0288), (27.0, 0.046), (28.0, 0.0345), (29.0, 0.046), (30.0, 0.0403), (31.0, 0.0403), (32.0, 0.0345), (33.0, 0.046), (34.0, 0.0403), (35.0, 0.0288), (36.0, 0.0403), (37.0, 0.0288), (38.0, 0.0345), (39.0, 0.0288), (40.0, 0.0288), (41.0, 0.023), (42.0, 0.0345), (43.0, 0.0403), (44.0, 0.023), (45.0, 0.0345), (46.0, 0.0173), (47.0, 0.046), (48.0, 0.0115), (49.0, 0.0345), (50.0, 0.023), (51.0, 0.0345), (52.0, 0.023), (53.0, 0.0403), (54.0, 0.023), (55.0, 0.046), (56.0, 0.0173), (57.0, 0.0403), (58.0, 0.0115), (59.0, 0.0345), (60.0, 0.0345)

Rel_increase_3 = GRAPH(Local_time_2)

(–110, 0.009), (–109, 0.003), (–108, 0.009), (–107, 0.003), (–106, 0.017), (–105, 0.009), (–104, 0.014), (–103, 0.029), (–102, 0.023), (–101, 0.02), (–100, 0.026), (–99.0, 0.026), (–98.0, 0.032), (–97.0, 0.04), (–96.0, 0.049), (–95.0, 0.037), (–94.0, 0.072), (–93.0, 0.023), (–92.0, 0.055), (–91.0, 0.055), (–90.0, 0.055), (–89.0, 0.052), (–88.0, 0.069), (–87.0, 0.083), (–86.0, 0.06), (–85.0, 0.06), (–84.0, 0.06), (–83.0, 0.095), (–82.0, 0.098), (–81.0, 0.069), (–80.0, 0.092), (–79.0, 0.083), (–78.0, 0.089), (–77.0, 0.101), (–76.0, 0.101), (–75.0, 0.109), (–74.0, 0.078), (–73.0, 0.127), (–72.0, 0.101), (–71.0, 0.115), (–70.0, 0.112), (–69.0, 0.135), (–68.0, 0.167), (–67.0, 0.132), (–66.0, 0.129), (–65.0, 0.138), (–64.0, 0.124), (–63.0, 0.15), (–62.0, 0.138), (–61.0, 0.144), (–60.0, 0.147), (–59.0, 0.147), (–58.0, 0.127), (–57.0, 0.184), (–56.0, 0.144), (–55.0, 0.144), (–54.0, 0.158), (–53.0, 0.155), (–52.0, 0.152), (–51.0, 0.144), (–50.0, 0.155), (–49.0, 0.141), (–48.0, 0.158), (–47.0, 0.144), (–46.0, 0.144), (–45.0, 0.161), (–44.0, 0.138), (–43.0, 0.144), (–42.0, 0.155), (–41.0, 0.141), (–40.0, 0.147), (–39.0, 0.121), (–38.0, 0.138), (–37.0, 0.138), (–36.0, 0.138), (–35.0, 0.129), (–34.0, 0.138), (–33.0, 0.138), (–32.0, 0.138), (–31.0, 0.135), (–30.0, 0.138), (–29.0, 0.141), (–28.0, 0.127), (–27.0, 0.141), (–26.0, 0.135), (–25.0, 0.098), (–24.0, 0.129), (–23.0, 0.129), (–22.0, 0.112), (–21.0, 0.135), (–20.0, 0.118), (–19.0, 0.106), (–18.0, 0.083), (–17.0, 0.104), (–16.0, 0.115), (–15.0, 0.132), (–14.0, 0.098), (–13.0, 0.109), (–12.0, 0.086), (–11.0, 0.092), (–10.0, 0.092), (–9.00, 0.086), (–8.00, 0.081), (–7.00, 0.086), (–6.00, 0.081), (–5.00, 0.081), (–4.00, 0.069), (–3.00, 0.063), (–2.00, 0.075), (–1.00, 0.086), (0.00, 0.069), (1.00, 0.063), (2.00, 0.075), (3.00, 0.081), (4.00, 0.075), (5.00, 0.075), (6.00, 0.081), (7.00, 0.063), (8.00, 0.063), (9.00, 0.069), (10.0, 0.058), (11.0, 0.058), (12.0, 0.052), (13.0, 0.052), (14.0, 0.058), (15.0, 0.052), (16.0, 0.058), (17.0, 0.052), (18.0, 0.052), (19.0, 0.052), (20.0, 0.046), (21.0, 0.052), (22.0, 0.04), (23.0, 0.046), (24.0, 0.046), (25.0, 0.046), (26.0, 0.029), (27.0, 0.046), (28.0, 0.035), (29.0, 0.046), (30.0, 0.04), (31.0, 0.04), (32.0, 0.035), (33.0, 0.046), (34.0, 0.04), (35.0, 0.029), (36.0, 0.04), (37.0, 0.029), (38.0, 0.035), (39.0, 0.029), (40.0, 0.029), (41.0, 0.023), (42.0, 0.035), (43.0, 0.04), (44.0, 0.023), (45.0, 0.035), (46.0, 0.017), (47.0, 0.046), (48.0, 0.012), (49.0, 0.035), (50.0, 0.023), (51.0, 0.035), (52.0, 0.023), (53.0, 0.04), (54.0, 0.023), (55.0, 0.046), (56.0, 0.017), (57.0, 0.04), (58.0, 0.012), (59.0, 0.035), (60.0, 0.035)

oscillator
NEG1(t) = NEG1(t – dt) + (inputneg – outputneg) * dt
INIT NEG1 = 0

INFLOWS:
inputneg = neginputfactor
OUTFLOWS:
outputneg = negoutputfactor*Poscoupling*POS1
POS1(t) = POS1(t – dt) + (inputpos – outputpos) * dt
INIT POS1 = 0

INFLOWS:
inputpos = posinputfactor*NEG1*Negcoupling
OUTFLOWS:
outputpos = posoutputfactor
Fishkairomoon = 1.0
Negcoupling = 1*couplingfactor*10
neginputfactor = 5.2*6
negoutputfactor = 0.52*6
Poscoupling = 1*Fishkairomoon*couplingfactor*10
posinputfactor = 0.52*6
posoutputfactor = 5.2*6

Swimming Behaviour
Swimming_Distance(t) = Swimming_Distance(t – dt) + (Swim_velocity) * dt
INIT Swimming_Distance = 0

INFLOWS:
Swim_velocity = IF POSNEGsum> = 1 then percentreactions* (Calibr3* (POSNEGsum + 1.63) + 0.85*POS1NEG1sum) else if POSNEGsum<–1 then percentreactions*(Calibr3*(POSNEGsum–1.63) + 0.85*POS1NEG1sum) ELSE 0.85*POS1NEG1sum
Calibr3 = acdeceleration?*PC/31.34

Not in a sector
couplingfactor(t) = couplingfactor(t – dt) + (input1 – output1) * dt
INIT couplingfactor = 0

INFLOWS:
input1 = IF teller = 0 then distribution else 0
OUTFLOWS:
output1 = If teller = 0 then couplingfactor/dt else 0
AttenuatxSvelo = Attenuation_Coeff*Swim_velocity
distribution = normal(0.8,0.4)
feedback = –1*AttenuatxSvelo*0.4
percentreactions = IF kairomone_conc = 0 THEN (2.93 + 542*Rel_increase2–761.7*Rel_increase2^{2})/100 ELSE 1. teller = counter (0,120)

Chapter 6
Light and Temperature

Contents

6.1 Introduction: The Relevance of "Biological Factors"

Species-specific relations between organisms and the physico-chemical properties of the environment are important in ecology. Yet the correlation of a biological property and a physicochemical factor is often poor. Some "holists" blame the reductionist method of analysis because the organism as a "whole" experiences an interactive combination of environmental factors. Not the unspecific environment but the species-specific "Wirkwelt" and "Merkwelt" is of importance. A more modern term like abiotic niche reflects that opinion. The notion is difficult to make operational. Even large data sets and sophisticated statistical analysis guarantee no biological understanding.

If a correlation is poor, a large variance of the biological component might be blamed or it is thought that an essential factor has been overlooked. Moore (1955) obtained poor correlations between the depth at which Siphonophora and Chaetognatha were found and the in situ temperatures and illuminations. He "suggests that a third factor, related to depth" (p. 181), although unknown, yet had to be included in the study.

J. Ringelberg, *Diel Vertical Migration of Zooplankton in Lakes and Oceans*,
DOI 10.1007/978-90-481-3093-1_6,

There is inconsistency between being convinced of species-specific relations and the practice of measuring physical and chemical units with instruments. It is not plausible that these units represent specific relations for all species because they describe the environment independent of the animal's experience. The "Merkwelt" of a migrating calanoid is different from the "Merkwelt" of a predating fish, although both are swimming in the same part of the ocean. For a very long time, absolute light intensity was considered important in the explanation of diel vertical migration, although morning descent or evening ascent did not correlate with light intensity. This was easily explained assuming that adaptation is an important aspect of light perception. Knowledge of some first principles of physiology comes in handy when studying the role of environmental factors in the ecological context. Light gets its importance from the way migrating zooplankton experience it and for this reason, the introduction of the physical factor "light" was postponed until the photobehaviour of copepods and cladocerans was discussed. The relative change in light intensity as a stimulus is not physics anymore but biology.[1] This is not a revolutionary new idea. Biological factors were introduced a long time ago, for example, the photometric systems[2] has used a.o. in studies of human vision. If the spectral sensitivity of an eye is known, the intensity of radiation can be expressed as a wavelength-dependent sensitivity distribution. A non-physical system of units like the lux is then created and we deal with a system for humans only. Deep-sea crustaceans have a different spectral sensitivity curve with a peak in the blue-green at 475 nm and an unusual high sensitivity in the near ultraviolet (Frank and Widder, 1996). For this reason, lux is better not used in studies of these animals because UV is outside the range on which the photometric system is based.

Using spectral sensitivity curves, photometric units were created for animal species too and one of the oldest is that for bees (*Apis*). Compared to humans, the eye of a bee has a higher sensitivity in the blue and the UVA part of the radiation spectrum. For *Mysis relicta*, the "mylux" was created based on the spectral sensitivity curve of this freshwater crustacean (Gal et al., 1999). Cohen and Forward (2002) showed that the marine copepod *Calanopia americana* reacts to radiation at wavelengths between 350 and 580 nm. They calculated wavelength-specific weighting factors and created the "copelux" unit. Cohen and Forward (2005) compared the copelux with two other systems of light units: PAR or photosynthetic active radiation (400–700 nm) and the quantal flux at 480 nm, which is the wavelength at which *C. americana* displays maximum photoresponses. They concluded that for migrating animals like *C. americana*, with a broad spectral sensitivity, using PAR even suffices reasonably well as the spectrum for measuring light intensity if photon units are used. This is reassuring because the spectral sensitivity curve of most animals is not known.

"Light" is a complex physical factor with several aspects that play a different role in the life of organisms. In Table 6.1 some of these are listed along with their relevant biological activities. The different physical modalities need to be kept in mind as the relevant part of undefined "light."

Table 6.1 Aspects of light as relevant for various biological phenomena

Physical modality	Biological phenomenon
Absolute intensity	Intensity of photosynthesis
	Threshold of visual systems
Polarisation pattern	Spatial orientation of animals
Spectral composition	Efficiency of photosynthesis and visual systems
	Colour vision
Rapid intensity changes	Shadow response sessile animals
	Stimulus for kinetic component of DVM (secondary phototaxis)
Diel intensity changes	Synchroniser of circadian rhythms
Annual day length changes	Synchroniser circannual rhythms
Spatial differences in intensity	Contrast orientation
	Vision
	Counter shading
	Leaf patterns
	Position choice sessile animals

6.2 Physical Aspects of Light in Water

6.2.1 The Air–Water Interface

The human eye is sensitive to wavelengths in the 400–700 nm range and this visible part of the radiation spectrum is called light. As an approximation, this band of wavelengths can be used in other organisms also, although some animals are sensitive to a slightly extended range, as the already mentioned deep-sea crustaceans.

Light, impinging upon a water surface, consists of direct sunlight and indirect light from the sky. The latter is the result of scattering when sunlight passes through the atmosphere. The spectral composition of both sources is different: the spectrum of skylight is shifted to the shorter wavelengths when compared to that of direct sunlight (Fig. 6.1). In other words, the sky is blue. The spectral composition is not constant throughout the day. With decreasing altitude of the sun its spectrum shifts towards the red, as is obvious when looking in the direction of the sun at sunrise and especially sunset. However, the total spectral composition of light, as received on a horizontal lake surface, is shifted towards the blue at the beginning and end of the day due to a decreased contribution of the sun to total radiation (Table 6.2). In Fig. 6.2, the more rapid decrease of intensity of long wavelengths when the sun sets is illustrated. Under water, this shift to shorter wavelengths is even more marked because the percentage of reflection of direct sunlight at the water surface increases rapidly and thus less radiation of longer wavelengths enters the lake. Occasionally, this phenomenon has been overlooked. Because *Daphnia* respond to red light with upward swimming, ascent migration in the evening was ascribed to the red sunset glow by Smith and Baylor (1953) and Baylor and Smith (1957), which is thus a

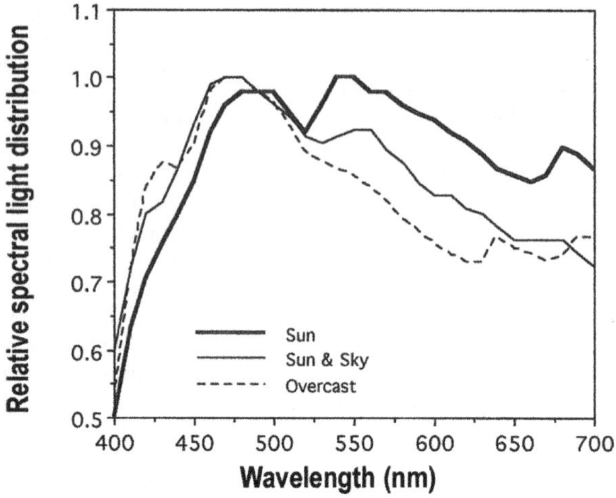

Fig. 6.1 Relative spectral light distribution of direct sunlight and total light on a clear day and when overcast. Averaged values between 9.30 and 15.30 h (based on modified data from Taylor and Kerr, 1941)

Table 6.2 Reflected of sunlight and skylight at a smooth water surface and as dependent on solar altitude

Solar elevation	5	10	20	30	40	50	60	70	80	90
Incident radiation	100	100	100	100	100	100	100	100	100	100
Contribution sun[a]	47	60	70	76	79	84	85	86	86	86
Contribution sky	53	40	30	24	21	16	15	14	14	14
Fresnel reflection %	69	35	15	6	4	3.2	2.5	2.2	2	1.9
Reflected of sunlight	32	21	10	4.6	3.2	2.7	2.1	1.9	1.7	1.6
Reflection % skylight[b]	17	15	13	11	9.3	8.0	7.4	7.1	6.6	6.1
Reflected of skylight	9.0	6.0	3.9	2.6	2.0	1.3	1.1	1.0	0.9	0.8
Total reflected %	41	27	14	7.2	5.2	4.0	3.2	2.9	2.6	2.4

[a]Whitney (1938); [b]Sauberer (1962).

wrong conclusion. The factors playing a role before light passes through the surface of lake or sea are listed in Table 6.2. From the total incident radiation, about 2–40% is reflected. Altitude of the sun and thus the sun's contribution to total radiation, especially the reflections of direct sunlight, are responsible factors. Waves are important too. Fresnel reflection holds for an ideal smooth surface and the data given in Sauberer (1962) are valid for a smooth lake surface on a clear day. If the sky is overcast, the average angle of incidence is about 45°, which results in a reflection of 4–5%. This percentage is approximately constant over the day. Whitney (1938) mentioned 7% for diffuse light and a moderately disturbed surface. In the presence of waves, reflection increases but the extent is variable and difficult to estimate. Waves vary in height, form and orientation with regard to the direction of the sun. According to Atkins and Poole (1933), a light wind over the surface of the sea increases reflection up to 17% and a strong wind up to 30%. For a lake, I guess that reflections in Table 6.2 must be multiplied by a factor of 1.5–2 for a midrange of solar altitudes and small waves.

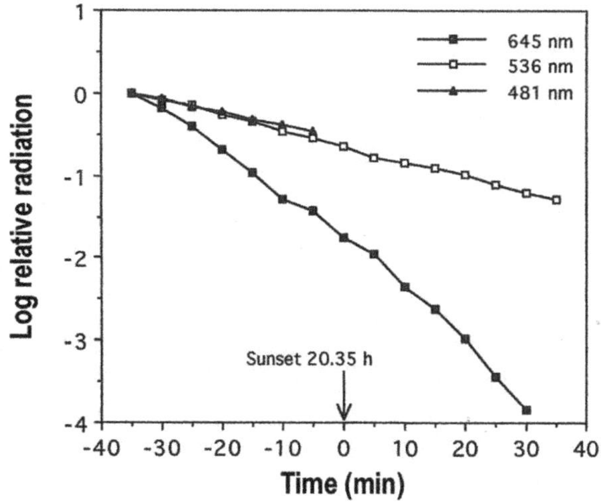

Fig. 6.2 Decrease in relative light intensity at three wavelengths around sunset. Measurements on a horizontal plane using narrow-band filters. A mackerel sky and a slight haze were present (Vrouwenpolder, The Netherlands, 21 May 1959)

6.2.2 The Angular Light Distribution

Light rays penetrating the water surface are diffracted towards the perpendicular axis (normal) according to the ratio sin i/sin $r = 1.333$ (i is the angle of incidence and r is the angle of refraction). For values of i between 0° and 90°, the angle of refraction r ranges from 0° to 48.6°. Thus a maximum angle of refraction is present. All light rays, coming directly from above the water surface, are incident at a point P within an inverted cone with an angle twice the angle of maximum refraction (Fig. 6.3), thus 97.2°. This holds for an ideal smooth water surface. This inverted cone of light should be present at all depths if internal scattering is absent. An observer, looking upwards underwater, will see a bright circle of light called Snell's circle or window.[3] However, various factors distort this simple illumination pattern. Waves disturb the

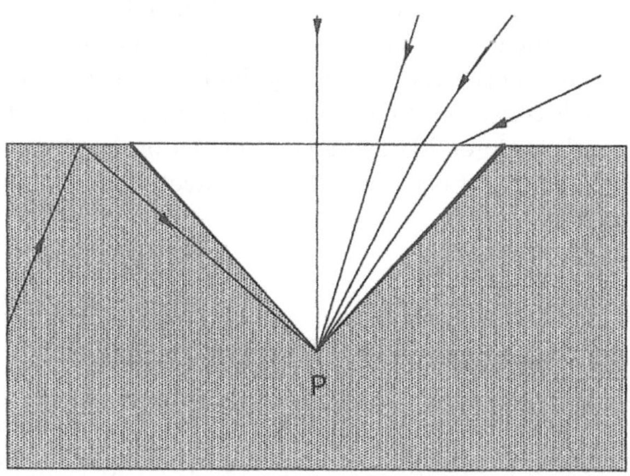

Fig. 6.3 Refraction of light rays towards the perpendicular axis upon entering the water surface. Angle of incidence is 0°, 20°, 40° and 60°. One light ray reaches the top P from outside the cone after refraction against the water surface (modified after Ringelberg, 1976)

ideal flat reflection surface and, consequently, light rays penetrate into the water at angles exceeding 48.6°. In addition, light rays that were scattered underwater will be incident at P from outside the cone. Because this light has travelled a longer path before reaching P than light rays inside the cone and thus must be of lower intensity due to heavier absorption, the sharpness of the window's edge is affected. Nevertheless, the boundary of Snell's window is a light–dark contrast and is even visible with simple eyes. Because the angular position is fixed in the vertical plane, it is a potential beacon for vertical plane orientation by pelagic animals (Chapter 7).

The angular light distribution (ALD) of the underwater light field has been studied by several investigators (Whitney, 1941; Sasaki et al., 1958, 1960, 1962). Near the surface, most of the light is coming from the direction of the sun. The two-dimensional light vector diagram (LVD), thus the graphical representation of the ALD, measured in a vertical plane including the sun, has an elongated shape (Fig. 6.4a, b). With increasing depth, the vector diagram becomes more symmetrical about the vertical axis. In the plane at right angles to that including the sun, the radiance distribution is nearly symmetrical about the normal, even near the surface.

Light vector diagrams are of various shapes but have relatively large angular differences in light intensity or contrasts between 30° and 60° to the normal in common. The magnitude of these differences depends on the angular distribution of the light impinging on the water surface but especially on scattering in the water column, thus on turbidity. Therefore, with increasing depth the contrasts fade away and become intensity gradients. However, LVDs also depend on the aperture of the used light measuring devices. With increasing aperture, the sharpness of the light–dark boundaries or contrasts decreases. This must be borne in mind when these contrasts, as presented in the literature, are considered beacons for orientation. Because measurements of ALD in lakes and oceans were done by physical oceanographers, apertures of eyes were not considered.

Refraction towards the perpendicular axis leads to shortening of the distance that a light ray travels to cover one vertical metre. The extent of shortening depends on the angle of refraction, thus on the angle of incidence at the water surface, which depends on the sun's altitude and the relative contribution of the sky to total radiation. Consequently, the mean path length of light rays covering the distance of one vertical metre changes with time of day and cloud cover (Fig. 6.5). When the sky is totally overcast and the influence of the sun is negligible, the mean path length converges to 118 cm. A maximum value of about 135 cm is found at a zenith angle of 75° and a contribution of the sky of 30%. This variable length of travelling over one vertical metre affects theoretically the attenuation of light with depth.

6.2.3 Downwelling Light Attenuation

Attenuation is the net reduction of light intensity with depth resulting from absorption and scattering. It is measured by taking readings of irradiance intensity (*I*) at

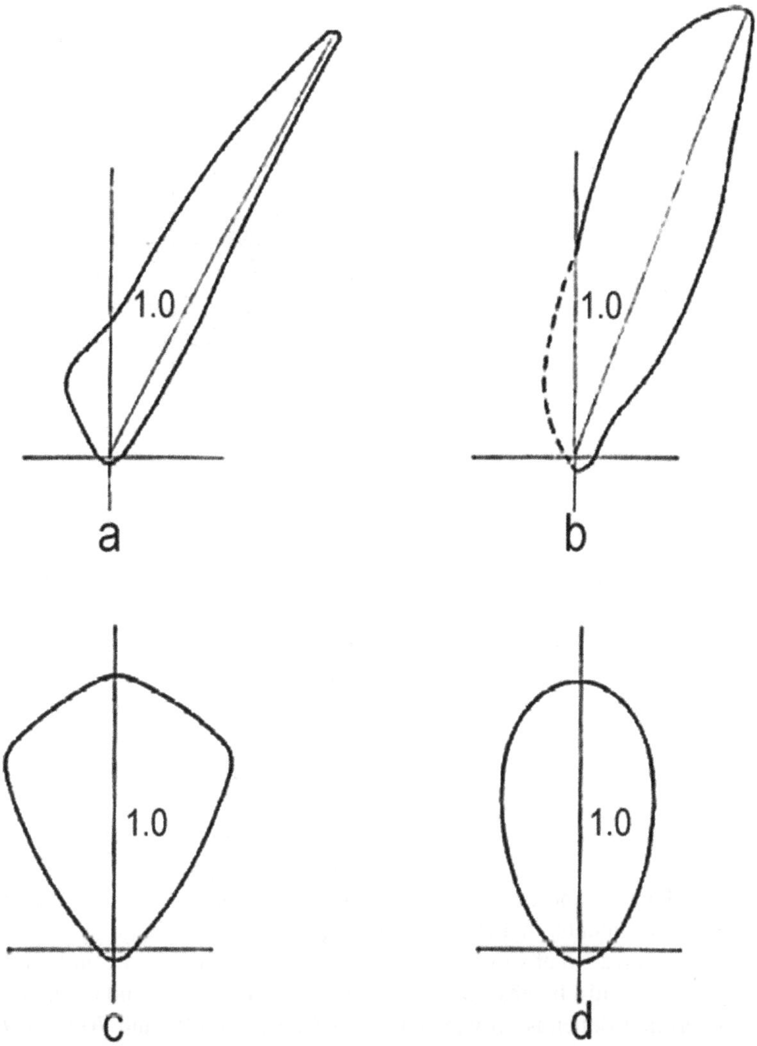

Fig. 6.4 Four light vector diagrams. (**a**) A plane through zenith, sun and the observation point P at a depth of 10 m; (**b**) the same plane but P at 30 m depth; (**c**) a plane perpendicular to the plane in (**a**) at a depth of 10 m; (**d**) as in (**c**) but P at 30 m depth (from Whitney (1941) in Ringelberg, 1976)

different depths (z) and the attenuation process can be approximated as the difference in the natural (Napierian) logarithm of light intensities over a distance of d metres:

$$\text{Ln}\, I(\text{d}) - \text{Ln}\, I(\text{o}) = K_{\text{d}} \times d$$

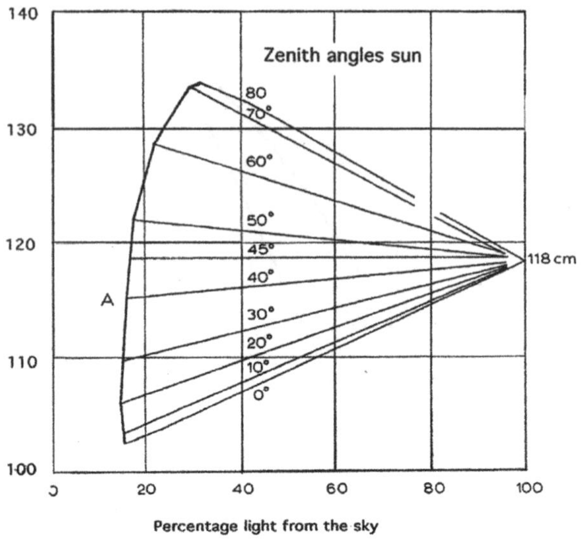

Fig. 6.5 Mean path length (*vertical axis*) of total radiation in water as a function of the zenith angle of the sun and the percentage of contribution of the sky. Curve A holds for a clear sky (after Whitney (1938) in Ringelberg, 1976)

The vertical attenuation coefficient for downwelling light K_d (also called the vertical extinction coefficient) is defined as the attenuation over a vertical distance of $d = 1$ m. The adjective "vertical" was introduced by Atkins and Poole (1933) to indicate that the measured decrease with depth is a net result of in situ measurements. As a true physical characteristic of water bodies, this vertical attenuation coefficient is incorrect because it is dependent on the mean path length of the light beams and thus on the time of day. On sunny days with clear skies and in clear lakes, the vertical attenuation coefficient might be found to increase from noon until sunset,[4] but the differences are small and negligible for biological purposes. Vertical attenuation differs largely for oceans, coastal waters and lakes. Empirical values for the 400–700 nm wavelength band are presented in Table 6.3 for a selection of marine and freshwater environments. In Table 6.4 some K_d values for UV radiation are given.

Attenuation is due to absorption by water, by dissolved organic carbon (DOM), especially humid acids, by absorption of particles such as algae and by internal scattering. Absorption depends on wavelength and between 600 and 700 nm, $K_d(\lambda)$is predominantly determined by water. Yellow-coloured humic acids are responsible for the rapidly increasing attenuation at shorter wavelengths. Far from land, in clear ocean water without dissolved organic material (DOM), attenuation also increases between 450 and 300 nm mainly by absorption of chlorophyll. In unproductive oligotrophic lakes, light below 10 m consists predominantly of wavelengths between 400 and 550 nm. In coastal waters and in mesotrophic and eutrophic lakes, the highest transmission is for green light (500–600 nm) and it has been suggested that herbivorous zooplankton use this wavelength to stay in patches of algae (Smith and Baylor, 1953; Stearns, 1975). Absorption by phytoplankton is maximum at about the absorption peaks of chlorophyll, 436 and 670 nm. In dystrophic lakes, with a very high concentration of humic substances, the spectrum shifts towards 650–700 nm (Fig. 6.6).

Table 6.3 Vertical attenuation coefficients for light of 400–700 nm wavelengths in some freshwater and marine environments. Due to pollution, the situation in a particular water body might have changed since the measurements were performed

Water body	K_d (m^{-1})	Reference
Open oceans		
Sargasso Sea	0.03	Tyler (1975)
Gulf stream	0.08	Tyler and Smith (1970)
Continental slope		
Off New York, upper 90 m	0.062–0.072	Clarke and Wertheim (1956)
Off New York down to 600 m	0.032–0.041	
Coastal waters		
Bjornafjord, Norway	0.15	Jerlov and Nygard (1969)
Gulf of California	0.17	Tyler and Smith (1970)
Oligotrophic lakes		
Canadian Arctic:		Markager and Vincent (2000)
Char lake	0.14	
Meretta lake	0.26	
North lake	0.29	
Crater lake (USA), 0–5 m	0.093	Tyler and Smith (1970)
Crater lake 300–600 m	0.017	Smith et al. (1973)
Lac Pavin (France)	0.15	Flik et al. (1973)
Lake Tahoe		
Lake Tanganyika	0.16	Heckey and Fee (1981)
Lake Moreno West	0.21	Alonso et al. (2004)
Lake Moreno East	0.29	
Mesotrophic lakes		
Lake Maarsseveen I (Neth.)	0.41–0.61	Swain et al. (1987)
Eutrophic lakes		
Lake Maarsseveen II (Neth.)	1–3.2	Swain et al. (1987)

6.2.4 Ultraviolet Radiation

UVR is subdivided into UV-B, ranging from 290 to 320 nm, and UV-A, extending from 320 to 400 nm. Attenuation in water increases rapidly from 400 to 300 nm (Fig. 6.7). Nevertheless, in oligotrophic lakes and oceans, UV-B penetrates sufficiently deep to have a harmful effect on organisms that are tens of metres down in the water column (Fig. 6.6, Kö and O_1, respectively). In clear oceanic water, it has even been suggested that near-UV might function in the vision of some deep-sea crustaceans to a depth of 300 m because these crustaceans have a pigment with a maximum spectral sensitivity at 390–400 nm (Frank and Widder, 1996).

Chromophoric or coloured dissolved organic matter (CDOM) (Gelbstoff; humic substances) is the main UVR-absorbing component of DOM, and DOC is a bulk measure of DOM. The concentration of CDOM and dissolved organic carbon (DOC) is often correlated. Production of CDOM is autochthonous originating

Table 6.4 Vertical attenuation coefficients for ultraviolet radiation in some freshwater and marine environments

Marine environment	Wavelength	K_d (m^{-1})	Reference
Gulf St. Laurence	300–400	1–0.3[a]	Kuhn et al. (1999)
Estuary St. L.		4–0.9[a]	
Bahamas			
Depth 75–150 m	380	0.066	Frank and Widder (1999)
150–200 m	380	0.041	
>200 m	380	0.028	
Lakes			
Canadian arctic:			
Char lake	320–400	0.55–0.2[a]	Markager and
Meretta lake	360–400	0.95–0.53[a]	Vincent (2000)
North lake	360–400	0.80–0.50[a]	
Lake Maarsseveen	280–320	9.03	De Lange (1999)
(the Netherlands)	320–400	3.68	
Lake Moreno West	320	0.52	Alonso et al. (2004)
Lake Moreno East	320	0.68	

[a]These ranges are part of an exponential decrease of which the given values are beginning and end.

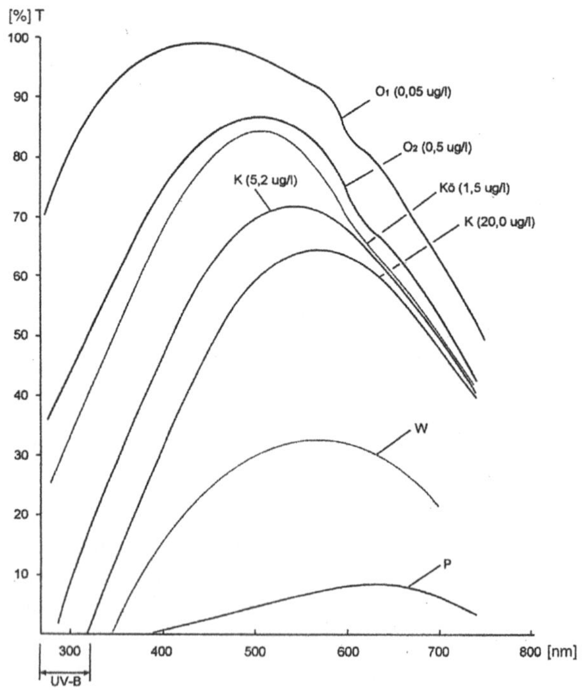

Fig. 6.6 Relative spectral energy transmission. O_1, open ocean water; O_2, coastal waters; Kö, Königssee (Germany), oligotrophic; K, Klostersee (Germany) eutrophic, with two different chlorophyll *a* concentrations; W, hypertrophic Waginger See; P, dystrophic Klein See. Chlorophyll *a* concentrations are presented between brackets (from Siebeck, personal communication)

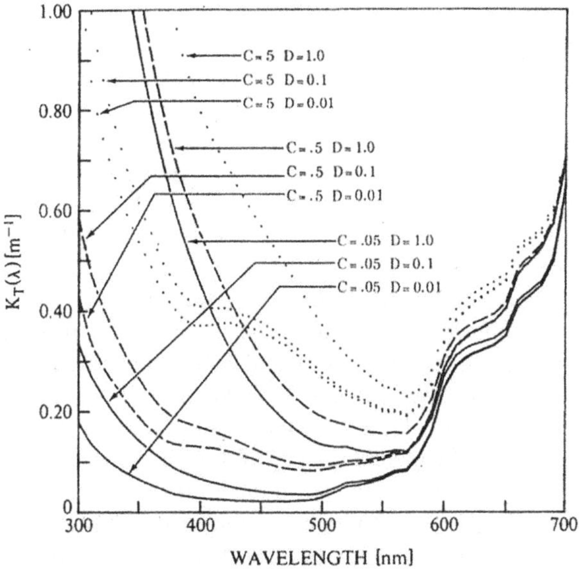

Fig. 6.7 Spectral attenuation coefficient K(g) for a range of chlorophyll (mg m^{-1}) and dissolved organic material (mg DOM l^{-1}) as calculated from a model developed by Baker and Smith (1982) (Reproduced by permission of the American Society of Limnology and Oceanography, Inc.)

from photosynthesis, but high amounts might also be washed into rivers and lakes from the surrounding watershed. Mineralisation of these organic substances is by photochemical reactions by which process the capacity to absorp UVR is lost.

In highly productive lakes or waters surrounded by peat, CDOM concentration is high and penetration of UVR is over a small distance only. Then the damaging effect is not a problem (Ringelberg and Flik, 1984; Siebeck, 1978). In the shallow lakes of the Netherlands, De Lange (1999) mentioned attenuation coefficients ranging from 9.1 to 70.7 m^{-1} for UV-B, from 3.7 to 31.1 m^{-1} for UV-A and from 0.8 to 6.4 m^{-1} for PAR. The 1% depth for UV-B varied from 51 to 6.5 cm. In lowland lakes, the role of chlorophyll *a* might be significant because of a high algal biomass. Nevertheless, also in the eutrophic shallow lakes of the Netherlands, De Lange (1999) found but a weak correlation between UVR attenuation and chlorophyll *a* but a better one with DOC. In lowland lakes, the concentration of DOC correlates in that case with CDOM and is often used as a predictor of attenuation (Morris et al., 1995). These authors studied attenuation in a large number of oligotrophic and mesotrophic lakes and published extensive lists of attenuations at different wavelengths, PAR included. Correlations with chlorophyll *a* and DOC were made but the best estimation of K_d was obtained with a power model based solely on DOC concentration. Considerable variation between lakes were found though, and they emphasised that calculation of K_d values from archival DOC concentrations can be done but provide rough estimates only. In alpine and subarctic lakes, DOC is less useful and CDOM explains most of the attenuation (Laurion et al., 1997, 2000). The most transparent lakes and small alpine waters are found on exposed rock, and Laurion et al. (2000) mention K_d305 values of 0.23–3.46 m^{-1} and K_d380 values ranging from 0.05 to 1.20 m^{-1}. Corresponding 1% depths are 20–1.3 m and 92–3.8 m. In Estany Redo, a lake in the

Pyranees, total depth of the epilimnion (0–14 m) was exposed to relative intensities higher than 1% of the surface UV-B at 305 nm. Although in alpine lakes, chlorophyll *a* is not a good predictor for the attenuation of UV-B, there are exceptions. In these lakes, the algal biomass often increases with depth (Tilzer and Schartz, 1976) and then attenuation of UVR becomes a function of chlorophyll *a*. Also DOC is not the best factor to correlate attenuation with in alpine lakes. Some oligotrophic lowland lakes are sufficiently clear to realise deep UV penetration, for example, the Patagonian Moreno lakes with 1% depths of 7–9 m at 320 nm (Alonso et al., 2004). These lakes are of interest because fish go deep and DVM of zooplankton is an enigma (see Chapter 9 for a discussion).

In oligotrophic ocean water, UVR may penetrate to a considerable depth. Measurements between 75 and 150 m made by Frank and Widder (1996) near the Bahamas showed a $K_d380 = 0.066\ m^{-1}$. If we assume the same attenuation coefficient (probably larger) for the upper 75 m, the 1% value for this wavelength should be at the considerable depth of 140 m. Penetration of UV radiation in near-shore water is lower, of course. In the Gulf of St. Laurence the maximum measured 1% depth for UV was 15 m, but in the estuary it was only 5 m (see Table 6.4).

Several papers were published with models to calculate attenuation of UV radiation as a function of environmental factors. Some of them are mentioned in Table 6.5, together with the used model parameters. In Fig. 6.7, the result of a model by Baker and Smith (1982) for oceanic situations is presented.

6.3 The "Biological Factor" Light

Considering the role of the physical aspects of underwater radiation in DVM, the domain of " biological factors" is entered. Three aspects of changes in light intensity during dawn and dusk are presented in Figs. 6.8 and 6.9. Measurements were done in Lake Maarsseveen on 29/30 May when the yearly returning period of migration had just started. The highest changes in absolute light intensity occurred after sunrise and before sunset. By that time the migrations of *Daphnia* and *Eudiaptomus* were over or were yet to begin, respectively. Absolute intensity and absolute changes are

Table 6.5 Environmental parameters used by different authors in models to calculate UV radiation attenuation

	Chlorophyll *a*	DOC	CDOM
Lowland regions	1, 8	1, 2, 4, 8	
Alpine/Arctic	2	4	2, 3, 7
Oceans	5	5	
Coastal waters	6	6	

1. De Lange (1999); 2. Laurion et al. (2000); 3. Laurion et al. (1997); 4. Morris et al. (1995); 5. Baker and Smith (1982); 6. Kuhn et al. (1999); 7. Markager and Vincent (2000); 8. Smith and Baker (1981).

Fig. 6.8 Aspects of the light intensity increase around sunrise on 30 May 1990 as measured 0.3 m below the water surface in Lake Maarsseveen (The Netherlands). See text for discussion

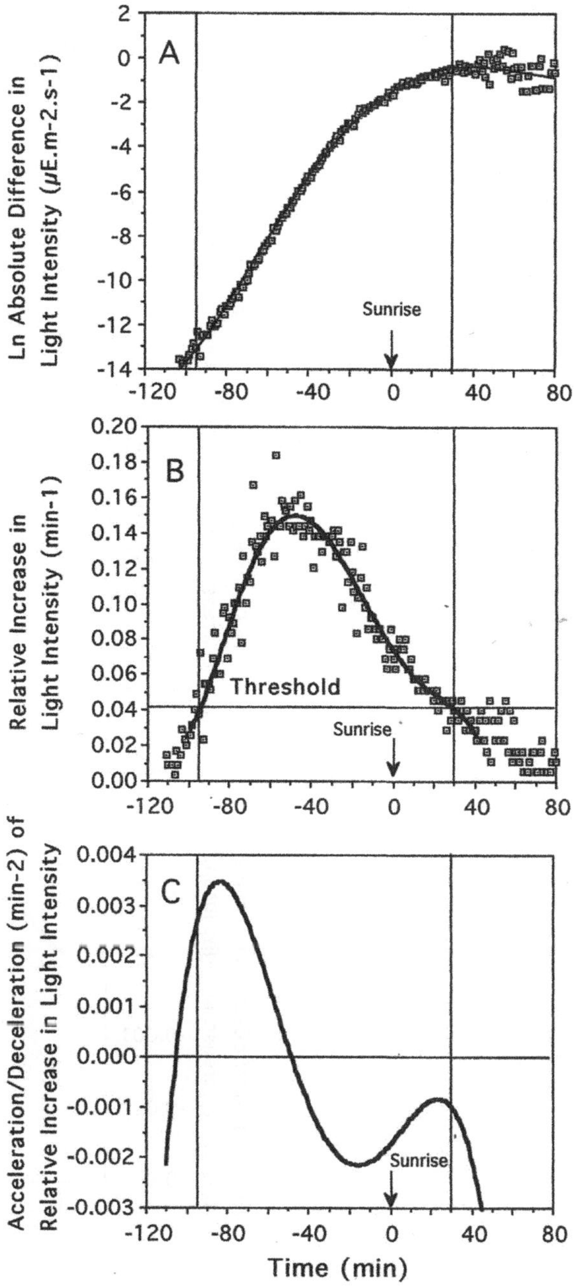

Fig. 6.9 Aspects of the light intensity increase around sunset on 29 May 1990 as measured 0.3 m below the water surface in Lake Maarsseveen (The Netherlands). See text for discussion

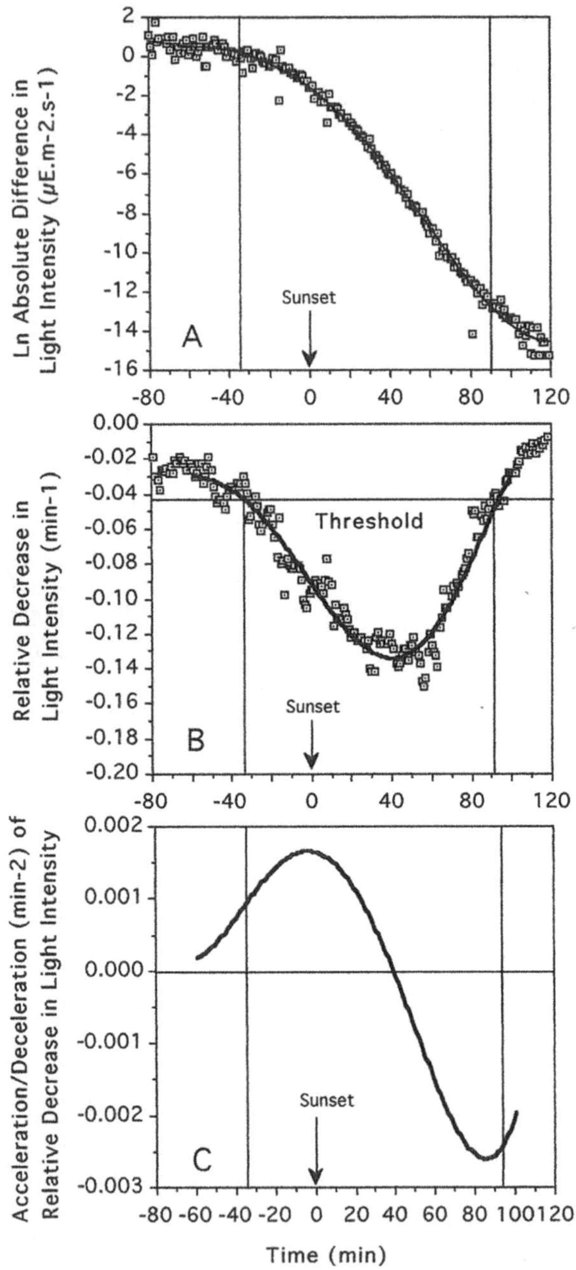

physical aspects and the biological importance is restricted to light acclimatisation. On the other hand, a relative change in light intensity is physics and physiology combined, and high values co-occur with extensive vertical migrations. As discussed in Section 3.2, sensory physiology provides the arguments to make absolute changes relative. In fact, ΔI has to be divided by a *function* of the acclimation intensity because $\Delta I/I$ is not constant over the large range of light intensities that occurs in nature (see Fig. 3.4). Since this function is unknown for continuous light intensity changes, dividing by the ambient intensity is an approximation of the true "biological factor".

Dawn and dusk are mirror images. The pattern of the relative increases and decreases is asymmetrical, at least in June. In the morning, the relative increase expands rapidly until the maximum is reached and then decreases slowly to become subthreshold shortly after sunrise. The afternoon starts with a slow increase and the maximum relative decrease occurs after sunset. A biological consequence of the mirror image of dawn and dusk is that dark-acclimated zooplankton start DVM in the morning and light-acclimated animals in the afternoon. If a relation as in Fig. 3.4 also holds for continuous changes in light intensity, the threshold for phototaxis should be lower in the morning than in the afternoon. Of course, the animals are deeper down in the water column in the afternoon, but the difference with a dark-adapted animal just after dawn might be large enough to be of influence. To illustrate this, the afternoon intensity on 31 May, just before ascent migration started, was 0.35 and 0.0039 $\mu E\ m^{-2}\ s^{-1}$ at a depth of 10 and 20 m, respectively. At dawn, the next morning, at 3 m depth, light intensity was only $1.65 \times 10^{-5}\ \mu E\ m^{-2}\ s^{-1}$.

A migration caused by relative changes in light intensity is a well-adapted strategy to avoid visual predators. At dawn, the stimulus is high and the threshold low and thus the sensitivity is high, which makes the rapid descent possible at very low light intensities before visual predation by fish becomes an important danger. In the afternoon, circumstances work the other way round, leading to a slow upward swimming before sunset, which thereafter accelerates. The maximum stimulus occurs 40 min after sunset. In Lake Maarsseveen, *Daphnia* arrive in the upper part of the epilimnion 1 h after sunset. Juvenile fish is "waiting" for them and light intensity although low is still high enough for predation during a short period.

This pattern of upward and downward migration is enhanced by accelerations in the rate of relative changes (Figs. 6.8c and 6.9c) if fish kairomones are present (Section 4.5). As soon as the threshold for the swimming response is surpassed at dawn, acceleration becomes maximum and downward swimming is enhanced.

In the afternoon, maximum acceleration does not occur until after sunset when the zooplankton is still below the danger zone.

The way light intensity changes around sunset and sunrise depends on the time of the year and the latitude on earth. North of the Tropic of Cancer, the longest twilight and the longest period of supra-threshold changes happen in June, the shortest twilight is in December and 21 March and 23 September are in between. On the other hand, the lowest maximum value of the relative change in light intensity is in June. The highest maximum is at both equinoxes and 21 December is in between. In

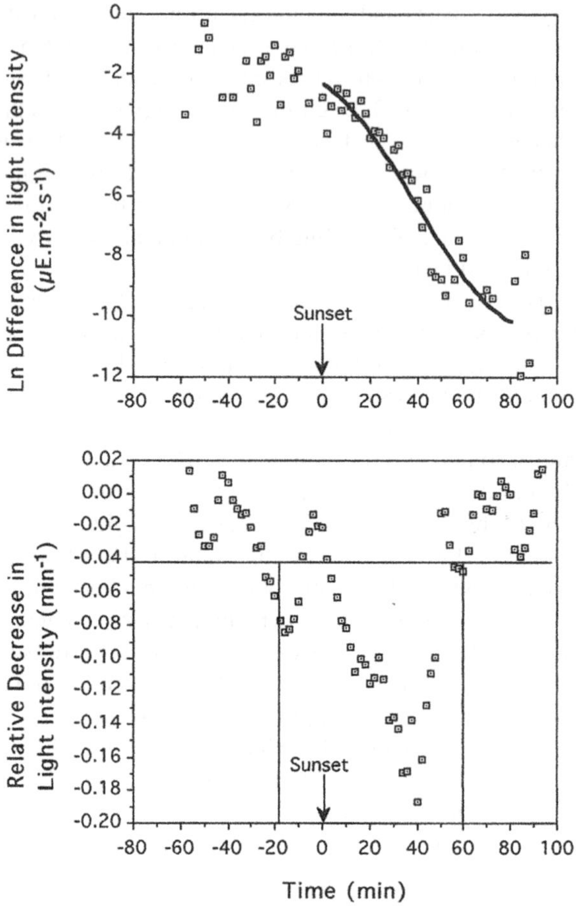

Fig. 6.10 Light intensity changes around sunset as measured on 20 October 1988 below the water surface in Lake Maarsseveen. The *horizontal line* indicates the threshold; between the *vertical lines* the estimated supra-threshold period is situated

Fig. 6.10, the light change of sunset on 20 October 1988 illustrates the shorter supra-threshold period (less than 80 min) and the higher maximum (about 0.19 min^{-1}) as compared to June. The influence of clouds is also apparent from this figure: light intensity is irregular and the beginning of the supra-threshold period cannot be estimated precisely.

Individual clouds passing before the sun in a clear sky cause large changes in light intensity (Fig. 6.11). Within the observation interval of 12 s, light intensity dropped for more than 80% and the original intensity was restored just as fast. A swimming response might be elicited. In the laboratory, responses to instantaneous changes of comparable magnitude last for less than half a minute (Ringelberg, 1964). Therefore, swimming distances caused by these clouds will be small. When a clear sky gradually becomes overcast, the relative decrease in light intensity might be above the threshold for a longer period and the upward movement will be more extensive.

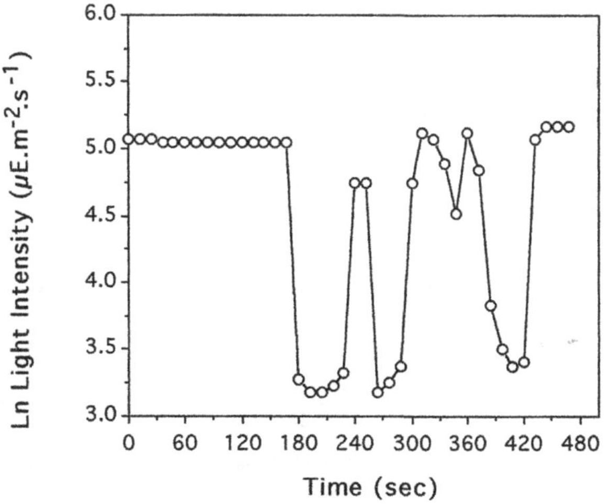

Fig. 6.11 Changes in light intensity caused by small clouds obscuring the sun in a clear sky

Angular light distributions (ALD) are physical properties of underwater light. The contrasts around the angle of maximum refraction might be thought too feeble to be of much use as a beacon in vertical plane orientation (see Chapter 7), but light sense organs possess contrast enhancing mechanisms. The light–dark boundaries at a depth of 10 m in Fig. 6.4c are certainly strong enough to be used for orientation. At 30 m, the pronounced contrasts have disappeared (Fig. 6.4d), however. A problem for orientation might also be the pronounced presence of the sun, concealing the contrasts of Snell's window as in Fig. 4a, b. Since the altitude of the sun changes continuously, the position of the sun might not be a very useful beacon for orientation in the vertical plane. Many terrestrial animals use the sun for horizontal direction finding and compensate for its movement across the sky by using an internal clock. For vertical plane orientation the sun could be used too but such mechanism has never been described.

In the laboratory and in nature, a strong light in dark surroundings attracts animals. Examples are the swimming of fish to a lamp at night (light fisheries), the movement of insects to a light in the dark ("Kommt geflogen und gebannt bist du Schmetterling verbrannt", Goethe) or a migrating bird to an old-fashioned lighthouse. This "trapping effect", as Verheijen (1958, 1985) called it, disappears if additional light sources realise a more natural ALD.[5] On land, sufficient environmental scattering is present for normal orientation and the sun is ignored (sun compass orientation excepted). Under water, normal orientation should be possible in planes where the high intensity of the sun is absent.

The light vector diagram in Fig. 6.12 of the plane, perpendicular to the one including the sun, is symmetrical with a smooth decrease in angular light intensity. At this depth of 60 m, nothing seems to be left of the light–dark boundaries of Snell's window and a beacon function seems impossible. However, if the angular intensity differences are made relative, the characteristic high values near the position

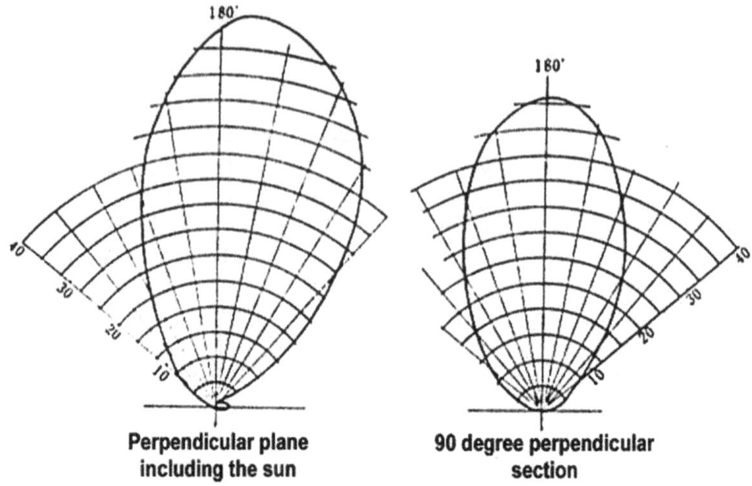

Fig. 6.12 Light vector diagram of ALD in perpendicular planes at a depth of 60 m in Sagami Bay (Japan). Aperture of the measuring unit was 10°. *Left* figure represents the plane including the sun; *right* figure represents the plane perpendicular to the previous one (from Sasaki et al., 1960)

of the contrasts of Snell's window become visible again (Fig. 6.13). With contrast reinforcement, called lateral inhibition,[6] a beacon function is probably realised. The mechanism of contrast orientation in *Daphnia* was studied extensively, but in Chapter 7 it will be argued that calanoid copepods also use contrasts. The mobile eyes of Euphausiacea are able to follow a light source (Land, 1980) and probably scan the vertical plane for contrasts to be used in vertical plane orientation. The biological function of the physical ALD becomes apparent if transformed into relative angular light distributions and the analogy with relative changes in time becomes even greater if it is realised that these spatial differences become temporal differences when the scanning eye moves over the angular gradient.

6.4 Temperature

Temperature is biologically effective as a constant factor or as a change per unit time. Constant temperatures exert influence without sensory perception and determine the velocity of vegetative physiological functions in poikilotherm animals. Ambient temperatures determine metabolism including all functional aspects dependent on metabolism, such as development times, generation length and life history characteristics in general. Swimming velocity too can be temperature dependent. The effect of constant temperature is not always easy to separate from the effect of temperature changes perceived by sense organs. A criterion for the first effect might be that the observed quantitative changes of a phenomenon are a function of absolute temperature. Many examples can be found in the literature.

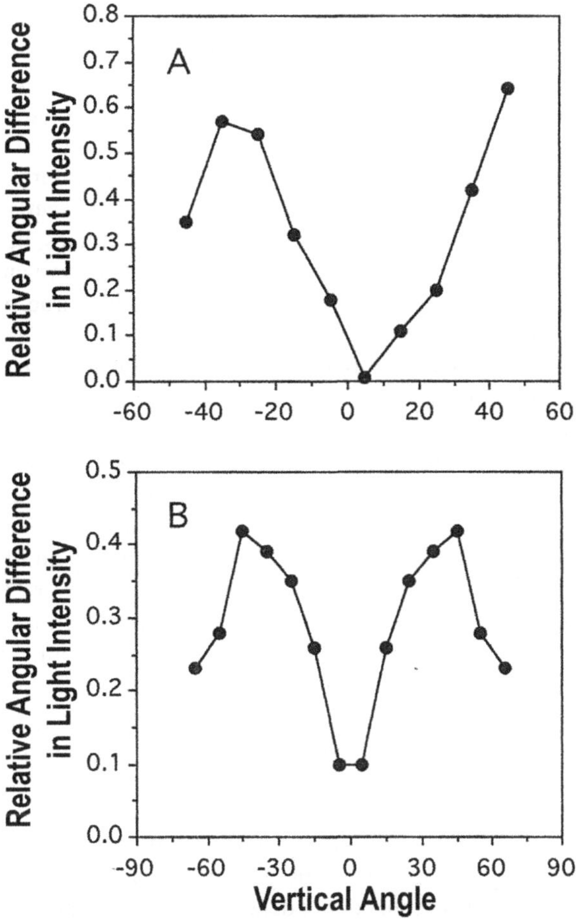

Fig. 6.13 Based on the light vector diagrams of the previous figure, the relative difference in angular light intensity $(\Delta I)/0.5(I_v + I_{(v+1)})$ was calculated: (**a**) for the vertical plane including the sun; (**b**) the plane perpendicular to the previous one

A sensory perception of a temperature change per unit time ($\Delta T/\Delta t$) causes an alteration of behaviour. The response occurs fast and can be species specific. The change in behaviour might be quantitative but is often qualitative. Since sensory perception is involved, a threshold must be surpassed. For the understanding of the behaviour of migrating zooplankton passing through the temperature gradient of the metalimnion, knowledge of the effect of this modality of temperature is important. However, insufficient experimental quantitative data are available and conclusions must be tentative. We start with an example to illustrate how difficult it might be to decide whether sensory perception of temperature changes is really involved or not.

Gerritsen (1982) raised and decreased the temperature of a small water column 0.3°C min^{-1} and 0.2°C min^{-1}, respectively, and observed that *D. magna* drifted to a higher or lower depth. The upward drift resulted from longer periods with strong strokes (see Section 2.2). The downward drift occurred because these periods became shorter. Gerritsen (1982) considered this a qualitative modification of a part of behaviour and not a physiological rate response. Moreover, the altered swimming was a linear function of the rate of temperature change and not of absolute temperature. It was suggested that the reaction could possibly reinforce DVM: encountering increasing or decreasing temperatures in a thermocline, swimming would be enhanced upwards as well as downwards. Generally, it is thought that a temperature gradient inhibits, and not facilitates, behaviour. Perhaps, temperature slightly disturbed the balance between the positive and negative neural oscillator at the base of normal swimming. If the latter is true, what aspect of temperature is involved?

In the experiments of Calaban and Makarewicz (1982), a "metalimnion" was realised in a tube with temperature differences of 7, 12 and 17°C covering a distance of 10 cm. Sunrise and sunset were "simulated" by "light-on" and "light-off". In the absence of a temperature gradient, *D. magna* moved downwards when the light was switched off and swam upwards upon light-on. These reactions are opposite to most phototactic responses caused by light intensity changes, but no explanation was offered. The "reverse migrations" over the length of the tube also occurred in the presence of a gradient of 7°C. If the gradients were 12 and 17°C, the "migrations" were confined to the "epilimnion". Obviously, these stronger temperature differences acted as a barrier. Experiments in a tube with temperature gradients were also performed by Loose and Dawidowicz (1994). They were discussed in Section 4.4. Phototactic swimming responses were enhanced by increased concentrations of fish kairomones, which led to more animals swimming through the temperature barrier.

In terms of a stimulus, swimming through a temperature gradient is complicated. A gradient can be defined as a temperature change per depth ($\Delta T/\Delta d$). Swimming through the gradient creates a temperature change per time ($\Delta T/\Delta t$). If perceived, a biological factor has emerged. A response is possible if a threshold is surpassed. Whether this occurs or not depends on the steepness of the gradient and on the swimming velocity. In the mentioned experiments, swimming is initiated by an increase or decrease in light intensity ($\Delta I/\Delta t$) and, obviously, the self-induced temperature stimulus counteracts the effect of the light stimulus. A large variety of combinations in temperature gradients and light intensity changes are possible and thus of stimuli. The results must be a variety of reaction patterns. Effects of these combinations of light and temperature have not been studied systematically.

Temperature has for a long time figured in the explanation of vertical distributions and migrations of zooplankton in lakes and oceans (Cushing, 1951). Deep lakes stratify in summer. The upper water layer or epilimnion has a high, nearly constant temperature; in the deep part or hypolimnion also, the temperature is constant but low; and in the metalimnion in between, the temperature declines rapidly with depth (Fig. 6.14a). The depth of the thermocline[7] is variable and depends among other factors on the surface of the lake and the predominant direction of wind in relation to its longest axis. The depth of the stratum of steep thermal change in oceans is much deeper and about 80–100 m. Due to internal waves, the depth of a

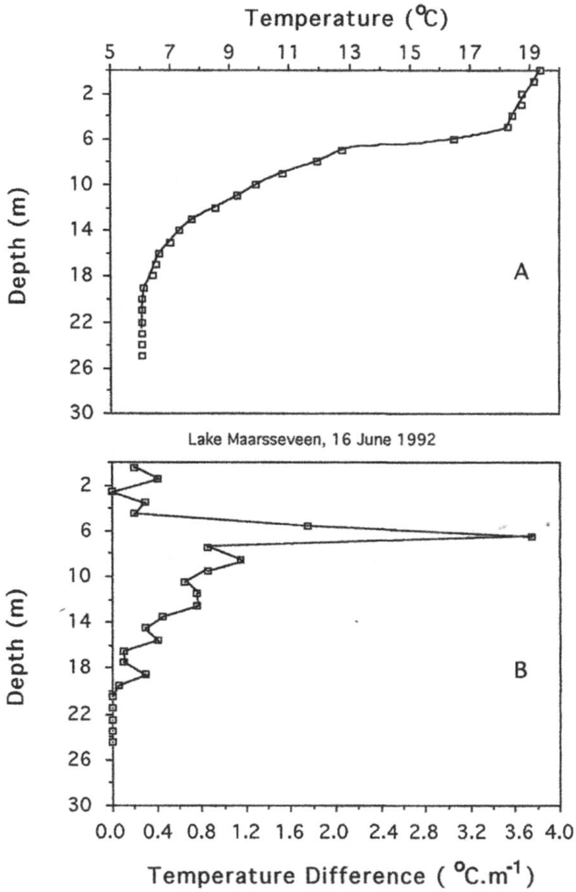

Fig. 6.14 Temperature stratification in Lake Maarsseveen. (**a**) the decline of absolute temperature with depth; (**b**) the temperature change ($\Delta T/\Delta d$) with depth. The steepest change in temperature or thermocline is at 7 m

thermocline might change in a rather short time. Angel (1968) reports of a thermocline in the Atlantic Ocean off the Moroccan coast to rise occasionally for about 20 m. This effect of internal waves or seiches is also present in lakes. The amplitude depends on the surface area. In the small Lake Maarsseveen amplitude is a few metres only (Kersting, 1981), but in a large lake, like Lake Constance, Stich (1985) measured 5–10 m. For the same lake, Schröder (1962) found water layers with a strong temperature gradient shifting over a distance of 30 m in a few days.

Considering the discussion, it will not be surprising that the reaction of zooplankton, coming across a discontinuity layer, is very variable. The following examples for the marine copepod, *Calanus finmarchicus* and the freshwater cladoceran, *Daphnia*, are derived from Cushing (1951). Clarke (1934a) found that *C. finmarchicus* moved through a discontinuity layer with a temperature difference of 6°C over George's Bank, while in the North Sea the species remained below the layer (Savage, 1926). According to Farran (1947), adults and copepodites might behave differently, the first migrating through a temperature gradient while the latter remaining below it during the day and the night. Adult *Calanus* seemed to be rather insensitive because Esterly (1911) found them to move from 9 to 17°C in the Pacific Ocean. Also euphausiid species react very differently to the presence of discontinuity layers. *Euphausia krohni* and *Thysanopoda aequalis* performed extensive vertical migrations in the western Mediterranean Sea and did not seem to be influenced by temperature (Andersen and Sardou, 1992). On the other hand, Roger (1971) reported that *T. aequalis* remained above or in the upper part of a discontinuity layer in the Pacific Ocean. These anecdotal examples illustrate at least that temperature cannot be isolated from other factors, especially those causing a species to migrate.

Variable behaviour is also present in lakes. *Daphnia longispina* crossed a pronounced temperature gradient in Lake Lucerne (Worthington, 1931), as it did in German lakes where the temperature in the metalimnion dropped from 17 to 10°C. On the other hand, Langford (1938) found *Daphnia* to remain below a weak gradient. Several authors reported accumulations of zooplankton individuals in steep temperature gradients, in marine environments (Banse, 1964; Miller, 1970) as well as in lakes (Schröder, 1962). Zooplankton do not always gather in temperature gradients, for example, Angel (1968) found that euphausiids, decapods and ostracods avoided the discontinuity layer in the Atlantic Ocean off the Moroccan coast.

Cushing (1951) found it "difficult to generalise from such scattered evidence" (p. 165). Nevertheless, he made an attempt at a general interpretation. Species were thought to be acclimated to different temperature ranges. Some "preferred" a narrow range of low temperatures, other species a range of higher temperatures. The first category tended to remain below and the second above a temperature gradient. The barrier effect would be most pronounced when the thermocline was near the limits of the temperature range of an organism. A thermocline in the centre of an organism's temperature range could be crossed more easily. In the literature many examples can be found of zooplankton species differentially distributed over depth and often temperature. The suggested temperature preferences are not supported by experimental evidence. I think that experienced temperature changes must be considered in the first place if we want to understand the behaviour of migrating animals when encountering a temperature discontinuity. A migrating plankton animal is forced to swim vertically by changes in light intensity. Thus driven in space–time, a complicated configuration of light–temperature stimuli arises.

If swimming is "spontaneous", that is, initiated by the central nervous system, the "expected" light changes (see discussion of the "Reafferenzprinzip" in Chapter 1) do not inhibit these spontaneous swimmings. If by internal seiches a water mass with zooplankton is shifted vertically fast enough, the change in light intensity is not "expected" and swimming in a counter-direction takes place. The principle is illustrated by interesting observations done by Schröder (1962) in Lake Constance. In Fig. 6.15 (upper picture), a slow downward displacement of a strong temperature gradient took *E. gracilis* with it. On the other hand, when a weak thermocline moved

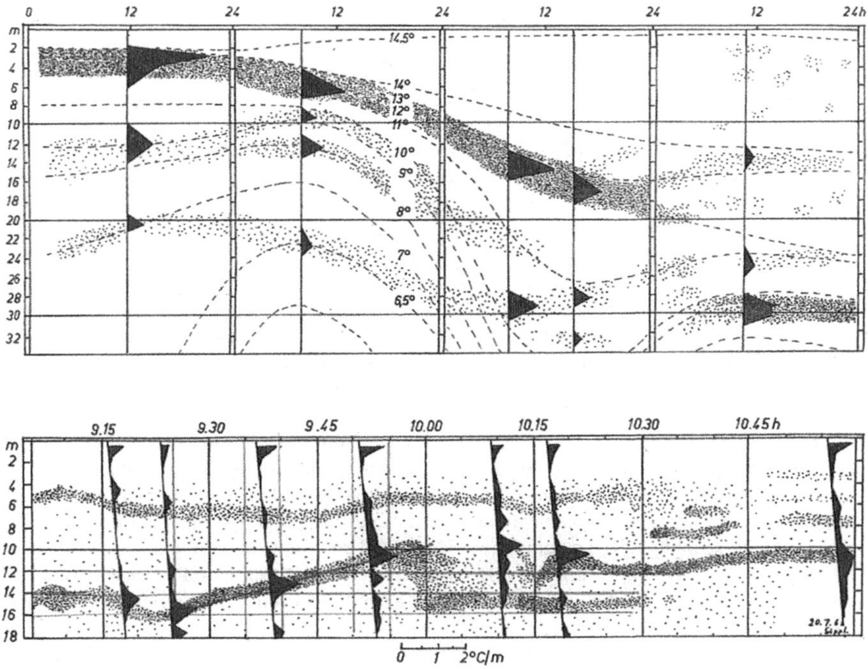

Fig. 6.15 Echo sounding of zooplankton in the Überlinger See (part of Lake Constance). *Dark bands* represent high densities of zooplankton (*Eudiaptomus* and *Daphnia* predominantly) made visible by ultrasound imaging. *Black triangles* are concentrations as determined by counting of individuals in samples taken in the water column. The *upper figure* is the result of a 4-day survey (18–21 October 1961). Zooplankton moved with the slowly descending temperature gradient. The *lower figure* illustrates the escape of zooplankton when internal seiches rapidly move a temperature gradient upwards. Combination of figures are from Schröder (1962) (Reproduced by permission of E. Scheizerbart'sche Verlagsbuchhandlung)

6 m upwards in 30 min, *E. gracilis* left the temperature gradient and swam downwards (lower picture). Schröder (1962) thought that the rapid light intensity increase was responsible and provided the stimulus that caused the reaction. Knowledge of the light changes, swimming velocities and temperature gradients is necessary to understand what occurs when migrating zooplankton enter a temperature gradient in a particular field situation.

Notes

1. A "biological factor" is not identical to a biotic factor such as competition or predation. A "biological factor" represents a composite of external and internal simple elements which, together, are of specific relevance for understanding a complex biological phenomenon. We cannot speak of a causal factor as we do with physical factors, although proceeding analysis of the "biological factor" and the complex biological phenomenon will reveal causal relations.

A nice example of how "biological factors" can lead to deeper insight is the distribution of invertebrates over brackish inland waters in the Netherlands. For 8 years, Heerebout (1970) determined chlorinity and estimated the number of invertebrate species in 19 ponds. The number of species correlated weakly with median chlorinity ($R^2 = 0.567$) and not at all with the amplitude of the fluctuations. This might have been considered the influence of other factors to be measured in a successive study, but Heerebout thought otherwise and made a clever shot. Each pond was positioned in a graph by median chlorinity and amplitude of fluctuation. Then, he calculated the distance of a line originating on the abscis at the chlorinity of 16.6‰ (= seawater, constant concentration) towards the position of a pond. Finally, the number of species was correlated with these vector lengths. The result was amazing: species numbers were close to a straight line ($R^2 = 0.929$). With increasing distance from the stabile seawater situation, the number of species decreases. It is essential to measure the vector length from the seawater point. If, for example, vector length is calculated from 8‰ on the abscis, the correlation is poor ($R^2 = 0.296$). Obviously, we deal with a "biological factor" and not a simple combination of two aspects of a chemical factor. Many brackish water species have a "recent" evolutionary origin in the sea and this historical aspect was taken into account.

In the given example, the vector is a complex with a.o. salt concentration (salinity, chlorinity) as a part. This salinity is a causal factor for osmoregulation of the individual species but neither salinity nor the mechanism of osmoregulation can explain the composition of a particular community. Due to a suitable combination of parts, a property emerged through which a deeper understanding of the communities in brackish environments evolved.

2. The photometric system of illumination is based on the relative spectral sensitivity of a light-adapted human eye. Lux is not a physical but a biological unit and the human photometric system differs from that for bees (*Apis*) a.o. because the compound eye of a bee has a higher sensitivity in the blue and the UV part of the radiation spectrum.
3. Snellius or Snel, 1580–1626, Dutch mathematician and physicist, professor at the University of Leiden. Cited as Snell in Anglo-Saxon literature.
4. Spectrophotometric transmittance or absorbance values of natural water are, from a physical point of view, better characteristics of optical properties. Due to the near absence of scattering of a parallel light beam in a cuvette, the obtained values are always lower than those of the vertical attenuation coefficient, measured in the field.
5. Verheyen placed a few-metres-long cardboard box over a beehive. The box had a diameter of about a metre and the inside was black. If this box was directed towards the sun, bees shot like bullets out of the box and then started to make the usual orientation flights. It was remarkable that returning bees were never able to re-enter the box and thus could not return to the hive. Obviously, inside the box normal orientation was impossible due to the absence of a "normal" ALD.
6. Lateral inhibition has first been demonstrated for the compound eye of *Limulus* (see Miller et al., 1961; Ratliff et al., 1966). Axons of each eccentric cell of an ommatidium (see Section 7.2) give off branches in what is called the lateral plexus that make inhibitory synapses on axons of neighbouring ommatidia. These connections become fewer as the distance from each axon increases. Hence, the interaction becomes weaker between more distant neighbours. This results in the accentuation of the contrast of a border between dark and light.
7. Hutchinson (1957) defines the thermocline as the plane of maximum rate of decrease of temperature with respect to depth ($\Delta T/\Delta d$). In the older literature, any outspoken change in temperature over depth is often called "thermocline". I will use the older English term "discontinuity layer" in that case.

Chapter 7
Optical Orientations

Contents

7.1 Introduction

Why vision in an optical empty environment? Why orientation in a vast homogenous pelagic environment by animals having a small action radius only? For a *Calanus* in the middle of the ocean, even for a *Daphnia* in the middle of a lake, it is without sense to swim 10 m to the left or to the right or in whatever horizontal direction. On the other hand, orientation in the vertical plane is of paramount importance, otherwise there would be no diel vertical migrations. In the literature of the first half of the last century, geotaxis was thought as an important mechanism to distinguish between up and down. Geotaxis was, however, in experiments difficult to distinguish from phototaxis and the two have never been properly separated. Therefore, the role of geotaxis in vertical plane orientation became questionable (Cushing, 1951). Another environmental factor must be available for finding the vertical direction in the pelagic environment and downwelling light is a reliable one making optical orientation plausible. The following experiment with different angular light distributions illustrates this.

A *Daphnia* placed in a small glass sphere and illuminated from all directions with light of equal intensity (Fig. 7.1) is completely disoriented. It makes somersaults or

J. Ringelberg, *Diel Vertical Migration of Zooplankton in Lakes and Oceans*,
DOI 10.1007/978-90-481-3093-1_7, © Springer Science+Business Media B.V. 2010

Fig. 7.1 Device used to study the orientation of *Daphnia* in different angular light distributions (ALD). The animals were placed in a small sphere within another sphere of opal glass, suspended by strings and illuminated by 14 incandescent light bulbs regularly distributed round the opal sphere. By varying the distances of the light sources to the sphere and by changing resistances in series with each light, a particular ALD could be produced. With binoculars the animals were observed through a round opening in the opal glass sphere (Ringelberg, 1964)

lies on the bottom. Somersaults also occur when a daphnid is suddenly illuminated by strong light. Obviously it is "blinded", which means that optical orientation is impossible. If the homogenous illumination is sufficiently changed by increasing the light incident from above (Fig. 7.2), normal swimming returns.

These experiments demonstrate that body posture is maintained by means of optical orientation. In the first half of the last century, several investigators studied oriented behaviour of diverse zooplankton species and found that body posture and

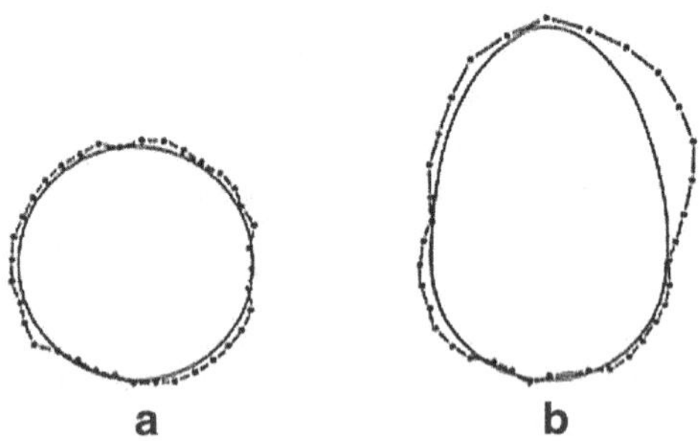

Fig. 7.2 Light vector diagram of an ALD with ratio $I_o/I_{180} = 1$ (homogenous) (**a**) and with ratio 2 (**b**). In (**b**), the upper half of the diagram is part of an ellipse with the long axis four times that of the radius of the bottom half circle (Ringelberg, 1964). The dotted line represents the measurement of the realised ALD.

the direction of swimming was determined by the direction of incident light (Radle, 1901; Ewald, 1910a, b, c; Hess, 1910; Von Frisch and Kupelwieser, 1913; Harris, 1953).

More recently, the importance of the angular light distribution (ALD) was demonstrated in calanoid copepods in a fine series of experiments by Siebeck (1980).

How body axis orientation in Cladocera and Copepoda is achieved will be discussed in this chapter, but first the structure of the optical sense organs of representatives of these taxonomic groups will be shortly described.

7.2 The Structure of Optical Sense Organs of Cladocerans and Copepods

The spatial orientation of zooplankton is difficult to understand without sufficient knowledge of the structure of the two important types of eye: the compound eye and the nauplius eye. Most or nearly all information of the compound eye is available from studies in *Daphnia*. Information about the nauplius eye comes from cyclopoid copepods, not from calanoids, which would have been better since the latter species are predominantly pelagic. The cladocerans have also pigment cups that are called nauplius eyes but their function is insubordinate to that of compound eyes.

7.2.1 The Compound Eye of Daphnia

Of the several types of light sensitive organs in zooplankton, the compound eye of *Daphnia* has been studied best. The single organ is kidney-shaped or roughly spherical with a flattened, centripetal side where the optic nerve parts for the lobus opticus.

The compound eye lies in the median plane close to the rostral point of the head (Fig. 7.3). Originally, two bilaterally located embryonic eyes fused into one during development. The eye is suspended by a membrane within a cavity filled with haemocoel. The membrane radiates from the equator of the spherical eye and is attached to the hypodermis of the carapace (Klotzsche, 1913; Downing, 1974). If a puncture is made in the carapace in front of the eye, it is squeezed through the hole until it rests on the carapace (Klotzsche, 1913; Schulz, 1928; personal observations). This means that the membrane is circularly attached realising a closed chamber in front thus, if punctured, enabling the haemocoel to push the eye from behind to the outside. Three muscle pairs are arranged symmetrically on each side of the eye making rotations possible (Leder, 1915; Consi et al., 1987). One pair is inserted at the dorsal side of the eye, one pair laterally and the third pair at the ventral side. Each pair of three muscles has one single point of attachment on each side of the eye. The dorsal and ventral muscles act antagonistically and generate rotations in

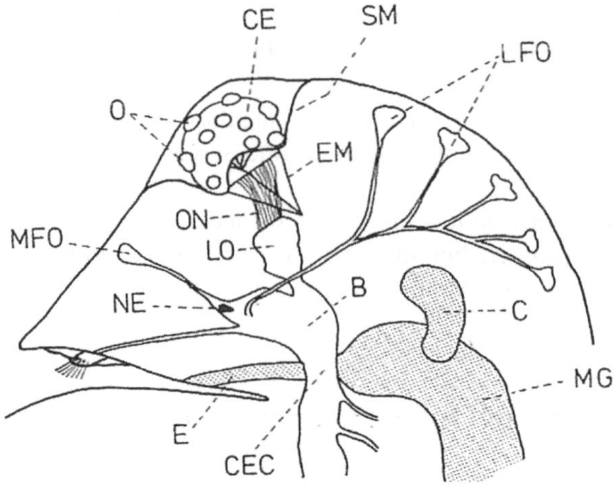

Fig. 7.3 Drawing of the possible optic sense organs in *Daphnia*. The figure was combined from data presented by Gicklhorn (1931), Klotzsche (1913), Sterba (1957) and Downing (1974). B = brain; C = caecum; CE = compound eye; CEC = circumesophageal commissure; E = oesophagus; EM = a set of eye muscles; LO = lobus opticus; LFO = lateral frontal organ with nerve; MFO = medial or ventral frontal organ; MG = midgut; O = ommatidia; NE = nauplius eye; ON = ommatidial nerves; SM = suspensory membrane

the median plane of the eye, while the lateral muscles produce side-to-side rotations. Downing (1974) compares the peculiar hydraulic suspension system with a ball-and-socket joint that also allows extensive rotations in many directions. The daphnid system is superior though, because it has better frictional characteristics and thus requires smaller muscular forces to rotate. A ball-and-socket joint would be impossible anyhow because of the protruding lenses of the ommatidia. A disadvantage of this suspension system is that rotation is limited by elastic stress in the membrane. Over what angle rotation is possible will be discussed in the next paragraph.

The eye of *Daphnia* has 22 ommatidia bisymmetrically arranged over both sides of the median line of the eye. Predatory cladocerans like *Leptodora kindtii* and *Polyphemus pediculus* have a much higher number of ommatidia, viz. 500 (Wolken, 1971) and 130 (Odselius and Nilsson, 1983), respectively. These eyes must have a higher resolution than the eye of the filter-feeding *Daphnia* and possibly play a role in prey recognition.

There is still some controversy about the precise arrangement of the ommatidia over the surface of the *Daphnia* eye and, consequently, about the angle between the optical axes of adjacent ommatidia. From histological preparations, Heberdey and Kupka (1941) derived an angle of 55° but Frost (1975) concluded 38° from similar observations. The latter value was also mentioned by Young and Downing (1976), who considered 55° too large because the normals of a polyhedron with 20 faces are only 41.8° apart and the compound eye consists of two more ommatidia. Moreover, there is a gap where the optic nerve leaves the eye. Assuming 23 sides,[1] they arrived at a value of 38° which agrees, as they wrote, with the angle estimated by Frost (1975).

Their reasoning can be criticised, however. The eye is not a regular polyhedron because it is composed of two symmetrical halves. Adjacent pairs of ommatidia on the midplane, where the embryonic eyes fused, lie closer together. There are six midplane ommatidia[2] and thus we deal with less than 22 equally spaced units. On each half of the eye, the ommatidia are arranged in trigonal arrays with angles of 54° as Young and Downing (1976) calculated. The exception is the double row of ommatidia straddling the sagittal midline and the angle between a pair of adjacent midplane ommatidia might be only 38°, which is the angle Frost (1975) probably measured. I will return to these ommatidial angles again when the measured angles of body and eye axes in oriented *Daphnia* are discussed. In Fig. 7.4 an idealised topography of one half of the eye and a three-dimensional sketch is presented. These figures play a role in the discussion of the orientation mechanism.

There is a reasonable close agreement about the optical properties of an ommatidium (Table 7.1), but some controversies exist and need attention. For example, Baylor and Hazen (1962) found that the form of the crystalline cone can be different at different locations within an eye as well as in eyes of different individuals. Another controversy is about blind gaps between ommatidia: are they large (Heberdey and Kupka, 1941), small or even absent (Young and Downing, 1976).

An ommatidium is the elementary anatomical and functional unit of compound eyes. In Cladocera, it is of the apposition type[3] as is usual in diurnal crustaceans with predominantly photopic vision inhabiting the littoral or the upper pelagic zone.

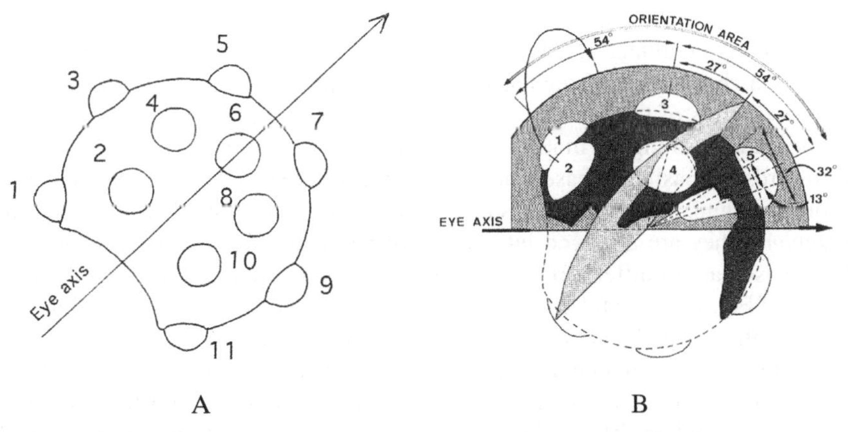

Fig. 7.4 Positions of the crystalline lenses over the compound eye of *Daphnia*. (**a**) Left half of the eye with numbered ommatidia. The stereographic projection of the non-midline lenses is not correct. The numbers are those used by Ringelberg et al. (1975) and correspond with lettering as used by Consi et al. (1990) in the following way: 1 = J, 2 = G, 3 = D, 4 = E, 5 = A, 6 = F, 7 = B, 8 = I, 9 = C, 10 = H and 11 = K; (**b**) a three-dimensional sketch of the orientation area is given as published by Ringelberg et al. (1975) (Reproduced by permission of Koninklijke Brill NV)

Table 7.1 Some optical properties of an ommatidium of the compound eye of *Daphnia*

Authors	Receptive field angle	Central zones
Heberdey and Kupka (1941)	45°	13°
Young and Downing (1976)	35°	10°
Frost (1975)		<15°
Consi et al. (1990)	40°	10°

A detailed description of the ultrastructure of the compound eye of *Daphnia* as revealed by electron microscopy can be found in Guldner and Wolff (1970).

In Fig. 7.5 an ommatidium is sketched. Most conspicuous is the crystalline cone (CC), which is secreted by and still forms a part of four cone cells (C). These are partly enveloped by covering cells (CoC) and a basilar membrane covers all. Enclosing the cone cells are seven neurosensory cells, the retinular cells (RC), arranged radially around the optic axis of the ommatidium. An eighth cell differs in size and is placed more proximal and eccentric relative to the ordinary retinular cells (Fig. 7.5a). In the centre of the retinular cells, along the optical axis of the ommatidium, is the rhabdom (R), which is a translucent, refractive cylindrical element. Dense black screening pigments in the retinular cells isolate the rhabdom. Smith and Macagno (1990) thought that the pigments do not move within the photoreceptor cells but Cellier et al. (1998) found that the black granula concentrated before sunset more distally in the retinular cells, thus near to the crystal cone, and more proximal around the rhabdom after sunset. They thought that a circadian rhythm was responsible for these migrations but additional observations are necessary to test the observations and the explaining hypothesis.

The rhabdom is apparently produced by the seven retinular cells (Waterman, 1961) although Guldner and Wolff (1970) think that also the eighth retinular cell produces a rhabdomere. Irregular compositions probably indicate different functionality of parts of the eye. The rhabdomeres are fused into one rod-like structure with one side touching the inner end of the crystalline cone and the other end extending towards the basilar or basement membrane, which bounds the proximate end of the ommatidium. A rhabdomere is an ordered array of fine tubuli or microvillii and in *Daphnia* they are arranged into two perpendicular axial directions (Fig. 7.5b). The precise pattern differs from ommatidium to ommatidium. For example, in the scheme of Fig. 7.5b, the tubuli of retinular cells 1, 4 and 5 run in one direction while those of cells 3, 6 and 7 are oriented at right angles to the former. The rhabdomeres contain the visual pigments and are the locus of the primary photochemical processes of vision. Each retinular cell sends a separate axon in a posterior direction to the optic ganglion. The eight nerve fibers from an ommatidium form a bundle together with a glial process that extends to the first layer of the optic ganglion, the lamina. Connections (synapses) are made with five laminar neurons and the bundle then proceeds to the underlying neuropilae.

This set of eight plus five fibers is called an optic cartridge (Macagno, 1984) and thus 22 cartridges enter the optic ganglion. To a large extent each neuron maintains

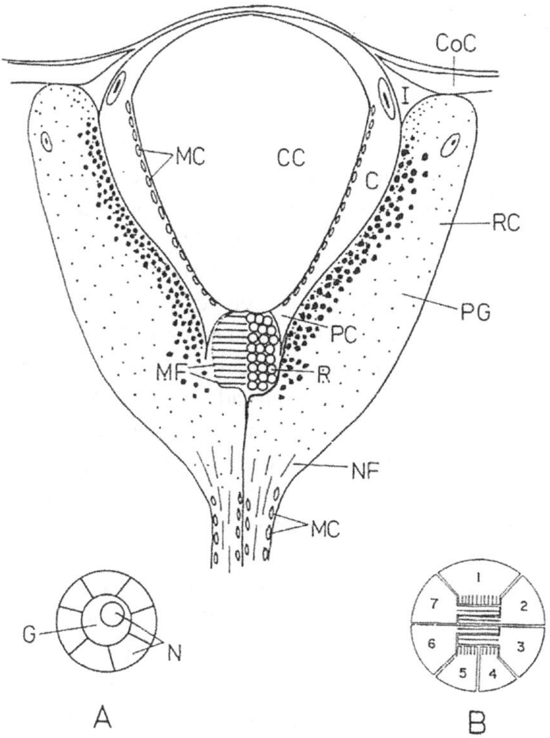

Fig. 7.5 Schematic reconstruction of a *Daphnia* ommatidium redrawn from (Ringelberg, 1987b), which was based on a figure in Waterman (1966b) and modified according to data from Guldner and Wolff (1970) and Gicklhorn (1931). (**a**) Cross-section of an ommatidial nerve with seven fibres, a central glia process enclosing the axon of the eighth photoreceptor cell (figure based on Guldner and Wolff, 1970); (**b**) Cross-section at mid-rhabdom level with the bipolar pattern of microvilli (from Waterman, 1966). C = cone cell; CC = crystal cone; CoC = covering cell; G = glia; I = intercellular space; MF = microvilli; MC = mitochondria; N = neurites; NF = neurofilaments; PC = process of cone cell; PG = pigment granula; R = rhabdom; RC = retinula cell

its individual character. Especially the axon of photoreceptor cell number 8 which has a special position along the optic axis is recognisable within the glia of the ommatidial nerve (Fig. 7.5a). The eight axons of the photoreceptor cells terminate in the anterior neuropilae, whereas the five laminar processes continue to grow into a more posterior area where synaptic connections are made with other neuronal cells.

In a series of interesting papers, Macagno and co-workers studied the embryonic development of the optic system of *D. magna* (Macagno, 1977, 1984; Consi et al., 1987). The information might be thought too detailed for an ecologist, even for one interested in behaviour. On the other hand, the effects of sub-lethal broadband UV radiation reported in Chapter 9 certainly has ecological relevance, and for good understanding some knowledge about the embryogenesis of the eye is needed. In 29- to 30-h-old embryos, small groups of embryonic photoreceptor cells of a particular ommatidium were deleted or delayed in maturation by UV. At this early stage, only few retinular cells have already sent an axon to the undeveloped laminar cells, some have made short axons and some nothing at all. The most lateral ommatidia are the first to develop. Some results of the experiments by Macagno (1984) can be summarised as follows. First, when a photoreceptor cell is radiated by UV before its axon has reached a laminar cell, this cell does not develop and dies. The making of contact by an axon is conditional for laminar development. Second, when irradiation results in partial deletions and delays of ommatidial development, the formation of aberrant synoptic connections might follow as, for example, more than one ommatidium may connect to one optic cartridge. Third, damage of the glia cells that bridge the ommatidia and the lamina layer and normally guide the axon to the lamina primordia makes that the photoreceptor axons fail to reach the lamina. Embryos thus irradiated might develop into adults (Macagno, 1977) but the development of the eye and the organisation of the lobus opticus is disturbed. Therefore, abnormal photobehaviour must be expected. *Daphnia* with eggs or embryos in the brood chamber are often found deeper in the water column and part of the adaptive significance of this behaviour might be prevention of UV damage to the developing embryo.

The eye muscles are innervated by different sets of motor neurons that are found dorsally in the supra-oesophageal ganglion. Two motor neurons innervate each dorsal and ventral muscle and one motor neuron innervates each lateral muscle. The two fibers of the dorsal and of the ventral muscles insert at somewhat different positions on the eye and are innervated by a different neuron. This suggests that these two fibers can operate independently which make fine adjustments of the dorso-ventral eye movements possible (Consi et al., 1987). For orientation of the compound eye to contrasts, a slightly lateral rotation is necessary (see Section 7.3).

7.2.2 The Nauplius Eye

This kind of eye is found in some form or other in many larval and adult Crustacea. It is present in *Daphnia* but seems to have an insubordinate function compared to the compound eye because extirpation does not change behaviour appreciable (Schulz, 1928). In *Daphnia*, the nauplius eye might be able to measure light intensity. This speculation is based on my observations that in a homogenous angular light distribution, thus with equal light intensities from all directions, a daphnid cannot maintain a normal body position but makes somersaults or lies on the bottom. However, below a certain light intensity the normal body posture is resumed. I concluded that a "dark" orientation mechanism is present which is inhibited at high light intensities, even if optical orientation is impossible (Ringelberg, 1964). Extirpation of the compound eye leads to the same somersaults, provided light intensity is sufficiently high (personal observation). As in the homogenous ALD information necessary for orientation is absent but again normal body posture is resumed at low light intensities. Therefore, even in the absence of the compound eye light intensity can be assessed. In addition, Harris and Mason (1956) found that *D. magna* without compound eyes swam in different light intensities with different swimming velocities (photokinesis). We must be careful, however. These results are no proof for light measuring via the nauplius eye: *Daphnia* is transparent and other organs might be involved.

The nauplius or median eye, as it is also called, is the most important light sensitive organ in copepods, adults included. It is considered a simple eye, a non-imaging, light gathering organ without properties for optical orientation in three-dimensional space guided by the direction of incident light (Fahrenbach, 1964). However, looking at Fig. 7.6, this nauplius eye seems to be rather complicated for a mere light-gathering organ. As will be discussed in Section 7.3, body axis orientation

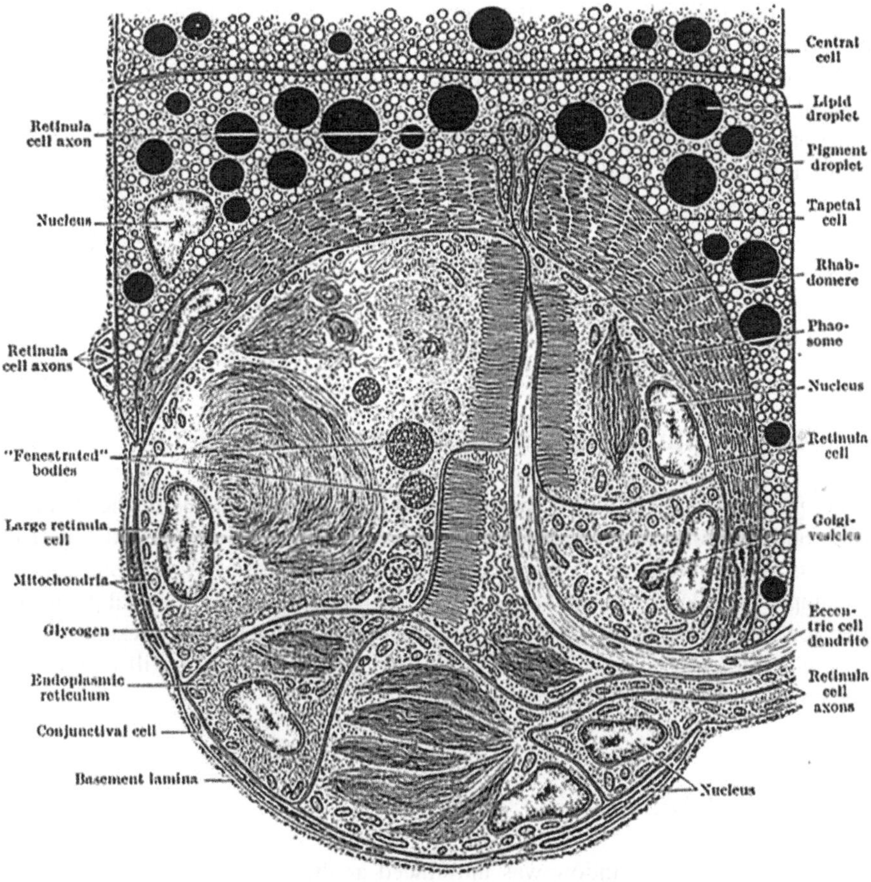

Fig. 7.6 Representation of the left dorsal ocellus of the eye of *Macrocyclops albidus* based on studies by means of the electron microscope. The anterior end of the animal is to the left, the protocerebrum is at the right margin of the picture. Retinula cells are shown, with the afferent axons and the efferent nerve (from Fahrenbach, 1964)

and directed swimming by copepods is possible and the nauplius eye is involved (Siebeck, 1980).

The nauplius eye of *Macrocyclops albidus* is composed of two dorsal ocelli and one ventral ocellus situated in the front part of the cephalothorax (Fahrenbach, 1964). The two superior ocelli point in a dorso-lateral direction and the ventral ocellus is directed towards the rostrum. Pigmented central cells partly enclose the ocelli and in well-fed animals they contain many orange-coloured droplets, giving the eye its colour which disappears when the animal is starved. In each ocellus two tapetal cells form a cup with retinula cells. These tapetal cells are filled with flat platelets consisting of refractile crystals forming a reflecting layer. Nine light receptor or retinula cells are present in the dorsal ocelli and five in the ventral one. Each ocellus contains one retinula cell that is considerable larger than the other cells. As in the compound eye of the Cladocera, all retinula cells contain a rhabdomere consisting of microvilli. The orientation of these microvilli is at right angles to the direction of incident light.

Afferent and efferent nerves innervate the eye. Each retinula cell gives off an axon in the direction of the protocerebral neuropile. The efferent innervation of each dorsal ocellus is by a single nerve fiber coming from the brain. The function of the afferent nerves is easily imagined as sending information to the central nervous system but the function of the efferent nerves is less clear, although they might initiate movements of the eyes of calanoid copepods. According to Radl (1901) the eyes of cyclopoids can not be moved. The article of Fahrenbach (1964) is a solid piece of work and has to be consulted for details on the fine structure of this complicated nauplius eye.

7.3 The Orientation of *Daphnia* in the Vertical Plane

In the previous paragraph the anatomy of the two types of eyes in the dominant planktonic animals was extensively discussed. Although the compound eye or the nauplius eye is sufficiently sophisticated to be used for image formation, they are not simple light detectors but have an orientation function in the vertical plane. The need for vertical plane orientation is not confined to *Daphnia* but present in many pelagic crustaceans in freshwater and the marine environment. I will now discuss how this is achieved in *Daphnia*.

7.3.1 Pinned Animals

In Chapter 6, Snell's window was introduced as the base of an inverted cone of downwelling light rays that were coming directly from above the water surface (Fig. 6.3). All other rays, impinging in the top of the cone (the place of a measuring apparatus or an eye), travelled over a larger distance through the water and are more heavily absorbed. This results in a large difference in light intensity or

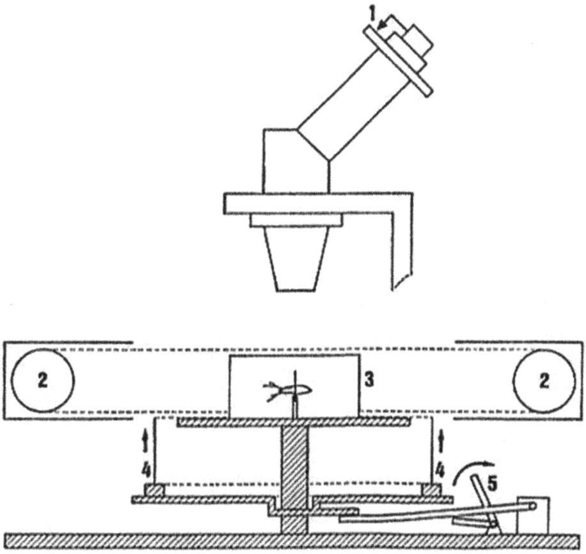

Fig. 7.7 The apparatus used to study body and eye orientations in pinned *D. magna*. (1) Binocular microscope with graduated arc and a measuring device in the eye piece, (2) circular fluorescent tube, (3) perspex box with a daphnid on a small needle placed in a glass capillary filled with wax, (4) black screen that can be moved upwards and downwards by means of levers (5) (from Ringelberg et al., 1975; Reproduced by permission of Koninklijke Brill NV)

contrast making up the edge of the window at an angle of 49.8° to the normal. I hypothesised that the eye of cladocerans (and also copepods) is well equipped to perceive these light–dark boundaries. To test this, animals were kept in place in the apparatus presented in Fig. 7.7 (Ringelberg, 1964).

An individual of *D. magna* was fixed horizontally by means of a small needle through the distal end of the brood chamber such that the animal was free to rotate around the needle. The animal was positioned in the centre of a round perspex box, which was placed in the centre of a circular fluorescent tube. A black screen could be lifted in order to screen off the light from the tube for 270°. In that way the screen edges produced two contrasts imitating the borders of Snell's window[4] in the median plane of the animal. In Fig. 7.8, the angular light distribution (ALD) in the centre of the apparatus is compared with an ALD, as measured in the ocean at 20 m depth (Whitney, 1941).

With the screen down, animals rotated steadily in either a backward or a forward direction but stopped as soon as part of the light was screened off. With a few strokes of the antennae the body axis was turned towards the lighted area or towards the dark area (Fig. 7.9) and then the obviously oriented *Daphnia* made vigorous beatings with the second antennae. The particular body position was maintained for some time and could be measured. The two different positions can be called positively phototactic and negatively phototactic orientations, respectively. Although swimming towards or away from the light is not possible in this pinned situation, the antennal beatings indicate that this would have occurred if the needle had not prevented this. If now either the dorsal or the ventral edge of the screen was shifted over some distance, the animals directed towards the lighted area followed the dorsal edge; those turned towards the dark followed the ventral edge. It was concluded

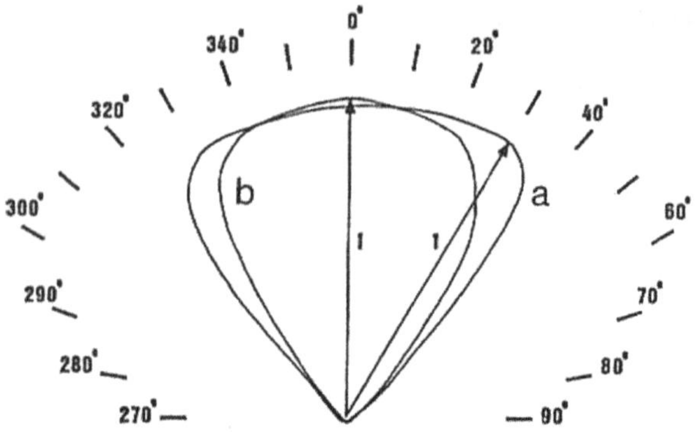

Fig. 7.8 (**a**) The light vector diagram of an angular light distribution (ALD) measured at 20 m depth in the ocean by Whitney (1941). Measurements in a plane perpendicular to a plane including the sun. (**b**) The light vector diagram as measured in the apparatus, shown in Fig. 7.7, after a screen of 270° was lifted (from Ringelberg et al., 1975)

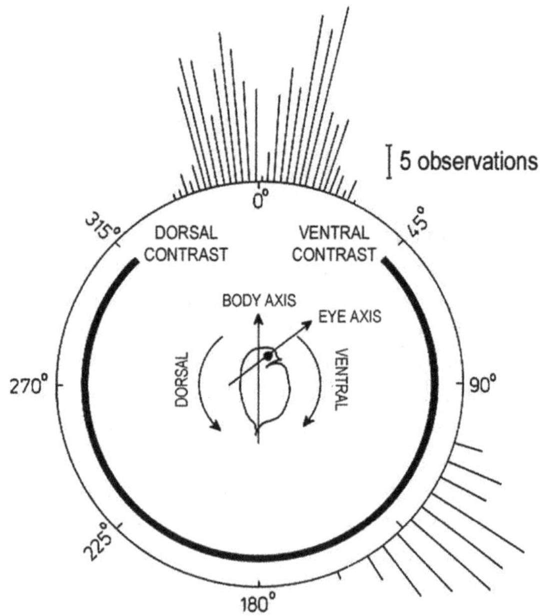

Fig. 7.9 Terminology as used with the experiments with pinned *Daphnia*. The results of body axis orientations towards the lighted area (positive phototaxis) and towards the dark (negative phototaxis) are also presented (combined data from Ringelberg (1964) and Ringelberg et al. (1975) as presented by the last authors; Reproduced by permission of Koninklijke Brill NV)

that optical orientation to the screen edges was involved and in both cases the edge at the dorsal side of the animal was the beacon for this orientation. One amazing fact became clear: an individual turning dorsally unoriented in the homogeneous circular light field always turned the body towards the lighted area when the screen was up, while animals previously rotating over ventral, turned towards the dark area (Ringelberg, 1964; Van Gool, 1999). Obviously, the "intention" to orient to the light or the dark is present before orientation becomes possible! For an individual animal this "internal motivation" – a term borrowed from ethology (Baerends, 1957) – is not always the same but depends on internal and external circumstances. For example, Ringelberg (1964) described how a positive female started to moult on the needle and had to be removed because shedding of the carapace was obstructed. After a few days of swimming freely, the animal was tested again and now proved to be negatively phototactic. The examples of orientation given by Van Gool (1999) are good illustrations of the "internal motivation". With a similar apparatus, he studied orientation of some of the "persistently" positively and negatively phototactic *D. magna* selected by De Meester (1991, 1995) (see also Chapter 2). In Fig. 7.10, orientations of three of these strange *D. magna* are presented in the presence and absence of fish kairomone.

The two persistently positively phototactic *Daphnia* (A and B) behaved differently when kairomone was added: A did not change orientation, but B became negatively phototactic. Negative phototaxis is usually seen in the presence of fish kairomone. An external factor changed the "internal motivation" in daphnid B but not in A, suggesting that the strength of this motivation to be positively or negatively phototactic can be variable. From the beginning individual C had a strong tendency to turn to the dark and this negative phototaxis did, of course, not alter when fish kairomone was added. The model of oscillating swimming discussed in Section 5.2 indicates how we can imagine that "motivation" has a neural base. If the two neural oscillators are out of balance, primarily positive or negative phototaxis results. A problem with kairomone experiments is that the used concentration is unknown which means that the power to induce a negative motivation might have been less in animal A than in B (Van Gool, 1999).

The positively phototactic orientations, presented in Fig. 7.9, are bimodally distributed. Averages are $\overline{m}_1 = 13°$ and $\overline{m}_2 = 349°$. Probably, the orientations at the left hand side of 0° do not occur in nature because the centre of body gravity will prevent this. A *Daphnia* resting horizontally on the small capillar is not troubled by gravity and this angle of orientation can become visible. Anyhow, the results indicate that two different orientations to the contrasts and subsequent body positions are possible. When the results of orientations of the eye axis are presented the discussion will be continued.

7.3.2 Free-Swimming Daphnia

Of interest is a comparison of body axis orientation in pinned daphnids and body postures of free-swimming animals. Ewald (1910b) mentioned an angle of 20° with the perpendicular for *D. pulex* and Fraenkel and Gunn (1961) of 20–30° for *D. longispina*. Scuba diving in Lake Constance, Schröder (1962) estimated angles of 20–25° also for *D. longispina*.[5] He also observed that these animals swam not perpendicular but obliquely down. This compares well with the body positions of negatively phototactic orientations presented in Fig. 7.9. Within the confines of experimental tubes, rapid downward swimming caused by increases in light intensity also occurs in zigzag paths or in spirals. Measuring body postures in free-swimming animals is difficult but a modest attempt was made by Van Gool (1999). He placed the apparatus of Fig. 7.7 in a vertical position and took photographs of the swimming animals. The majority of individual daphnids swam from left to right and vice versa. With differently sized screens, contrasts were presented at different angular distances. As long as both contrasts made the same angle with the vertical, body axes were the same in either direction of the swimming. With contrasts at 315° and 45°, thus comparable to those of Snell's window, body axis orientation was on average 26.1° (s.e. = 0.69). This agrees with those mentioned for free-swimming animals by Schröder (1962), but is more oblique than in pinned animals (see Figs. 7.9 and 7.10). With each stroke of the second antennae, the body

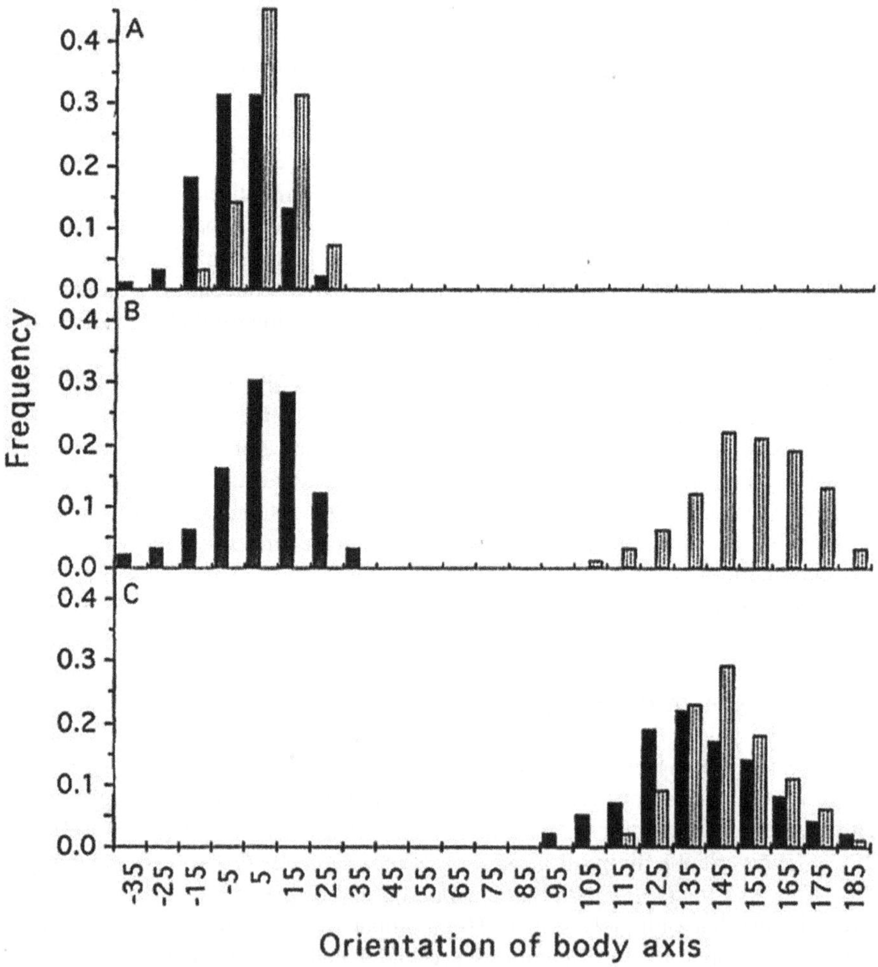

Fig. 7.10 The orientation of the body axis in relation to a screen with edges at 45° and 315° (see Fig. 7.9). A. clone $P_1$3260; B. clone $P_1$3285; C. clone $P_1$3296 as provided by Dr. L. De Meester, Leuven. The *black bars* represent frequencies of orientations in the absence of fish kairomones and the *open bars* orientations with fish kairomones present (from Van Gool, 1999; Reproduced by permission of E. Scheizerbart'sche Verlagsbuchhandlung)

of a swimming daphnid tilts forward which might explain the difference. When left and right contrasts were at 70°, the body axis turned to nearly the same angle and thus swimming was in a more horizontal position. The shore of ponds and lakes alters Snell's window and shore-directed contrasts might be at a variety of angles. Pond-dwelling species like *D. magna* and *D. pulex* possibly have more flexible orientations, although lake-dwelling daphnids are also able to swim in a horizontal position. This occurs during the so-called "shore-avoidance" reaction (Siebeck, 1969) when orientation is related to the uplifted, shore-directed contrast. In large vessels, *D. magna* often swims horizontally with the dorsal side upwards to and from opposite walls or, if the vessel is circular, in a crowded carrousel near to the wall. Therefore, the horizontal swimming of *D. magna*, mentioned by Harris (1953) and Harris and Wolfe (1956), is certainly part of the behavioural repertoire.

In the experiments with free-swimming animals (Van Gool, 1999), the two contrasts did not function independently. If the left and the right contrast were located at different angles, for example, 315° and 70°, respectively, swimming to the left was at a slightly larger angle (on average 32°) than the expected 26.1°. Swimming to the right was at a slightly smaller angle than expected (55° instead of 66°). An ANOVA confirmed the interaction ($P < 0.001$) (Van Gool, 1999).

7.3.3 The Mechanism of Contrast Orientation

The orientation function of the compound eye in *Daphnia* has been studied for a long time (Radl, 1903; Hess, 1910; Ewald, 1910a, b, c; Von Frisch and Kupelwieser, 1913). It was found that the eye followed the movement of a small light bundle but quantifications were not made. The importance of the light as a contrast was not recognised and no connection with Snell's window was made. A detailed description of body and eye axis in relation to two contrasts is given in Fig. 7.11. An individual of *D. magna* was placed in the apparatus shown in Fig. 7.7 either completely fixed with the head directed to the centre of the lighted part (C_L), or in such a way that rotation round the needle was possible. In the first case, only eye rotations could be measured; in the second case, oriented body positions were also measured. With C_L at 45° – and thus a dorsal contrast at 0° and a ventral contrast at 90° – no turning of the eye occurred.[6] This is why the origin of absis and ordinate in Fig. 7.11 was placed in this point. With the muscles of the eye relaxed, the eye axis makes an angle of about 45° with the body axis: this is the basic rest position .

Lifting the screen in an either dorsally or ventrally shifted position caused rotation of the eye as if to correct for the changed position of the screen. The correction is not complete, however, since the "compensation" angle is smaller than the angle over which the screen is shifted. This results in Fig. 7.11 in lines either above or below the diagonal line. Such residual values are characteristic of cybernetic steering mechanisms. The two function lines extend over distances of about 45–140° and 45–310°, or 95° in a ventral direction and in a dorsal direction. Because of the uncertainty in the precise rest position and the error in measurement, we can assume

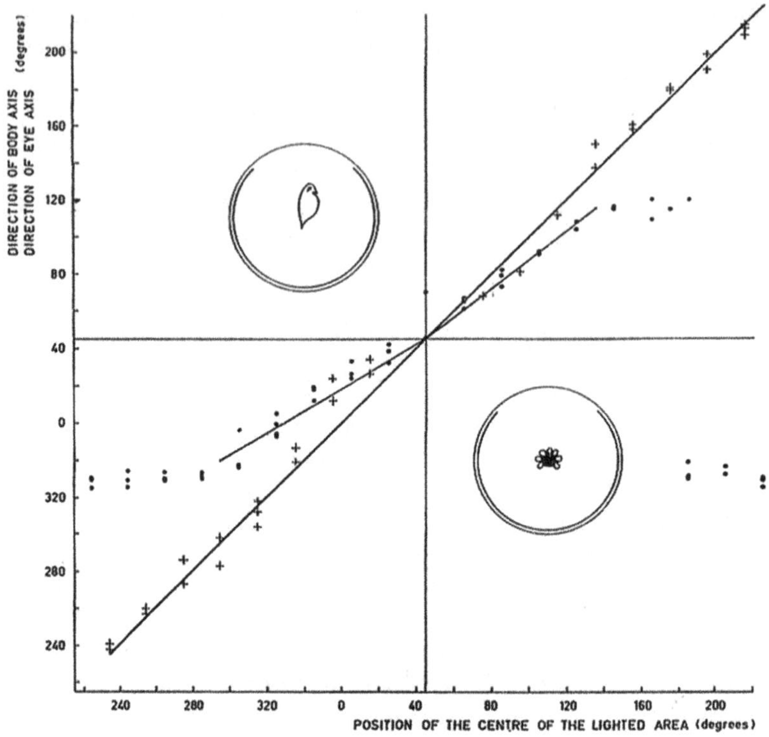

Fig. 7.11 The direction of the eye axis (*dots*) to different positions of a screen of 270°, when a *D. magna* is fixed in the point 0° position. The crosses give the direction of the body axis of the same animal to different positions of this screen when the animal is set free to rotate round the needle. The insets give oriented positions of body and eye with regard to the dorsal edge of the screen (from Ringelberg, 1964)

that eye rotations are possible over equal distances in both directions with a total action radius of about 190°. When the centre of the lighted area C_L exceeds 180°, the eye attains no stable oriented position but rocks alternatively between a ventral and dorsal direction. Eye rotations are anatomically constrained and both contrasts are outside the part of the eye where ommatidia with an orientation function are positioned (see further on).

Of interest is the position of the body axis in a freely rotating animal. Outside the region of 310–140°, thus outside the range of possible rotations of the eye in a fixed animal (see Fig. 7.11), the angle of the body axis is located on the diagonal line. The body seems to turn into a position comparable to the one occupied when C_L is at 45° and the eye is at rest. It is strange, however, that within the range of possible eye rotations, the body follows the eye, although lagging behind. An explanation might be that the body position follows the orientation of the eye and is determined by the eye.

The orientation experiments were repeated by Ringelberg et al. (1975) but now the body of *D. magna* was fixed and the screen was lifted in different positions. The eye turned in the usual way and after it stopped it was noted whether the eye reacted to a small displacement of the dorsal, the ventral or to both screen edges. The results were comparable with previous observations, although the position of the relaxed eye with regard to the body axis seemed to be shifted ventrally. Consequently, the angular rotation distance in the ventral and dorsal direction was different but the total action radius had remained the same, viz. 190° again. Another important difference with respect to the animals, used by Ringelberg (1964), was that both the dorsal and the ventral contrast seemed to be used for orientation of the body axis. The freely swimming *D. magna* in the experiments of Van Gool (1999) also used both contrasts, as was mentioned previously.

A summary of the results is given in Fig. 7.12 to illustrate that both contrasts are used in eye orientations. The figure also indicates which part of the compound eye is used for orientation. For example, in Fig. 7.12c and d, a shift of either contrast causes a turning of the eye. In both cases, the contrasts can be projected on the dorsal part of the eye axis (drawn line with arrow). However, if the screen is lifted with the contrasts in a slightly more dorsal position with regard to the eye axis, as in Fig. 7.12e, the eye only reacted to one of the contrasts. Obviously, the other one was outside the orientation range of the eye. It was concluded that only a part dorsal of the eye axis is used in contrast orientation. One boundary is between 10° and 14° (see Fig. 7.12b and c) and the other boundary is between 120° and 134° (see Fig. 7.12d and e). The total extension of what was dubbed the "orientation area" of the compound eye of *Daphnia* is about 106–124°.

Functional specialisations of particular eye regions are well documented in other arthropods (Consi et al., 1990). These areas might be anatomically visible as flattened arrays of ommatidia with small interommatidial angles. Also the compound eye of the cladoceran *P. pediculus* has anatomically distinct regions in its single, fused eye (Nilsson and Odselius, 1983; Odselius and Nilsson, 1983). In *Daphnia* no distinct array of ommatidia is visible. Therefore, a functional specialisation seems to precede an anatomical adaptation.

We can now hypothesise how "contrast orientation"[7] operates in *Daphnia*. The area used for orientation is located dorsal from the eye axis and extends over an angular distance between 106° and 124°. The optical axes of the median plane[8] ommatidia 1, 3 and 5 are within these limits and this also holds for the lateral ommatidia 2 and 4 on both sides of this plane (Fig. 7.4). Which ommatidia are used in contrast orientation can as yet only be deduced from the experimental results. The bimodality of positively phototactic orientations in Fig. 7.9 suggests that two different fixations of the dorsal contrast are possible. Assuming that the angle between the axis of an oriented eye and the body axis is constant the average mutual distance of 24° between these two fixations suggests that a median and a transversal ommatidium are used because the optical axes of these ommatidia are 27° apart. Ringelberg et al. (1975) found that alternative negatively phototactic body orientations also are close to 27° apart ($\bar{m} = 137$, $s_m = 2$; $\bar{m} = 109$, $s_m = 3$; Total $N = 417$). To use a lateral ommatidium in orientation, tangential rotations of the eye are necessary.

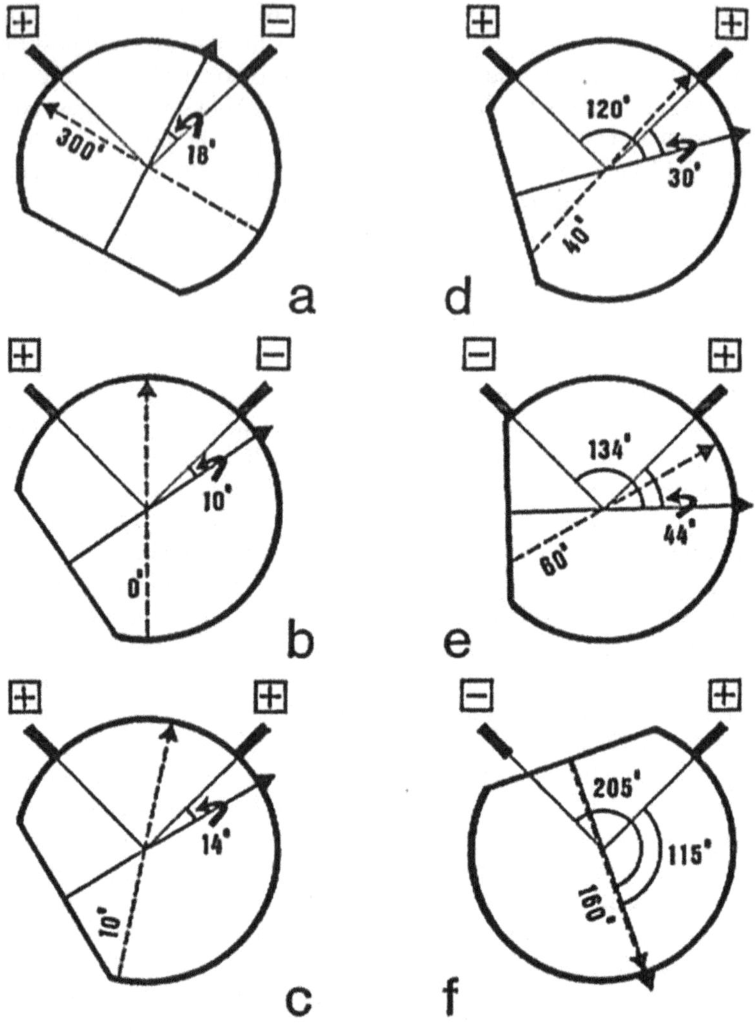

Fig. 7.12 Projections of the dorsal and ventral screen edges (*black bars*) on the compound eye in different positions of orientation. *Broken lines* with arrow indicates the direction of the body axis; the *continuous lines* with arrow that of the eye axis. The plus or minus sign indicates presence or absence of a reaction to a shift of the particular contrast (from Ringelberg et al., 1975; Reproduced by permission of Koninklijke Brill NV)

An angle of tangential rotation of 35° has been mentioned (Eckert, 1935), which enables ommatidia 2 or 4 to play a role (see also the discussion in Notes 7–12). The eye is constantly in tremor and makes saccadic movements (Consi et al., 1987). Frost (1975) and Harris (1953) have suggested that the function might be to prevent adaptation, as in humans. Ommatidium 3 is probably the most important one for orientation because it is at the position of the "null region" of no eye rotations in the experiments of Consi et al. (1990) (see Note 6). These authors stated that the eye and the body system may always attempt to point the null region at a strong contrast boundary. However, in some animals both contrasts play a role, therefore more than one ommatidium must be involved at the same time.

We must now consider how *Daphnia* might use this mechanism of orientation in nature. Below the water surface, Snell's window provides contrasts for orientation to maintain a more or less upward body posture during swimming. Contrast orientation is also used in diel vertical migration. An increase in light intensity causes a ventral turning of the compound eye which is in the direction of the contrast at the ventral side of the animal.[9] Ifthis contrast is then fixed by ommatidia of the dorsal orientation area and the body axis is turned ventrally to the eye axis, as was observed in the experiments, the body is directed obliquely down. The increased activity of the antennae, as occurring in a fixed animal, suggests a swimming response and the result is swimming downwards obliquely. I have already mentioned that Schröder (1962) found *Daphnia* to swim obliquely down in nature. A decrease in light intensity causes the eye to rotate dorsally and the contrast at the dorsal side is fixed. The body axis is kept dorsally from the eye axis, which means a nearly vertical position in the water. Upward swimming follows.

Positively and negatively phototactic orientations are thus highly comparable, although different contrasts are used. The difference in position of the body axis with regard to the eye axis in both ways of orientation makes that the ommatidia ventral of the eye axis are turned to the dark, away from the cone of light in negative phototaxis, while these ommatidia are illuminated by the cone of light in positively phototactic orientations. Whether these differences play a role in the distinction between the two different orientation modes or whether they are merely a consequence of eye and body positions is unknown.

In nature two sources of optical information can be distinguished: contrasts and "light as such" (Verheijen, 1986). Both play a role in orientation. The positively and negatively photo-orientations of *Daphnia* are examples. In the vector field of light directions a contrast is fixed and, subsequently, by using the scalar field of light intensities, the body axis is turned such that ommatidia of the non-orientation area perceive a high or a low light intensity, respectively. Normally, the vector field and the scalar field are spatially related but in the experiments of Clarke (1930), with a convergent bundle, this natural relationship was severed which caused negatively phototactic *Daphnia* to move into increasingly higher light intensity.

The observations by Van Gool (1999) with pinned *D. magna* showed that fish kairomones stimulated the ventral turning of the compound eye. With the coupled, obliquely downward directed body axis, this would lead to downward swimming, which is of adaptive significance. Interesting from a physiological point of view is that the "internal motivation" to do so is present before eye orientation is displayed. Obviously, the direction of turning of the eye is under efferent, central control.

The picture of contrast orientation given above is incomplete, as is illustrated by differences in "rest" positions of the eye. Ringelberg (1964) found that with muscles relaxed the eye axis makes an angle of 40–45° with the body axis. Van Gool

(1999), however, found 25° for the persistently positively phototactic clone P 3285 and 50° for the persistently negative clone P 3296. Both clones belong to the aberrantly behaving *D. magna* isolated from populations of "normal" animals by De Meester (1991, 1993a,b,c). "Normal" *Daphnia* turn the eye axis in these directions as responses to light intensity decreases and increases, respectively. The origin of the aberrant rest positions is unknown. The fusion of the two eyes during embryogenesis might be a delicate process and prone to aberrations. Macagno (1977, 1984) described how abnormal synaptic connections in the visual system were formed after UV radiation, although the animals developed into adults. Variability in number of fibres per eye muscle was also observed, even within clones (Consi et al., 1987). Whether also the embryonic development of the eye muscles is sensitive to UV, for example resulting in different insertion points, is unknown. A study of development and behaviour of the strange *D. magna* must clear the issue.

The model of contrast orientation, presented here, is almost exclusively based on research by Ringelberg and co-workers. Therefore, it is important that comparable experiments, with a different objective in mind and by different methods, were done by another research group. Consi et al. (1990) probably learned of contrast orientation in a late stage of research because a different terminology was developed but this makes their research all the more independent. Consi et al. (1990) stimulated a fixed *D. magna* with a small stripe of light[10] impinging from different angles on the eye and measured the resulting eye rotations. The light was also moved over the median plane and the pursuing ("tracking") of the eye was measured. I will discuss some relevant results.

If illuminated from different angles, the compound eye rotated and kept the attained position for the duration of the stimulus. If the light felt on ommatidium D (number 3, see Fig. 7.4 for the conversion of letters in numbers) in the middle of the dorsal part of the compound eye, no rotation occurred. The authors called this the "null" region of the eye. Obviously, rotation of the eye was unnecessary because the eye was already in the oriented position. Stimulating ventrally from this "null" region evoked a ventral turning, stimulating dorsally caused a dorsal rotation of the eye. In both cases the eye turned to bring the light stimulus into the direction of ommatidium 3 and the extent of rotation depended on the distance of the stimulating light beam to that ommatidium 3. For beams incident more ventrally than 30° from the eye axis, an initial ventral rotation occurred but the eye oscillated irregularly. This phenomenon was also described by Ringelberg et al. (1975). An eye that is unable to rotate far enough to pick up a contrast starts with maximum ventral and dorsal rotations alternately.

The eye also followed the stripe of light, when it moved slowly back and forth over the dorsal half of the eye. Such following of a moving light spot was already described by Von Frisch and Kupelwieser (1913). The light was followed over a distance of about 100°. This was called the "tracking zone" by Consi et al. (1990). An identical part was called the orientation area by Ringelberg (1964) and Ringelberg et al. (1975). The angular extension is close to the 106–124° proposed by these authors.[11] Consi et al. (1990) concluded that fixation of the stripe of light is possible with the three median ommatidia in this "receptive" field, numbered 1, 3 and

5 in Fig. 7.4. No role was assigned to the laterally positioned ommatidia 2 and 4. Ringelberg et al. (1975) thought it necessary to include these ommatidia in order to explain the bimodel body axis orientations. Consi et al. (1990) discussed their results in terms of contrast orientation and Snell's window while referring to the publication of Ringelberg et al. (1975).

7.4 Optical Orientation in Other Planktonic Animals

It might be expected that contrasts of Snell's window are also used by other zooplankton species, for example, copepods. The eye of copepods consists of three ocelli (Section 7.2) and might be considered simple, but is too complicated for mere light perception. The ocelli of calanoids are able to make small movements, which is important for contrast orientation. The observations on "shore avoidance" in lakes and relevant experiments on directed swimming by Siebeck (1965, 1968, 1969, 1980) revealed that optical orientation is involved.

In early morning, planktonic animals that have drifted into the littoral zone swim away from the shore into the direction of the open water (Fig. 7.13). Siebeck suspended a flat cylinder with 16 partitions, close to the wall, in shallow water of the littoral zone of the Lunzer Untersee (Austria). After some time, animals that had been liberated in the centre had swum into the partitions and were counted. Independent of the place around the lake, the resultant swimming direction was always away from the shore. The angular light distribution (ALD) near the shore is different from that of the open water. In the direction of the shore light intensity is lower and the light–dark boundary of Snell's window is "elevated". The ALD near the shore is bisymmetric with the plane of symmetry directed from shore to open water. Siebeck suggested that zooplankton used this plane of symmetry to swim towards the lighted part of the open water: "the copepods apparently orient themselves in a plane of symmetrical stimuli, which they find in the plane of symmetry of the angular light distribution" (Siebeck, 1969, p. 839). Probably, he thought of an equal illumination of the left and the right side of the optical sense organ. This reminds one of the so-called "tropotaxis" (Fraenkel and Gunn, 1961), which was a supposed type of orientation in the first part of the last century.[12] However, more is needed for "shore avoidance": animals must also turn the body axis more or less horizontally in order to swim away from the shore. Ringelberg (1969)suggested that orientation to the "elevated" shore-directed contrast would tilt the body axis horizontally (see Fig. 7.13) making swimming away from the shore possible, at least in *Daphnia*.

For the calanoid *Mixodiaptomus laciniatus* orientation was worked out more precisely (Siebeck, 1980). A white translucent plexiglass dome was placed over a large circular swimming arena. The dome was blackened except for an oval segment that was uniformly illuminated. Inside the arena the ALDs were different depending on the location (Fig. 7.14). All light vector diagrams were bisymmetrical except at one location where the vector diagram could be divided by two perpendicular vertical

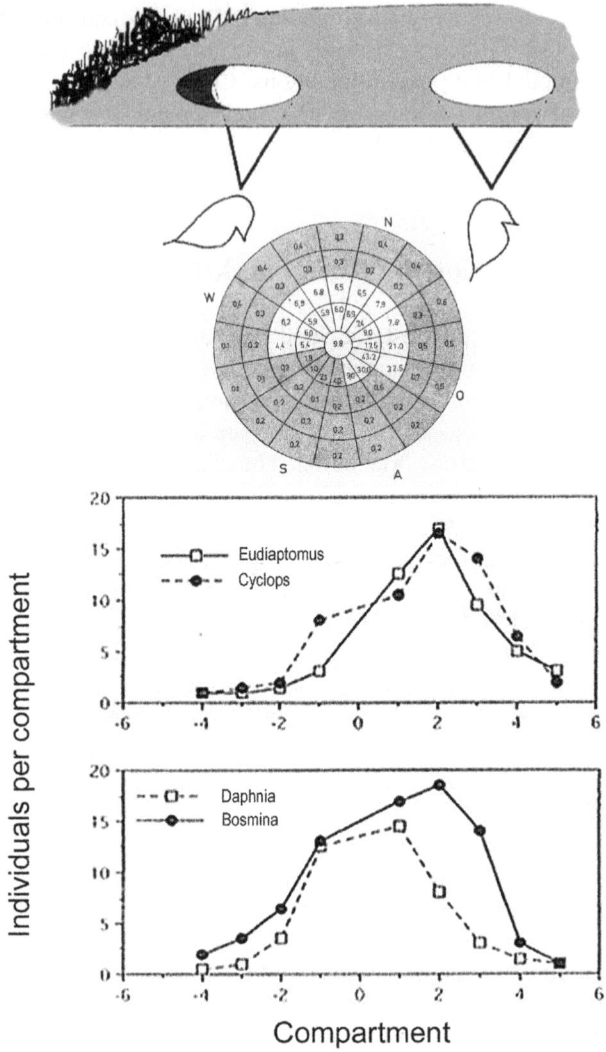

Fig. 7.13 Shore avoidance. In the upper part a simple representation of Snell's circle near the surface of a lake in the vicinity of the shore and out in the open water is given. Under water *Daphnia* orient to the elevation of the shore-directed contrast with a more horizontal body posture (free after Siebeck, 1980). The circular light intensity distribution of the middle figure was measured near the shore of the Lunzer Untersee (Austria) (Siebeck, 1968, Fig. 33). North is about the direction towards the open water. The lower figures give the number of animals in the compartment of a flat circular cylinder, placed in the littoral zone at the place where the light intensity measurements were made. These figures are composed of those in Fig. 33 of Siebeck (1968). The zero compartment corresponds to the direction North (N) of the middle figure; West (W) is number –4 and East (O) is +4

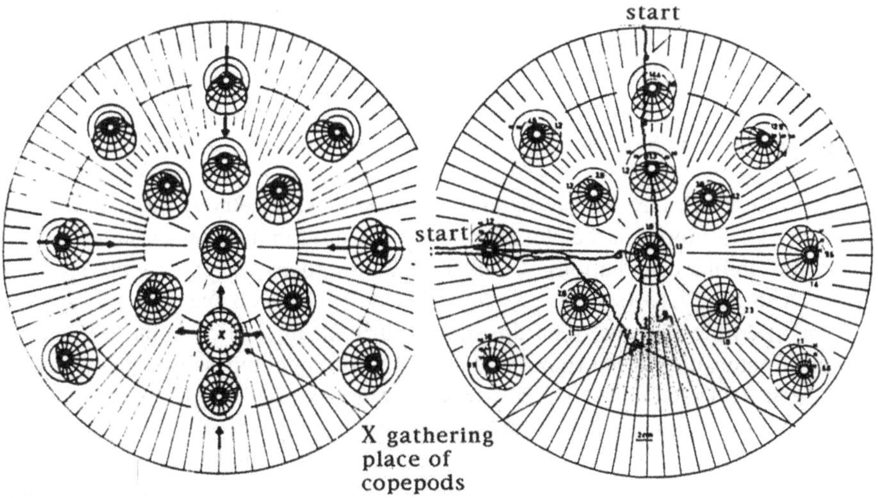

Fig. 7.14 Light vector diagrams of several ALDs under the large dome (radius = 85 cm) used by Siebeck (1980) to study orientation behaviour in *Mixodiaptomus laciniatus*. The transparent swimming arena had a radius of 45 cm and was placed under the centre of the dome. At X, the ALD is circular and comparable to that of the open water and at this point the copepods gather. Some swimming tracks are illustrated (from Siebeck, 1980)

planes of symmetry. All individuals gathered at this place (indicated by X), independent of the starting position. Swimming paths in Fig. 7.14 show that the initial bearing need not be in the direction of the final destination but takes place within the plane of symmetry of that particular place.

Since the ALDs change during swimming and, thus also the direction of the symmetry planes, the animals alter directions accordingly until X is reached. No "preferred" direction exists at this place and *Mixodiaptomus* concentrate at this place.

Siebeck (1980) does not refer to contrast orientation but I think his explanation of the swimming directions makes use of it. A model copepod with two pairs of ocelli was introduced (actually only three ocelli are present) and all four were thought to be fixed on the edges of Snell's window. In the pelagic area this would result in a vertical body position with no "preferred" direction of swimming. Near the littoral, however, the shore-directed contrast is shifted in a vertical direction and Siebeck suggested that one ocellus of a pair is aimed to this elevated contrast. This is possible only if the body axis is tilted into a more horizontal position. The symmetrical illumination of the two pairs of ocelli is now disturbed. The other ocellus of the pair remains somehow fixed to the contrast of Snell's window directed to the open water and in this way symmetrical stimulation of the two opposing ocelli is re-established. The two other ocelli remain directed left and right to the edges of Snell's window. This model seems to be a hybrid between swimming in the plane of symmetry and contrast orientation. A problem is that the eye of a copepod consists of only three ocelli but Siebeck (1980) solved this stating: "it can be shown that there are sensory cells in the unpaired pigment cup which are in a common plane with sensory cells from one of the paired pigment cups. It is therefore possible to compare the brightness in two planes that are perpendicular to each other" (p. 35).

I tried to determine to what extent the plane of symmetry correlated with the direction of swimming in the plexiglass carousel, used in the Lunzer Untersee. The data of the circular distributions of four zooplankton species together with the ALDs are given in Fig. 7.13 (see Fig. 33 in Siebeck, 1968). As a measure of "symmetry" was used the ratio between the sum of all light intensities at the left and the sum of all intensities at the right hand side of a certain compass direction. Siebeck gives these light intensities as measured under angles of 30° and 50° with the vertical. If the ratio equals 1, symmetry of the angular light distribution is complete and under Siebeck's hypothesis all animals should be found in the compartment of that particular compass direction. With increasing ratios, the number of animals ought to decrease. Therefore, a negative correlation is expected between the ratio of these light intensities and the number of individuals in a corresponding compass direction. It proved that for none of the species the correlation was significant in the 30° light field (Table 7.2). Moreover, the regression coefficients were positive, not negative. For the 50° fields, significant regression coefficients were found for *Cyclops tatricus* and *Bosmina longispina*, not for *Eudiaptomus gracilis* and *D. longispina*. In all cases, R^2 was small thus the symmetry of the light field does not contribute importantly to the direction of swimming. As an alternative hypothesis, the importance of the contrast realised by the "elevation of the horizon" in a particular compass direction was compared with the number of individuals in the compartment of that compass direction. Light vector intensity values in the direction away from the shore at 30° and 50° were divided by the intensity at these angles in the same plane but in the direction towards the shore. With Snell's window in the open water, the ratio would be 1. Near the shore and depending on the extent of the "elevation of the horizon" the ratio would be larger than 1. If orientation occurred to the contrast of the elevation, it was expected that the number of individuals in the compartments directed to the open water would be a function of the strength of the contrast, thus the value of the ratio. The result of the regression analysis is given in Table 7.3. For all plankton species, studied by Siebeck (1968), a significant regression was found.

I therefore concluded that orientation to the contrast provided by the "elevation of the horizon" is of more influence on the direction of swimming than the plane of symmetry. In Fig. 7.15 the number of all zooplankton individuals in the diverse compartments is compared with the contrast ratio. A weaker contrast is accompanied by a smaller number of individuals swimming in that plane. The gradual decrease in numbers left and right from the maximum can be ascribed to a decreasing contrast ratio resulting from a gradual disappearing effect of the "elevation of the horizon". The ratio in the plane parallel to the shore is near 1, as it is for all directions in Snell's

Table 7.2 Results of regressions of the radial distribution of some zooplankton species and the ratio between the sum of light intensity left and right (symmetry) from a compass direction; *a* is regression coefficient. Calculations based on Fig. 33 in Siebeck (1968)

Zooplankton species	Comparison of left–right light intensity at 30°	Comparison of left–right light intensity at 50°
Eudiaptomus gracilis	$a = 5.35$; $P = 0.603$; $R^2 = -0.096$	$a = -4.89$; $P = 0.063$; $R^2 = 0.327$
Cyclops tatricus [a]	$a = 3.47$; $P = 0.015$; $R^2 = 0.535$	$A = -1.15$; $P = 0.011$; $R^2 = 0.578$
Daphnia longispina	$a = 15.9$; $P = 0.149$; $R^2 = 0.169$	$a = -4.66$; $P = 0.140$; $R^2 = 0.182$
Bosmina longispina	$a = 12.16$; $P = 0.146$; $R^2 = 0.158$	$a = -5.00$; $P = 0.024$; $R^2 = 0.475$

[a]For cyclops small numbers were present.

Table 7.3 Regressions of the radial distribution of some zooplankton species and the mean ratio of light intensities at 30° and 50° at both sides of the perpendicular in a particular vertical plane; *a* is regression coefficient. A ratio of 1 indicates complete symmetry in the angular light distribution as present in Snell's window of the open water

Species	*a*	*P* "model"	R^2
Eudiaptomus gracilis	2.065	0.021	0.494
Cyclops tatricus [a]	0.380	0.028	0.454
Daphnia longispina	2.563	0.008	0.603
Bosmina longispina	2.246	0.001	0.804

[a]For cyclops small numbers were present.

window of the open water. Thus no horizontal body position and no swimming parallel to the shore is expected as was actually found by Siebeck (1968). I concluded that the studied freshwater zooplankton species are able to orient to contrasts of the angular light distribution as realised by Snell's window.

Optical orientation in the important group of marine zooplankton, the euphausiids, was studied by Land (1980). *Euphausia* have well-developed compound eyes that are, just as in *Daphnia*, capable of rotation in the sagittal plane in reaction to a moving light source. The eye rotations are accompanied by up or down movements of the tail: it sweeps upwards when the eye rotates in a caudal direction and downwards when the eye rotates forwards, which suggests that the steering of the body is linked to eye movements. Land (1980) showed that the eye tracks a moving light but lags behind in relation with the position of this stimulus. If the stimulating light is in front of the animal, the eye turns slightly posterior. If it is behind, the eye is turned anterior. Lagging behind the shifting light source can also be concluded from the course of the two lines in Fig. 7.16: they are not parallel to one another nor to the (not drawn) diagonal line. The trajectories of the four individuals, studied by Land (1980), were rather irregular probably because the light source in the tracking experiments was moved from front to the tail of the animal and back again, which would

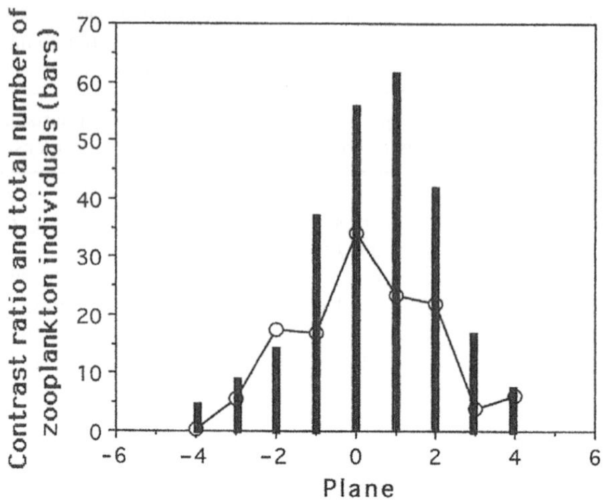

Fig. 7.15 A comparison of contrasts in shore-directed planes (*black bars*) and the number of zooplankton individuals (*open circles*) moving into the lake-directed compartment of that plane. Compartment 0 is North–South directed and more or less perpendicular to the shore line. With an "elevation of the horizon" the value of the ratio is larger than 1, a ratio parallel to the shore is, as in all directions in the open water, about one. Data were derived from Fig. 33 in Siebeck (1968). See also text

lead to a different extent of lagging. When all data from the four individuals were averaged, the regular curve of Fig. 7.16 was obtained. This figure allows for another interpretation than the one proposed by Land (1980). For example, this author thinks that the "uncertainty of the eye position" when the light is between 90° and 162° (the horizontal part in Fig. 7.16) is due to the "discrepancy between artificial and natural stimuli" (p. 258). Land is focused on the tracking function, although he realised that "in reality the euphausiid eye would not be attempting to track a small light source but would use the downwelling light for orientation". Inwhat way this is done is not mentioned.[13] I suggest that the constant position of the eye when the stimulus light moves between 90° and 162° indicates that orientation is complete and that no additional adjustment through rotation is necessary. Outside this traject, the eye has to rotate to keep the "orientation area" directed towards the light spot. The highest precision is at about 125° for the light angle and eye angle (angles with regard to the body axis). If the light source impinges on the eye from an angle outside the 90–162° range, Land (1980) found that the tail flips in an appropriate direction to bring the body axis back into the oriented position. The mechanism is comparable to what we have seen in *Daphnia* (compare Figs. 7.11 and 7.16). Both figures suggest that a cybernetic steering mechanism is involved which is not amazing because control mechanisms are universal. If the light source is outside the anatomically limited tracking range, the eye flips over 180°, as the compound eye of *Daphnia* does.

A light source as the one used by Land (1980) cannot easily be compared with the natural underwater light distribution. Nor are the experimental data sufficient to get

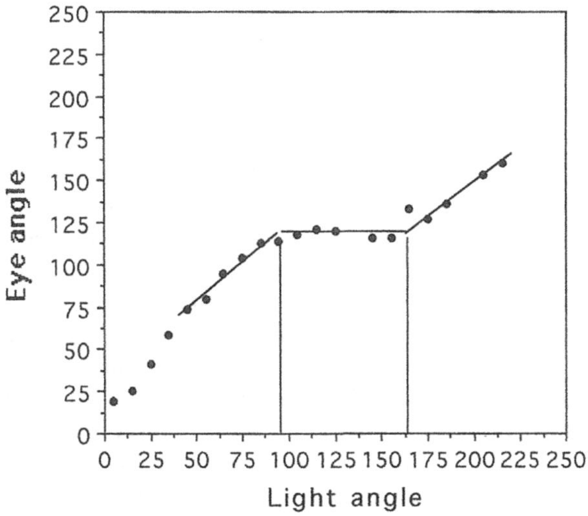

Fig. 7.16 The position of the eye as a function of the position of a light source in *Nematoscelis atlantica* (Euphausiacea). Angles (degrees) are in relation to the body axis. Data were derived from Fig. 2 in Land (1980). Measurements within class intervals of 10° were taken together and each point is a (mean) value of 1–4 eye angles

a clear picture of the orientation of these euphausiids. Considering a "rest" position of 125° between eye and body axis, an oblique body posture might be predicted. This seems to be the normal body posture in free-swimming animals, indeed (see Plate 3).

Price (1989) observed free-swimming euphausiids in the tower tank at Dalhousie (Canada). Individuals of *Thysanousa raschii* oriented relative to the vertical axis. They swam upwards at an angle of 57.4° ± 28.6° with the vertical. Orientation changed into a more horizontal direction within an algal patch and sinking stopped almost entirely. This suggests that food (as kairomones do in *Daphnia*) influences the angle of contrast orientation .

To determine the angle of swimming during evening ascent migration, De Robertis et al. (2003) used high-resolution, multi-sonar techniques. During the phase of fast upward swimming, the average angle with the horizontal was 12.5° ± 37.6° and 18.9° ± 38.7° on 11 and 12 August 1996, respectively. On average this is 74° with the vertical, which is not a very efficient direction of swimming for the rapid evening ascent. During the early and late slow phases, the angle could be directed even downwards. This is strange, of course. I think that at a certain moment only part of the individuals in the population really moved upwards, while other individuals did not. The large standard deviation points in this direction. Also the observation "that most observed swimming speeds were substantially lower than the upper bound imposed by the tracking procedure" (p. 892) corroborates this. Migration is discontinuous and interrupted by phases of no swimming as predicted by phototactic swimming in *Daphnia*. After a rapid upward response to a relative decrease in light intensity, the stimulus becomes zero and might even shortly change into an increase. If so, the positively upward swimming would stop or might change into a short downward movement. Interrupted swimming provides an explanation too for the relevant observation by Heywood (1996). She describes that individual euphausiids in the leading edge of a migrating sound scattering layer occasionally moved back to the centre of the layer. I will return to this phenomenon in Chapter 11 to explain how SSL can migrate more rapidly than an isolume.

Different species of euphausiids have different types of eyes and this seems to be correlated with depth distribution. A detailed study of the optical orientation in these important pelagic crustaceans would be interesting.

Tracking experiments reveal physiological aspects of an eye but are difficult to apply to optical orientation in the natural environment. As Consi et al. (1990) did, Land (1980) also tried to extrapolate tracking data to oriented behaviour in a pelagic environment. Although a point source of light is an excellent stimulus for tracking experiments, it lacks essential features of the natural angular light distribution which might have prevented to think of contrasts as the real beacons for orientation. Land (1980) wrote: "if we accept that what the eye is 'trying' to do is to track the direction of the downwelling light ... then it is evident that the 4° light-guide used here is not very appropriate as a stimulus" (p. 257). The ALD at 50 m depth in the ocean is not a small light source and "in reality the euphausiid eye would not be attempting to track a small light source, but rather the centre of a rather wide patch of light whose brightness falls of rather gradually with increasing angle" (p. 258). How to orient to the centre other than by fixation of the delimiting boundary or contrast? Contrasts are stimuli for oriented vision. Also a small light source is defined by the intensity difference with the darker surroundings as is a small white spot on a black background or a black spot on a white background. The universal aspect which makes things visible (apart from colour) are light–dark boundaries or contrasts. Low resolution eyes like the compound eye of *Daphnia* and even the nauplius eye of calanoid copepods are able to use light–dark boundaries for vertical plane orientation of the body axis. When resolution increases, image formation becomes an extension of contrast perception. Nevertheless, the more simple function of vertical plane orientation is maintained. The compound eye of predator euphausiids and predator cladocerans seems to be examples of this twofold function.

Notes

1. One additional ommatidium to fill the gap would be impossible because the eye consists of two embryological halves. The assumption seems to be chosen in order to get the same angle as Frost (1975).
2. Consi et al. (1990) say four, but I think they forgot the two pairs bordering the centripetal gap. See Fig. 7.4, the numbers 1 = J and 11 = K
3. In apposition eyes the crystalline cone and the rest of the ommatidium is enclosed in pigment, such that only light rays passing the cone can reach the photosensitive part of that ommatidium. In the superposition eye, light rays incident on different ommatidia are collected on retinular cells of a ommatidia.
4. In the first experiments by Ringelberg (1964) this angle was chosen instead of 97.8°, as actually found in Snell's window. In all subsequent experiments by him and co-workers the angle of 90° was maintained.

5. It is possible that this *D. longispina* in Lake Constance was actually *D. hyalina* or *D. galeata*.
6. Moving a small light bundle around the eye, Consi et al. (1990) found that with the light impinging on a similar small dorsal part of the eye, no "tracking" (that is, following the light with an eye rotation) was evoked. They called this part the "null region" because of the absence of this "tracking" reaction. The name is appropriate although misleading because it suggests insensitivity of that part of the eye. Obviously, the authors did not recognise that rotation is unnecessary because the eye is already oriented with a light in this position.
7. It is not intended to introduce a new kind of vision with this term. Apart from colour vision, also image formation is based on contrasts. Optical resolution of a *Daphnia* eye with only 22 ommatidia is small and images cannot be produced. This is no problem to a filter-feeding organism in an optical empty environment. With considerable more ommatidia, as in the eyes of predators like *Holopedium* and *Leptodora*, crude imaging of prey might be possible. We do not have much information about these eyes. *Daphnia* eyes with "contrast orientation" might be considered at the beginning of a long route towards the evolution of multifaceted compound eyes with image forming abilities.
8. Consi et al. (1990) say two ommatidia but they probably forgot number 1 in Fig. 7.4.
9. The contrast of Snell's window is a continuous circle, of course. A dorsal or a ventral contrast is always with regard to the animal's position.
10. A small light might also be considered a contrast. Von Frisch and Kupelwieser (1913) used a small light bundle and Ringelberg (1964) argued that orientation was comparable to that with contrasts.
11. Consi et al. (1990) recognised the likeness of their results and those of Ringelberg et al. (1975): "in addition, the area of the eye which became fixated at the contrast boundary was a dorsal region which was roughly the same as the tracking zone which we have identified" (p. 419). Moreover they write that based on their results, it can be proposed "that the two contrast boundaries on either side of Snell's window may provide a dual stimulus, and these cause the eye to remain stationary when their effects on the oculomotor system are balanced" (p. 419). They have, however, not carefully enough studied Ringelberg (1964) and Ringelberg et al. (1975) when they write that a daphnid will orient according to the dorsal light reaction.
12. The orientation mechanism, called "tropotaxis" (Fraenkel and Gunn, 1961), was based on the movements of various species along the diagonal path in the direction of two light sources that were placed in front of an animal. It was thought that a comparison of light intensities, falling on the left and right eye, was made and that a bilateral balance was maintained while moving. Eckert (1935) and Alverdes (1924) performed this so-called "Zwei Lichterversuch" with *Daphnia*. Depending on the angular distance between the lamp on the left and right (horizontal), swimming was in the median plane, until the lamps were seen under an angle of 90°. Then the diagonal path was abandoned and the animals moved randomly to the left or the right lamp. If light intensity of one of the light sources was made higher, more animals moved to the brighter one. These results do not prove unambiguously that moving is within a plane of symmetrical brightness. Other explanations are possible. For example, Von Buddenbrock (1952) thought that the lamps were used as orientation points. On reaching the "Entscheidungspunkt" (90°) the two lamps could not fall simultaneously on the "Fixierraum der Augen". The results of experiments of Eckert (1935) are in agreement with this hypothesis, although the author interpreted them in favour of tropotaxis.

    Experimental evidence for tropotaxis, thus a bilateral balance of light intensities at the left and right hand side of the median plane of animals, also *Daphnia*, is poor (Wehner, 1981).
13. We have to assume that "orientation to downwelling light" is, in fact, orientation to contrasts if we try to apply the *Daphnia* model. The eye of Euphausiacea is sufficiently complicated to suppose that both contrasts of Snell's circle can be used. If both are used in equal importance, "light coming from above" represents such orientation. As in *Daphnia*, we have to assume that in upward and downward orientation the angle between body axis and eye axis is fixed.

Plate 2 Adult female of *Calanus pacificus* (total length about 3.3 mm) from the Main Basin of Puget Sound, USA. (Photo by Kathy Newell; Courtesy of Bruce Frost, University of Washington, Seattle.)

Chapter 8
Considerations Before Going into the Field

Contents

8.1 General Introduction

If the depth of a zooplankton population in daytime differs significantly from the depth at night, the population is said to migrate. It has become customary to speak of *diel* vertical migration, thus of a phenomenon involving a 24-h cycle. In the older literature, the term *diurnal* was used, indicating that downward and upward swimming occurred in daytime. Since light was (and still is) considered important as a causal factor, the term diurnal was obvious. Sometimes migrations were called *nocturnal* if a single maximum at the surface was observed sometime between sunset and sunrise (Hutchinson, 1967). Also the term *twilight* migrations was used if it was thought that the vertical movements occurred at dawn and around sunset. I think a classification of migration patterns is difficult to make, hard to apply and not useful. The classifications suggest something about the underlying mechanism, for example, that the initiating stimulus in diurnal migrations operates during the daytime light period or during the night in nocturnal migrations but nearly always information from the field was not detailed enough and about the stimulus nothing was known. So the neutral term *diel* has come to be used but the experience from the past makes it clear that more information must be obtained than is possible with a few samples per day if DVM is to be understood.

Because the distribution of individuals over depth can be irregular, the depth of a population is not simple to estimate. For a mere demonstration of the presence of DVM, as was the objective of most of the earlier studies, statistical analysis is not necessary if migration is over a large distance with the daytime and night-time distributions far apart. But if this is not so and distributions have to be compared in a quantitative way either within or between species, a numerical summary is needed.

J. Ringelberg, *Diel Vertical Migration of Zooplankton in Lakes and Oceans*,
DOI 10.1007/978-90-481-3093-1_8,

As a measure of central tendency of a vertical distribution the mean population depth (MPD) can be used:

$$\mathrm{MPD} = \frac{\sum_{i=1}^{n} n_i \times d_i}{\sum_{i=1}^{n} n_i}$$

Depth (d_i) is weighted by the number of individuals (n_i) found at that depth. Because many vertical distributions are asymmetric or skewed, with the longer tail often pointed to greater depth, the arithmetic mean might be unduly influenced by a few individuals far from the centre. Logarithmic transformation of depth deals with this problem. If absolute light intensity is thought to play a role in determining the vertical distribution, as was the paradigmatic explanation for tens of years, the choice of logarithmic depth is justified because light intensity also changes logarithmically with depth. A log-normal transformation was successfully applied by Geller (1986) in his study of DVM in Lake Constance (Fig. 8.1). In Fig. 8.2, the expected linear-normal (open dots) and log-normal (black dots) frequency distributions are compared with the observed distributions (bars). The presented daytime distribution is reasonable well approached by a log-normal distribution as a high correlation ($R^2 = 0.918$) between observed and expected values demonstrates, while without transformation the correlation is poor ($R^2 = 0.564$). The night-time distribution is

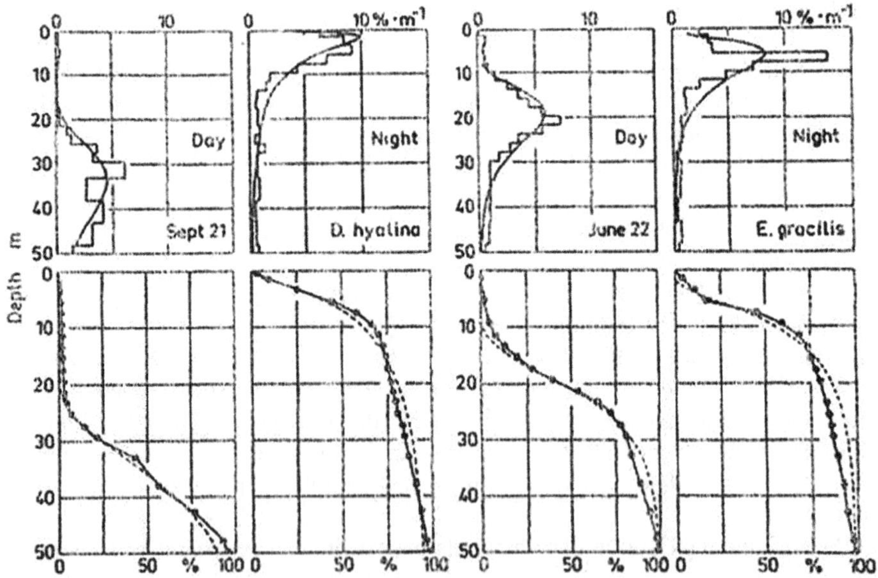

Fig. 8.1 *Upper panel*: relative frequencies as percentage of total population per metre, the observed distributions are shown by histogram, the approximated log-normal distributions by a line. *Lower panel*: corresponding cumulative frequency distribution curves; the observed values are shown by *dots* connected by a full line, approximated log-normal distribution by a *broken line* (after Geller, 1986; Reproduced by permission of E. Scheizerbart'sche Verlagsbuchhandlung)

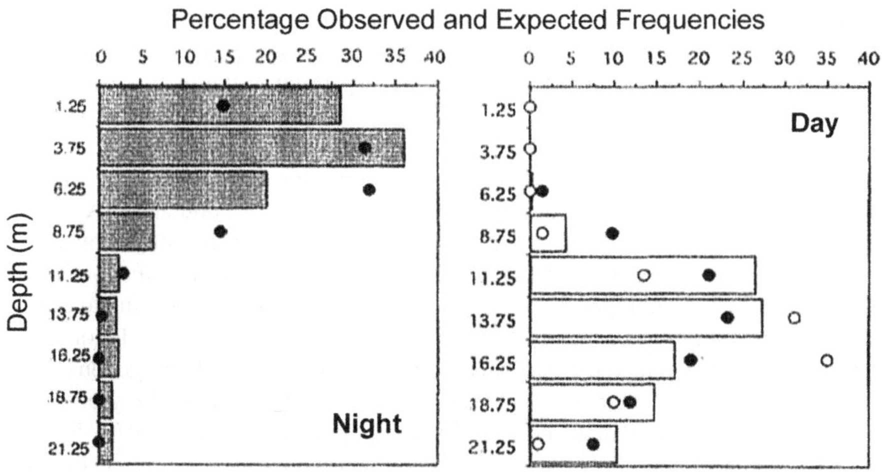

Fig. 8.2 A night-time and daytime vertical distribution of *D. galeata x hyalina* in Lake Maarsseveen on 10 June 1992. *Bars*: observed distributions. *Black dots*: log-normal frequency distributions. *Open dots*: linear-normal frequency distributions

also better described by a log-transformed distribution ($R^2 = 0.723$) and the linear frequency distribution (not presented) fits badly ($R^2 = 0.341$). Upward migrating animals tend to stop on approaching the water surface, which easily leads to a skewed distribution. The relatively few individuals, remaining at greater depth during the night, draw the average to a deeper depth layer. A back-transformed mean of a logarithmically transformed depth repairs this effect to a certain extent. For example, the arithmetic mean is 4.97 m for the night-time distribution in Fig. 8.2, while the geometric mean is higher up in the column at 3.66 m.

Strongly tailed distributions do not conform to standard probability models. Another descriptive statistic, the median, deals with skewness because the frequency distribution is divided into two halves, thus with 50% of the individuals on either side of the median. For the given night-time distribution, the median moves even closer to the surface, at 2.9 m. The disadvantage of this measure of central tendency is that a simple use of statistical tests is not possible. Medians were used by Hart and Allanson (1976) to compare vertical distributions with changes in the depths of an isolume during morning descent and evening rise.

The number of sampling depths is often too small to calculate the best fitting frequency distribution. Geller (1986) had 20 depths at his disposal, which is an imposing number for a DVM study. Nine depths as in Fig. 8.2 are not impressive but taking large samples simultaneously, which is necessary if the velocity of migration has to be estimated, is much work. Using linear and multiple regression, Ringelberg and Flik (1994) studied the effect of absolute light intensity and relative changes in light intensity on the displacement velocity of a migrating *Daphnia* population. The velocity was calculated from the difference in vertical distributions sampled with a time interval of 30 min. The distributions had to be characterised by one quantity and the difference in mean depth of the population was used. There must be a reason for an elaborate statistical presentation. Although Geller (1986) made one of the best statistical presentations, visual inspection of the figures led to most of his conclusions and without use of the theoretical distributions, means or standard deviations.

The calculation of an expected frequency distribution presupposes that the observed distribution consists of a homogeneous biological population. A skewed distribution can be repaired by a logarithmic transformation but might obscure a possible difference in genetic composition and, consequently, in behaviour of the individuals. Some genotypes do not migrate but stay near the surface in daytime or remain deep down during the night. Not only the *statistical* data population might be inhomogenous but also the *biological* population might be inhomogenous.

The vertical "displacement" of frequency distributions tells nothing about the movements of the individual members of the population. Pearre (1979) has discussed possible swimming behaviour of individuals and the type of vertical distributions one can expect. If individuals move synchronised, individual swimming speed equals displacement velocity of the population. If individual swimming is interrupted and not synchronised, with short alternate periods of downward and upward movements, the net changes in depth of the population do not reflect behaviour of individuals. We must also consider that the individual animal does not move precisely in a vertical plane and thus the vertical migration of a population is the vector sum of the components in the vertical plane. If upward and downward swimming of individuals is asynchronously distributed over the daily cycle, the centre of the population remains at a certain depth and the conclusion will be that no DVM occurred. This example represents an extreme case. If, however, all individuals occasionally swim upwards from about the same depth to another depth, the surface for instance, sampling will reveal a bimodal depth distribution. Especially so, if the stationary time at the extreme ends exceeds the transit time of swimming. Bimodal distributions were described in the literature but are no proof of asynchronous migrations of individuals. It is possible that only part of the population migrates while another part is stationary as, for example, when genetic differences in migration behaviour exist. Obviously, it is difficult to make definite conclusions about the behaviour of individuals from changes in population centres. This hampers the application of experimental results to the situation in the field. Explanations of DVM are based on assumed behaviour of individuals and hypotheses keep coming.

The study of DVM goes back for more than a century to Weismann (1877) who described migrations of zooplankton in Lake Constance (Germany). Since then, the number of publications has become very large. Generally, the described pattern is more or less the same, viz. a descent in early morning and an ascent in the evening, although reverse migrations have occasionally been observed (see review in Bayley, 1986). Large variations exist in the distance between the central depth in daytime and night-time. The amplitudes range from a few metres in lakes to several hundreds of metres in the oceans. Amplitudes differ between juveniles, males and females. Species behave differently within and between lakes or parts of the ocean, moreover at different times of the year. In other words, many values of the migration distance can be expected.

Differences also exist in the depth where morning descent starts. Especially in oceans, species migrate over different levels of the water column and the day depth of species A might be the night depth of species B. Separate layers of migration are less pronounced in lakes, although in deep lakes species may ascend from the hypolimnion into the metalimnion, while others leave the metalimnion for a near-surface depth in the evening.

A "unifying theory that explains the different patterns in all taxa" (Lampert, 1989, p. 25) will not easily be found if such unification is possible at all. Diversity in behaviour is the inescapable result of evolutionary development necessary for survival. Planktonic life is confined to the vertical plane. Inevitable problems of community life such as competition and predation must be solved in that two-dimensional space and DVM contributes importantly. In terrestrial biotopes, an important way to release competition is through habitat differentiation in a three-dimensional, highly structured space. The pelagial biotope is uniform in horizontal directions at least as far as the action radius of an individual plankton organism reaches and, compared to the terrestrial biotope, rather uniform in a vertical plane too. Therefore, habitat differentiation cannot be the universal solution. Only the oceans are deep enough to provide some vertical differentiation. For example, the composition of sinking food items changes with depth and can be responsible for food niche differentiation. The pelagial is without the everywhere available refuges to escape predators and only travelling in the vertical plane might offer a solution. To a large extent ecological knowledge about community structure and function originates from terrestrial research. However, pelagic environments are so different that this knowledge cannot simply be adopted for planktonic life and it is possible that for similar problems different solutions were found.

Variation in DVM originates at several levels of biological integration. Not much is known about genetic differences in migration behaviour, but examples of different genotypes with different migration patterns within or between populations are available. Within- and among-population genetic differences are certainly at the base of differences in migration amplitude in different lakes. For example, the parents and hybrids of *D. hyalina* and *D. galeata* in Lake Constance (Weider and Stich, 1992) migrate with considerable different amplitudes. In Lake Maarsseveen, the multilocus allozyme genotypes of these hybrids also differ in migration amplitude (Ringelberg et al., 2004). The parent species are probably genetically different from those in Lake Constance and local adaptation may be involved. From a set of phenotypes, short-term selection will lead to some "optimal" migration amplitude. Optimality is temporary, depending on the environmental situation, and a dominant phenotype is a passing favourite of the selection process. The current dominant type is replaced next year. This makes the notion "optimality" of limited value in a dynamic plankton world.

Different thresholds of photobehaviour or of sensitivities to predator kairomones may be the physiological cause of different amplitudes between and within species. Also during ontogenetic development the changing physiology must be at the base of different migration amplitudes of juveniles and adults. The often heard explanation is that juveniles are less predated upon by visual predators and do not need to perform large migrations. Although possibly true, this is an ultimate explanation and does no justice to the fascinating ontogenetic development in behaviour. Of which hardly anything is known.

The variations in vertical migration can be interpreted as quantitative differences of adaptive goals but it is insufficient to say that the ultimate reason for DVM is, for instance, the avoidance of visual predators. The intensity of the predation threat determines amplitude and predation pressure has to be quantified somehow. Other reasons might play a role at the same time. The ones proposed in the past such as a metabolic advantage (McLaren, 1974), interrupted grazing (McAllister, 1971) and starvation avoidance (Geller, 1986) were all rejected by Lampert (1993) in favour of the hypothesis of prevention of predation by visually hunting fish. The case is strong, certainly in lakes and especially for *Daphnia*, but other adaptive reasons must be considered too, especially for oceanic migrations.

In "normal" DVM, individuals move within 24 h from an upper part of the water column to a deeper part, and vice versa. We must suppose that in both parts circumstances are advantageous but also disadvantageous. If not, the animals would stay in the best of the two. In ecological jargon, costs and benefits are involved and the values of these conditional contributions to fitness determine whether migration is adaptive or not. Also the time spent in either part is of influence although fixed by the time of descent and ascent migration, amplitude, season and latitude. When comparing DVM in polar and tropical regions, for instance, these aspects are worthwhile to consider. Quantitative assessments of costs and benefits have not been made but qualitative reflections have provided ideas about the nature of migrations. For example, if visually hunting predators are present, the benefit of leaving the well-lit upper part of the water column is certainly a reduction of mortality by predation. The costs in the refuge consist of lower temperatures and lower food availability, both hampering birth rates. (A benefit of lower temperatures is an increased longevity and larger size.) If predation is always sufficiently large, and thus predictable, the "trait" DVM may become genetically fixed, or even disappear from the population because the individuals stay out of reach permanently.

Predation by 0+ fish fluctuates seasonally and the ratios between costs and benefits shift in time which makes that a conditional migration behaviour is more successful than a constitutive or fixed defence. When DVM is induced phenotypically, no costs have to be made in the absence of a reason for migration. A fast detection of a threat and a timely change in behaviour must be realised, of course. A physiological mechanism of sensory perception, integration of information and locomotor activation, in other words a sensor–actor chain, as discussed in Fig. 1.1, makes this possible. If the threat is by planktivorous fish, the cue is a kairomone, the role of which is now generally accepted as conditional for extensive, seasonal migrations in lakes (Lampert, 1989, 1993).

With a conditional migration, variable costs and benefits explain different amplitudes. When will the best amplitude be realised? According to the central neo-Darwinian paradigm, the fittest genotype is defined by having the highest relative reproductive success. The ratio between the risk of mortality by predation and the opportunity for reproduction must be such that the reproductive output is maximised. Application of this logic should enable the individual animal to choose the best possible migration amplitude. However, without evidence for a physiological mechanism to evaluate risk and opportunity, a claim like this is what Dennett (1995) would probably call a statement attached to a skyhook, instead of one that is fastened to a crane firmly fixed on the ground. Lampert (1989) writes that predation pressure on the one hand and food availability, together with the thermal gradient, on the other hand, should determine the pattern of DVM. He gives some examples in his review paper. For instance, if the costs of staying in deep water are small because differences in temperature and food concentration are small compared to those in the epilimnion, animals should stay down all the time. As an example, Lampert refers to very eutrophic Polish lakes. Animals with a genetically fixed migration do not need or are unable to make decisions and individuals can permit to be unaware of the environmental conditions. However, if migration is conditional, assessment of environmental conditions is necessary to make the decision to migrate or stay down or remain in the epilimnion. Hardly any suggestions for decision-making were made in the literature because it is difficult to study and (therefore?) not fashionable. Speculative predictions of adaptive behaviour have heuristic value but tend to stop instead of initiate research. Another example: if food conditions deep in the water column are very poor and near the surface insufficient for compensation during the short night-time feeding period, animals should not migrate but take the risk of being eaten by the predator. According to Lampert (1989), this situation was present in enclosures operated by Johnson and Jakobsen (1987). No migration occurred when food concentrations were low at all depths but when food was added to the upper part migration soon started. Why would *Daphnia* leave a food-rich surface to spent part of the time without food? This only makes sense if fish kairomones were also present but Johnson and Jakobsen (1987) do not mention this.[1] Only if we assume that a low concentration of kairomones in the water of the enclosures was present, the result can be understood and predicted by the decision-making mechanism, which was presented in Fig. 4.6.

Sometimes the formulation of ultimate aspects threatens to confuse causes and results, which leads to deceptive explanations. For example, large zooplankton, which are more easily detected by visually hunting fish predators, will perform more pronounced migrations than smaller animals (Lampert, 1993). Indeed, there are many examples that juvenile copepods and cladocerans are less deep during daytime than the larger adults. From an adaptive point of view such size-dependent vertical distributions can be expected. Consequently, the next question to ask is by what proximate mechanism differently sized individuals are able to migrate differently. De Meester et al. (1998) writes of "a deep-rooted relationship between day-depth preference (DVM behavior) and body size in zooplankton". "Preference" suggests an active choice and, thus, a recognition of depth layers with different properties relevant to the criterion of choice. The authors continue with: "for relatively large zooplankton, the daytime vertical distribution will be largely *determined by the changes in visual predation pressure* with depth, ..." (my italics). This "determined" points to a causal relation and, consequently, "changes in visual predation pressure" must be perceived by migrating individuals. In the view of De Meester et al. (1998), DVM is part of a general "depth selecting behaviour". But no indication for a physiological mechanism that makes depth selection possible exists. The danger is that suggestive formulations hamper research into those mechanisms that really explain how differently sized individuals migrate with different amplitudes. Let us look from a different angle at size-dependent vertical distributions. Suppose a prey population consists of migrating and non-migrating genotypes. Evidence for the existence of differently migrating sub-populations is available. Size selective predation in the lit upper part of the water column makes that the size distribution of the resident sub-population shifts to a smaller average size. In addition, a relatively high reproduction in the food-rich layer leads to a younger age distribution with smaller sized individuals. The predator kairomone in the water also results in a smaller size at first reproduction as life-history experiments have revealed. The combined results make that the non-migrating sub-population consists of small individuals. On the other hand, the migrating sub-population will tend to be composed of larger individuals. At low temperatures, individuals grow to a larger size than at high temperatures. Growth is not as fast as in the warmer upper layers and predation is less severe, thus the individuals live longer to attain that larger size. Therefore, visual predation and DVM both lead to daytime vertical distributions with smaller individuals in the upper part and larger individuals in the lower part of the column. This differential size distribution is a passive effect and not the result of an active choice of differently sized animals. With this reasoning, size-dependent differences in predation are not denied but I ask attention for detailed research and not to be satisfied with an ultimate reason that seems to explain a phenomenon.

In a subsequent review paper, Lampert (1993) made a list of predictions as consequence of the predator-avoidance hypothesis. The predictions are based on our being aware of costs and benefits, of predation risks, food availability and thermal gradients. We have, however, no idea what information a zooplankton individual has nor whether it is able to use it to make decisions. It is assumed that evolution must have moulded the animal such that adaptive behaviour is present. This may be right and without problem if behaviour is genetically fixed. If DVM is conditional, however, individuals are supposed to have flexible behaviour, to evaluate the environment and to make adequate decisions. Lampert's (1993) predictions have a qualitative character, but assessments and decisions must be quantitative. He was aware of this problem and proposed that future research must concentrate on the quantification of the interactions of predation pressure, food availability and temperature. Up till now, this has not happened. The narratives of evolution-oriented migration research are still predominantly qualitative and speculative and meanwhile I think it has come to a dead end. De Meester et al. (1998) wrote that costs and benefits are easy to measure and can be expressed in "units of commonly accepted fitness parameters such as the intrinsic rate of increase r". Why then has it hardly been done and is it so difficult to derive costs and benefits from descriptions in the literature.[2] If depths distributions, food concentrations and thermal gradients are described, costs in terms of fecundity and development time might be approximated. The estimation of benefits is more difficult, however, because the size of a fish population is mostly unknown and mortality by predation rarely estimated in the context of DVM.

In the literature of the last 30 years proximate aspects of DVM barely got attention. Obviously, it is taken for granted that zooplankton are able to migrate and that the presence of fish kairomones suffices to realise DVM. However, if it is predicted that descend must start early in the morning in order to avoid fish predators (Lampert, 1993), some other stimulus must trigger the animals because fish kairomones are present the whole day conveying no temporal cue. Nor do kairomones realise a directional cue and, thus, a steering stimulus for swimming in the vertical plane. In the Chapters 3 and 7, both aspects were discussed. Research in the field must be done to generate hypotheses and to test the validity of the experimentally found mechanisms.

Studies of DVM were often limited to a few days only. In few articles, the actual descent and ascent phases were described in sufficient details in order to make a connection with experimental data possible. Migrations were observed in a variety of lakes and oceans and locations range from polar regions to the tropics. If well-known lakes were studied, additional information is available, and I have chosen these lakes as starting points for analysis. Proximate and ultimate aspects are both important. Guidelines may facilitate the analysis of the literature. For example:

1. Is DVM present during the whole year or only during a seasonal period? This might be an indication of a constitutionally or a phenotypically induced migration, respectively. In both cases, it is interesting to find out the ultimate reason. If the migration is phenotypically induced, causes or rather the "set of necessary ecological conditions" must be investigated. The amplitude might change during the migration period which can be the result of a quantitative change in adaptive reason. For example, predator abundance might have changed. If so, what proximate cue makes animals aware of the change? Quite different factors may be involved, such as a developing temperature gradient that inhibits downward swimming thereby constraining amplitude.
2. A timely retreat into deeper parts of the lake in the morning is thought a necessary consequence of the predator avoidance hypothesis (Lampert, 1993). This is true but a descent migration at dawn does not necessarily exclude other reasons for DVM. If continuously changing light intensities of sufficient magnitude are needed to trigger downward swimming, migrations can only occur during dawn, whatever the ultimate reason is. Light intensity changes were not measured often, and certainly not at dawn when descent migration starts. The time of descent swimming can be used to estimate roughly the prevailing light changes. Better evidence for the role of light must come from correlations between displacement velocities and measured light intensity changes.
3. The evolutionary biologist is interested in the ratio of costs and benefits because it is a criterion for the outcome of selection. If different genotypes in a lake show differences in migration amplitude they are prone to a different risk of predation but also to a different food availability and different temperatures. The "optimal trade-off solution" is a fiction. Noncommittal statements do not contribute to

understanding.[3] The ultimate fitness criterion is progeny. The production of eggs and juveniles is primarily determined by metabolism, thus food availability and temperature. The latter is easy to measure, but food not only depends on quantity but also on quality. The quality of algae has been studied in the laboratory, but the incorporation of available information into field situations is a difficult task. Mortality by predation (and by other factors) is very difficult to estimate. Mortality depends on predator abundance times predation rate times available time. The abundance of 0^+ fish is of primary importance. Predation rates depend on visibility of the prey, thus, among others on light intensity. From a proximate as well as an ultimate point of view, light intensity measurements ought to accompany studies of diel vertical migration.

Migrations in lakes and oceans have to be compared for common aspects and specific differences. From the existing large literature, papers were selected devoted to extensively studied species such as *Daphnia* in lakes and Euphausiacea in oceans. Intensively studied lakes are important.

Notes

1. The enclosures of Johnson and Jakobsen (1987) were filled with water from a lake where several species of planktivorous fish were present.
2. Fitness is a characteristic of an individual or a genotype while the intrinsic rate of increase r is a population parameter. In addition, how to measure r in subpopulations with different migration amplitudes. As far as I know, it has never been done, and certainly r is not "relatively well studied" (De Meester et al., 1998).
3. Non-committal statements run like this: "to function adaptively, animals must have a 'norm of reaction' within which the acceptable degree of risk-taking is tailored to the environmental condition". Or "organisms balance the risk of predation from various sources against the opportunities of growth and reproduction in order to maximise reproductive output". It is better to start measuring the various items. "Meten is weten" ("to measure is to know") (Kamerlingh Onnes, Dutch physicist, Nobel Laurate). In the third edition of his important textbook on Limnology, Wetzel (2001) devotes a paragraph (p. 464) to "Buzz Words: The Plague of Fuzzy Concepts and Ambiguous Terminology". A few great scientists have founded evolutionary ecology, a large number of epigones have polluted the literature with "Just so stories" (Kipling, 1902), or with "compelling tale(s) based on scant evidence, which sometimes has the facile ring of reasoning after the fact" (Johnson in "Fire in the mind") Starting from beautiful figurative landscape paintings, Van Dongen gradually reached a limit state of simplicity. Imitators, unable to draw a tree, likewise fill a canvass with differently coloured squares.

Chapter 9
Diel Vertical Migration in Lakes

Contents

9.1 Introduction

Diel vertical migration of zooplankton in lakes has often been described and is obviously a common phenomenon. Most studies were confined to a few days, few sample times and depths. Therefore, most papers are rather anecdotal, lacking sufficient details to get insight into temporal development and underlying mechanisms. The few studies made during a greater part of a season have contributed more to our understanding. I therefore focussed on two detailed case studies: DVM in Lake Constance and in Lake Maarsseveen.

General limnological research has a long tradition in Lake Constance, and also DVM has been studied for a long time (Einsle, 1969; Elster, 1969; Stich and Lampert, 1981; Geller, 1986; Weider and Stich, 1992). Research in Lake Maarsseveen is of a more recent date. In 1975, a general limnological survey was

DOI 10.1007/978-90-481-3093-1_9,

started, soon followed by an analysis of the seasonal wax and wane of some dominant phytoplankton species. Zooplankton was sampled too, but no vertical migration was found to occur before 1988. However, in that year, large numbers of 0^+ perch (*Perca fluviatilis*) were observed for the first time in this recently excavated lake. During the crucial spring and summer months of the following 10 years, daytime and night-time samples were taken every week.

The two lakes are very different. Lake Constance, bordering Germany, Switzerland and Austria, is a large, deep, monomictic lake, originally oligotrophic, but eutrophication made it mesotrophic. The river Aar feeds the lake with melt water from the Swiss Alps, and the river Rhine leaving the lake goes north to the Netherlands and the North Sea. In Table 9.1, some characteristics of both lakes are listed. Lake Maarsseveen is in the centre of the Netherlands. Originally, it was a swamp where peat had been dug for centuries. In 1960, deep excavation started to obtain sand that was used to make solid bases for roadways and houses in the low-lying western part of the Netherlands, where the underground is soggy. The resulting lake has no surface inlet and one drainage ditch functions as outlet. Water is supplied predominantly by ground water seeping through the bottom and rain that falls on Pleistocene dunes east of the lake. Lake Maarsseveen is also mesotrophic. Its surface area is much smaller and the maximum depth of 30 m cannot be compared with the 252 m of Lake Constance. Nevertheless, both lakes harbour a comparable complex of *Daphnia* species, consisting of *D. galeata*, *D. hyalina* and a plethora of different hybrids of these two parent species (more details in Chapter 12) (Weider and Stich, 1992 for Lake Constance; Spaak and Hoekstra, 1993; Spaak and Ringelberg, 1997; Ringelberg et al., 2004 for Lake Maarsseveen).

Table 9.1 Comparison of Lake Constance and Lake Maarsseveen

Characteristics	Lake Constance[a]	Lake Maarsseveen[b]
Location	47°46′ N, 9°09′ E	52°08′ N, 5°05′ E
Surface area	476 km^2	70 ha
Maximum depth	252 m	30 m
Period of stratification	June–October	May–October
Depth of metalimnion	Not well defined	6–10 m
Thermocline	17–18 m	7 m
Secchi disk transparency	1.7–6.0 m	3.5–8.0 m
Attenuation coefficient	0.3–0.8 m^{-1}	0.4–0.6 m^{-1}

[a]Tilzer, 1983; Gaedke and Schweizer, 1993.
[b]Swain et al. (1987).

9.2 The Seasonal Character of Migration

Usually, DVM of *Daphnia* and *Eudiaptomus gracilis* starts in Lake Maarsseveen in the last week of May or the first week of June. The period of migration continues for 6–7 weeks and thus ends in the first half of July (Fig. 9.1). In Lake Constance, migration of *D. hyalina* and the hybrids starts a month later and was present on

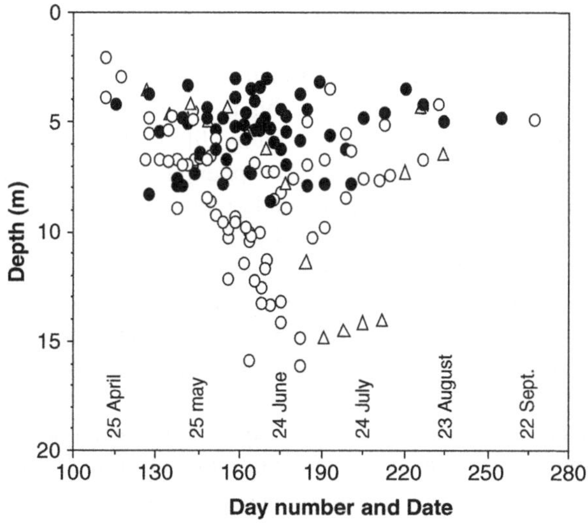

Fig. 9.1 The mean population depth of *D. galeata x hyalina* in Lake Maarsseveen. Data are from the years 1989 to 1999. *Open circles* represent the distribution around noon and the *black dots* around midnight. The exceptional year 1991 is represented by *triangles* and the daytime distributions only are given

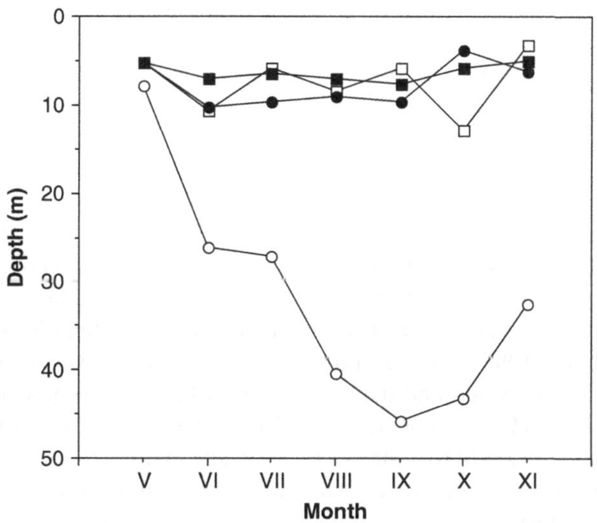

Fig. 9.2 The mean population depth of *D. galeata* (*squares*) and of *D. hyalina* (*dots*) in Lake Constance. Daytime is represented by *open symbols* and night-time by *black symbols*. The data were derived from Stich (1985)

15–16 June (Stich and Lampert, 1981; Stich, 1989) (Fig. 9.2). On 26–27 May 1982, Geller (1986) found no DVM, but a distinct one on 22–23 June of that year. DVM continues until November (Stich and Lampert, 1981; Geller, 1986). Other species, for example, *E. gracilis,* migrate during the same period as *Daphnia* in both lakes. Detailed descriptions can be found in Stich (1985) and Geller (1986).

Although it is obvious to suppose that factors responsible for periodic recurring phenomena must be periodic themselves, this is not necessarily so. The beginning and end of the phenomenon might be caused by different factors. Whether a factor is causal or just coinciding with the studied phenomenon is difficult to conclude from field data alone if no idea about the underlying physiological and behavioural mechanism is available. Perhaps this is why in the literature several factors have been held responsible for DVM. A striking example of this is the interpretation of DVM in Lake Constance. Stich and Lampert (1981) thought the presence of visually predating fish essential to its occurrence, while Geller (1986) rejected the role of fish and concluded that low food concentrations and temperature were responsible. Both Stich (1985) and Geller (1986) made extensive studies of DVM in Lake Constance with only a few years in between. The described migration patterns are similar too, so what arguments made such controversial interpretations possible?

It is common sense and now generally accepted that zooplankton prey migrate downwards in the morning to avoid visually hunting fish. Arguments in favour do not follow from a direct relation between prey and predator though, because these were hardly studied in the field. Arguments were derived from aspects of adaptations, for example, that large zooplankton go deeper than small ones or that the amplitude of migration is larger in clear lakes than in more turbid lakes (Lampert, 1993). Arguments also follow from laboratory experiments. For example, the extent of downward swimming of *Daphnia* in a cylinder depends on the concentration of kairomones, exuded by fish (Loose, 1993a). It is then taken for granted that migration amplitudes in lakes also must be determined by this kairomone concentration, although this concentration has never directly been estimated. Many arguments in favour of fish as an important factor are available, but precise knowledge is absent. The situation in a water column of a lake is more complicated than in the column of a small cylinder and the results cannot be simply applied to DVM in the field. For instance, although 0^+ perch (*P. fluviatilis*), producing kairomones in the laboratory, are abundantly present from the end of April in Lake Maarsseveen, it takes until the last days of May before DVM actually starts (Flik and Ringelberg, 1993; Ringelberg et al., 1997). This is rather strange because even fish larvae bring kairomones into the water, and daphnids react in the same way as to kairomones from larger fish (Van Gool and Ringelberg, 2002). After DVM has started, the amplitude gradually increases until a maximum is reached within a few weeks (Fig. 9.1). Thereafter, the amplitude decreases again until migration stops altogether. This pattern was observed for several years and holds for *Daphnia* as well as for *E. gracilis*. Thus, there is a well-defined, short migration period in Lake Maarsseveen.

A seasonal migration period was also observed in Lake Constance but migration begins about 1 month later and lasts until November (Stich and Lampert, 1981; Stich, 1989; Geller, 1986) (Fig. 9.2). This holds for *D. hyalina* and the hybrids *D. galeata x h*yalina (Weider and Stich, 1992) and for *E. gracilis* (Stich, 1985) but not for *D. galeata*, which does not migrate (or over a relatively short distance of 4.4 m only according to Geller, 1986). As in Lake Maarsseveen, the amplitude increases and decreases in the course of the period, but is much larger. For example, *D. hyalina* migrates over more than 30 m.

Few other lakes were studied long enough but, if so, a seasonal character of DVM was found too. In the Mondsee (Austria), *D. hyalina* and *Cyclops abyssorum praealpinus* start to migrate in the first half of June and migration extends, as in Lake Constance, to November (Nauwerck, 1993). Amplitudes increase gradually in summer and decrease in autumn. Other zooplankton species did not migrate, not even *E. gracilis*, although this species generally migrates, albeit in most cases not as extensive as *D. hyalina* (Angeli et al., 1995).

For the Oberer Arosasee a seasonal pattern of increasing and decreasing depth of *Daphnia* was described too but migration amplitude remained more or less constant (Winder et al., 2002). This pattern differs from those in the discussed lakes and is of special interest because fish and fish kairomones seemed to be of no influence. So what determined behaviour? An extensive discussion follows in Section 9.3.

We conclude that DVM of zooplankton in lakes might have a periodic, seasonal character. How can this be explained in causal (proximate) terms?

9.2.1 A Natural Experiment

In 1991, nature "performed" an experiment in Lake Maarsseveen. May and June were exceptionally cold, compared to other years, which led to a retarded development of the thermal stratification of the lake. In Fig. 9.3, four temperature-depth profiles for about the same dates in 1991 and 1992 illustrate the differences. At the end of May 1991, temperature of the upper 6 m was 12°C, while around the same date a year later, temperature was ca. 20°C. Even in the second week of June 1991, a metalimnion could hardly be recognised and the epilimnion had still a temperature of 13°C. The next year, stratification was well developed at this time and the temperature of the epilimnion exceeded 18°C. In 1991, it took until the first week of July before the temperature suddenly increased and stratification became comparable with other years.

The slow rise in temperature in 1991 strongly affected the development of phytoplankton and zooplankton density and growth of 0^+ fish. The migration period started nearly a month later (Fig. 9.4). A comparison of the postponed beginning of DVM and the development of biotic factors might indicate what factors are important for the start of migration.

The low temperatures in April and May inhibited the growth of algae. The small fraction (<25 μm) is important as food for *Daphnia* and the depth profile of chlorophyll *a* shows that the concentration, usual low in May (" clear water phase"), continued to be low in June. Growth started when the temperature rose at the end of the month. At the same time DVM started, as is apparent from the deep daytime distribution of adult *D. galeata x hyaline* in Fig. 9.5. One would expect that daphnids, when food concentrations in the epilimnion are low, will not migrate to deeper water layers with still lower concentrations of algae. Models, like those by Gabriel

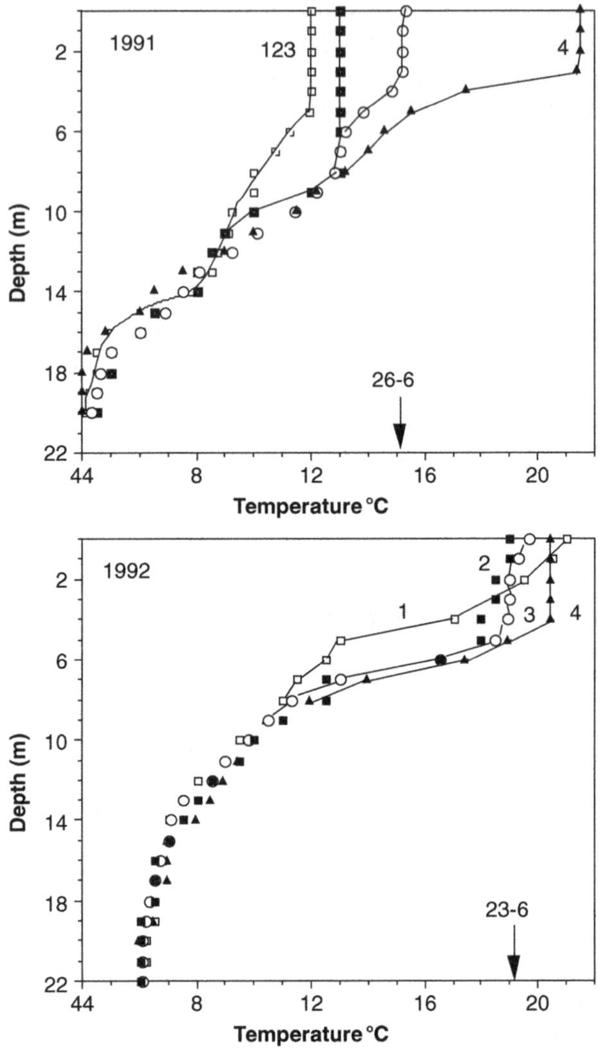

Fig. 9.3 The development of the temperature in Lake Maarsseveen during the summer of 1991 and 1992. The dates 23 May 1991 and 25 May 1992, respectively, are indicated by 1 and *open squares*; 13 June 1992 and 10 June 1992 by 2 and *black squares*; 26 June 1992 and 23 June 1992 by 3 and *open circles*; 10 July 1992 and 6 July 1992 by 4 and *black triangles*. The *arrows* point to the temperature in the epilimnion on nearly the same date in both years

and Thomas (1988) (see overview in Gabriel, 1993), show how the ratio between predation pressure and food concentration determines whether DVM is profitable or not. The decision-making mechanism, presented in Chapter 4, indicates what mechanism of behaviour is used by the animal to lead to the best possible strategy (Van Gool and Ringelberg, 2002). Obviously, food is an important factor but not a factor that in itself can cause migration. Food modifies the motivation to migrate. High food concentrations are also not necessary to maintain migration, as can be concluded from the still high algal concentration on 14 August 1991 (Fig. 9.5), when migration is already finished. Therefore, my interpretation of the information

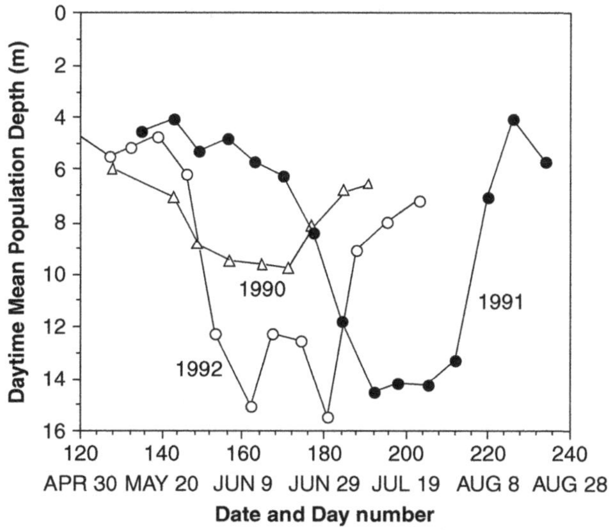

Fig. 9.4 To illustrate the retarded DVM in the exceptional year 1991, the mean population depth at noon of *D. galeata x hyalina* in Lake Maarsseveen is presented for three subsequent years (from Flik and Ringelberg, 1993; reproduced by permission of E. Scheizerbart'sche Verlagsbuchhandlung)

contained in Fig. 9.5 is that food concentration is one of the factors in the " set of necessary ecological conditions". In 1991, between the end of May and the beginning of July, algal food concentration was below the necessary level to contribute to the set. Food concentration was limiting indeed, as is corroborated by the rapid decrease in number of eggs in the brood chambers of *Daphnia* and the low fecundity, previous to the start of the DVM period in July (Fig. 9.6). The number of eggs increased rapidly after 11 July (day 192), more or less parallel to the increase in chlorophyll *a* (Flik and Ringelberg, 1993).

At the start of DVM, the 0^+ perch had grown to a mean length of > 15 mm (Fig. 9.7). This length is normally attained at the end of May (Flik and Ringelberg, 1993), but in 1991 growth was retarded by nearly a month. It is improbable that *Daphnia* individuals are triggered to swim down by a mere length of juvenile perch. At the body length of 15 mm, mouth gape of the perch has become large enough to eat adult daphnids. This makes this particular length of interest, because then predation on *Daphnia* may start. Such an anatomically defined transition point is not absolute because behaviour, as determined by previously eaten species, is of influence too. However, as is apparent from Fig. 9.7, the crucial mean length was reached nearly 17 days earlier and, being an average value, a considerable part of the perch population must have been able to predate on adult *Daphnia* already.

Nevertheless, migration had not started. This problem must be solved. It might be hypothesised that the concentration of fish kairomones in the lake has to be sufficiently high to start migration. Kairomone production is certainly positively

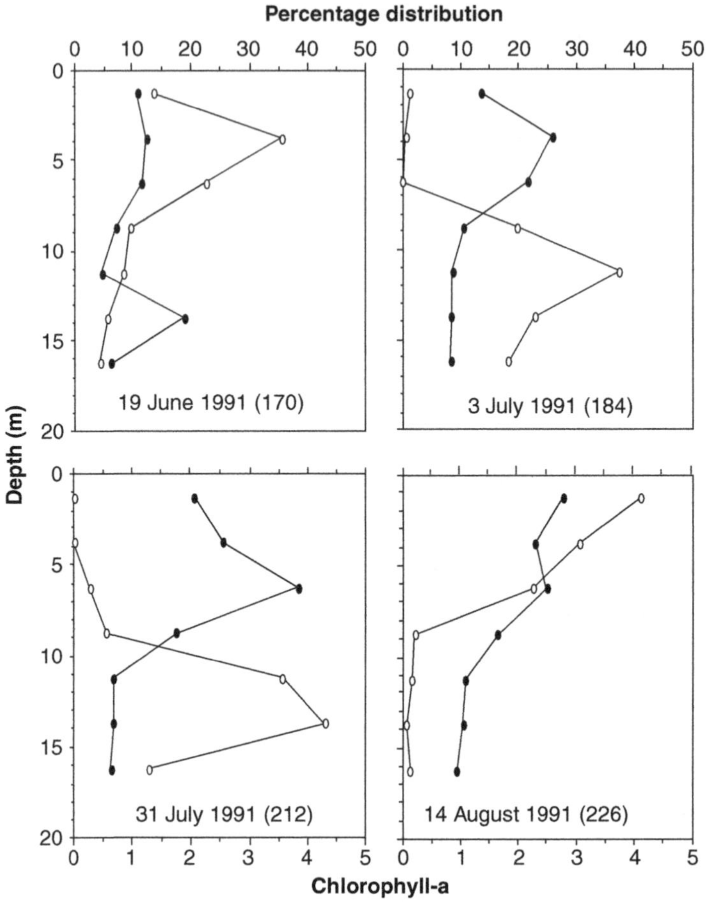

Fig. 9.5 Depth distribution of adult *D. galeata x hyalina* at noon (*open circles*) is compared with the chlorophyll *a* concentration (<25 mm size fraction) (*black dots*) on different dates of 1991. The deep daytime distribution of *Daphnia* on day number 184 and 212, indicating a migrating population, coincides with an increased chlorophyll *a* concentration. On day 226, the chlorophyll *a* concentration is still relatively high but DVM is over (modified after Flik and Ringelberg, 1993; reproduced by permission of E. Scheizerbart'sche Verlagsbuchhandlung)

correlated with fish biomass and thus with length (see further on in this chapter). This is no solution of the problem, however. If length did not change, biomass also must have been about constant for 17 days. Biomass might even have decreased during this period because a number of perch must have died. On first sight, also the factor kairomones does not explain why *Daphnia* did not migrate until the end of June.

However, the concentration of kairomones in the lake not only depends on production by fish but also on dilution in the water column.

Temperature stratification of the water column must, therefore, be of importance. On 26 June 1991, stratification was weak and irregular, while on 23 June 1992,

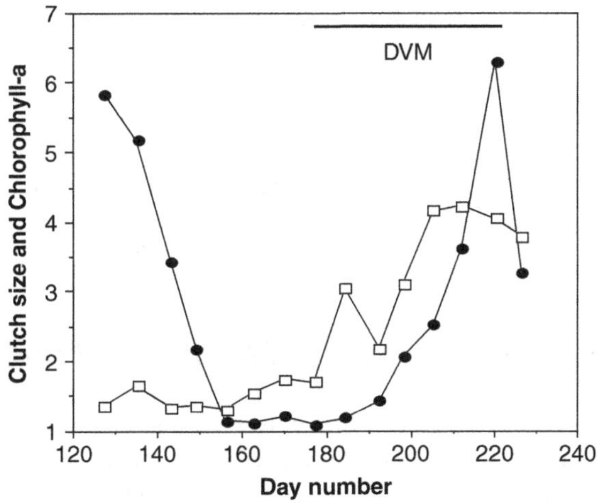

Fig. 9.6 Comparison of clutch size (*black dots*) and the total amount (μg × 5) of chlorophyll *a* (<25 μm size fraction) in a column (1 m^2 × 7.5 m) of the epilimnion of Lake Maarsseveen. The period of the retarded DVM is indicated by a *horizontal line* (from Flik and Ringelberg, 1993)

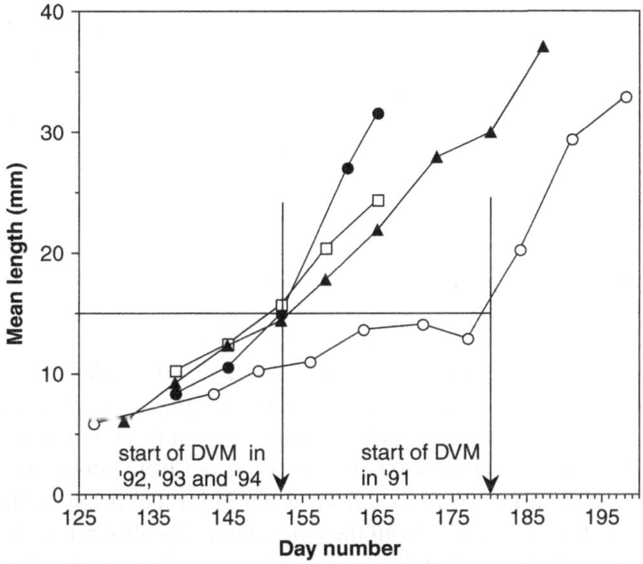

Fig. 9.7 The beginning of the period of DVM in Lake Maarsseveen in the "normal" years (1992, 1993 and 1994) and in the exceptionally cold spring and early summer of 1991. The beginning of migration, estimated as the first date with DVM minus 3 days (weekly observations), is compared with the mean total length of the 0^+ perch in the lake. The *horizontal* at 15 mm indicates about the size at which the perch start to feed on *Daphnia*

a strong thermocline already closed the epilimnion from the rest of the water column (Fig. 9.3). It took again 10 days before also in 1991 a strong temperature gradient prevented mixing over a larger depth than the upper 7 m to which the 0^+ perch are confined. If before that time the produced kairomones were diluted over a deeper part of the water column, the concentration could have been too low to make DVM possible. I think, we must certainly add kairomone concentration (and indirectly predators as producers) to the “set of necessary ecological conditions”. Temperature, stratification and the internal column stability influence the extent of dilution and thus concentration. Given a certain 0^+ fish biomass, these hydrological aspects might determine different kairomone concentrations. Nevertheless, they do not really belong to the “set of necessary ecological factors”.

9.2.2 The Beginning of the Period of Diel Vertical Migration

I suggested that a sufficiently *high* algal concentration was one conditional factor to make DVM possible. Geller (1986), instead, suggested that a *low* algal concentration initiated (the period) of vertical migration. He found that the onset of DVM coincided with a switch from a high level to a low level of food supply and he considered migration a strategy to diminish starvation. A physiological mechanism seemed to provide a basis for this hypothesis. If, with increasing temperatures, feeding increases more rapidly than respiration, an “energy bonus” could be obtained by zooplankton swimming from the cold hypolimnion into the warm and food-rich epilimnion. This energy surplus could then be used for growth and reproduction when metabolism is at a low rate in the hypolimnion. The evidence for a beneficial effect of fluctuating temperatures, as experienced by migrating zooplankton, is scarce and confusing. Orcutt and Porter (1983) found for *D. parvula* a higher rate of population increase (r) at temperatures alternating among 10 and 20°C compared with a constant 15°C. On the other hand, Manca et al. (1986), experimenting with *D. obtusa* and *D. hyalina*, found no positive effect of fluctuating temperatures. This conclusion was also reached by De Wit (1990) in life-history experiments with *D. galeata x hyalina* from Lake Maarsseveen. Independent of these controversial results, all authors concluded that a maximum fitness is achieved by animals that remain all the time in warm surface waters. The same conclusion was reached by Park et al. (2004). At equal temperatures *D. rosea* from subalpine Castle Lake (USA) grew faster on seston from a deep-water chlorophyll maximum compared to individuals cultured on algae from the epilimnion. The difference disappeared in cultures with temperatures of epilimnion and hypolimnion. At the alternating temperatures, as experienced by migrating animals, growth rates were intermediate. The authors concluded that in Castle Lake, the higher temperatures of the epilimnion override the higher food quality and quantity of the deep-water chlorophyll maximum.

Ghan et al. (1998) studied benefits and costs of vertical migration in the copepod *Skistodiaptomus oregonensis*. In the four lakes that were studied, two populations were migrating and two did not migrate. Only in the two lakes where pelagic three-spined sticklebacks (*Gasterosteus aculeatus*) were present, DVM occurred, which is consistent with a predator avoidance hypothesis. A foraging efficiency hypothesis did not explain migration because neither food abundance nor food distribution distinguished lakes where *S. oregonensis* migrated from lakes where they did not migrate. Migration and non-migration could neither be interpreted by a bioenergetic efficiency hypothesis nor by a thermal advantage hypothesis because the thermal structure of the four lakes was similar. Therefore it can be concluded that migration was the result of a trade-off between predator avoidance and resource acquisition.

The conclusion can be reached that DVM does not lead to a thermal demographic advantage. This makes Geller's "energy bonus" and "starvation avoidance" less probable propositions. The transition in Lake Constance of a high food concentration into a low one seems not to be a factor in determining migration.

Geller (1986) observed in Lake Constance that minimum temperatures at the lower boundary of the day depth distribution were nearly constant and species specific from June to September (however, not later in the year). He concluded "that the lower boundaries of the DVM amplitudes of the migrating zooplankton are controlled by the ambient temperatures" (p. 50). This conclusion seems to be inspired by his thermal advantage hypothesis. Zooplankton should stay at the lowest possible temperature to profit maximum from the bonus. Also Nauwerck (1993) concluded that migration in the Mondsee (Austria) led to constant daytime temperatures for *D. hyalina* and *C. abyssorum praealpinus*. As usual, vertical distributions were rather diffuse in both lakes and individuals of a species could be found over a wide range of temperatures. For Lake Constance, we now know that what was previously considered the migrating species, *D. hyalina*, includes hybrids of *D. galeata* and *D. hyalina*. The different genotypes distribute at daytime over a wide vertical range (Weider and Stich, 1992) and, thus, are found at different temperatures.

To what extent migration amplitude and temperature is a causal relation or a correlation of parallel seasonal developments only is unsure. Field evidence is often non-conclusive and experiments on the relation between behaviour and temperature are nearly absent. Flik and Vijverberg (2003) made likely that the seasonal slow descend of *D. pulicaria* is possibly initiated by the increasing temperature of the epilimnion. Also experimental results seem to indicate a relationship. If temperature in a cylinder was gradually raised from 10°C to above 16°C in 2 days, *D. pulicaria* changed from a near-surface distribution into a near-bottom distribution.

Why did Geller (1986) reject predation avoidance as ultimate reason and, consequently, have to create a new hypothesis? A few years earlier, Stich and Lampert (1981) had proposed this – by then already widely accepted – ultimate reason for a similar migration pattern in the same lake. In the first place, Geller argued that migration was not deep enough to reach a sufficiently low light intensity to make predation by visual predators impossible. Second, the largest of the *Daphnia* species, *D. galeata*, remained near the surface where size-selected predation is highest, while the smallest one, *D. hyalina,* migrated. This he considered "the reverse to be expected from strong fish predation" (p. 34). Although both arguments are valid, reasoning is incomplete because migration can be profitable, even if predation is not completely prevented. What holds is that benefits are larger than costs and both were not estimated. In general, large individuals go deeper than smaller ones, indeed, but again, the ratio between benefits and costs is criterion for selection of the traits. The results of growth and reproduction experiments with the two species, done by Stich and Lampert (1984), revealed that both species produce more offspring under the favourable conditions of the epilimnion. However, *D. hyalina* is more successful under the adverse conditions of migration (Stich and Lampert, 1984; Manca et al., 1986). Before the transition from an oligotrophic into a mesotrophic lake, *D. galeata* did not occur in Lake Constance. Also this indicates that this species needs a higher food concentration than *D. hyalina*. It seems the two species have evolved a different strategy to enlarge the chance of producing progeny. At the start of the migration period, food concentration is low and *D. galeata* would probably lose competition if it also migrated. Its higher birth rate near the surface must compensate for predation mortality. *D. galeata* succeeds, although the size of the adult population is slightly smaller than that of *D. hyalina*, notwithstanding the bad conditions of temperature and food experienced by the latter species.[1] I do not think that "predation prevention" has to be replaced by "starvation avoidance" as the ultimate reason, as Geller (1986) proposed. Nor do I think that food and temperature are primary factors, causing migration. Within the realm of proximate aspects, they play a secondary role as conditional factors.

The results of the "natural" experiment suggest that fish kairomone is an important component of the "set of necessary ecological conditions" . I also pointed out that dilution by vertical mixing of the water column can lead to a kairomone concentration too low for facilitating DVM. A concentration below the minimum might explain why no migration occurs, even if a high biomass of kairomone producing fish is present. This threshold hypothesis is difficult to test as long as the chemical nature of the kairomone is unknown and concentrations cannot be determined. Thus we have to consider indirect data. For example, an experiment by Loose (1993a) in the large, semi-natural Plöner plankton towers revealed that the day depth of migrating *D. galeata x hyaline* was a linear function of the "amount" of kairomone pumped into the water column. The data do not point to the existence of a threshold amount, however. The same holds for the photoresponses at different amounts of kairomone,[2] performed by Van Gool and Ringelberg (2002). So we must look for other possible explanations for the absence of DVM.

Small perch[3] (4 to 15–20 mm) eat rotifers, nauplii, copepodites and small cladocerans like *Bosmina* spp. (Jachner, 1991; Flik and Ringelberg, 1993; Wahl et al., 1993; Wang, 1994; see also review by Mehner and Thiel, 1999). The transition into 0^+ perch is characterised by a fully developed stomach and a larger mouth gape width. Predation on *Daphnia* starts when the critical total length (15–20 mm) is surpassed (Flik and Ringelberg, 1993; Wang, 1994; Wang and Eckmann, 1994). In Lake Maarsseveen, adult *Daphnia* are present in small numbers in the gut for the first time at the end of May or the beginning of June and heralds the beginning of DVM (compare Fig. 9.8a and b). For Lake Constance, Wang (1994) found that from late June (day number 180) to early August (day number 213) small and mid-sized (0.7–1.0 mm) daphnids were eaten. At that time, the perch were also of the size-class of 15–20 mm (Wang and Eckmann, 1994). Compared to Lake Maarsseveen, the important shift to predation on *Daphnia* seems to be a month later in this lake. Spawning and hatching is also later and the growth rate is lower. In Table 9.2 relevant data are listed. It can be concluded that in both lakes the start of predation

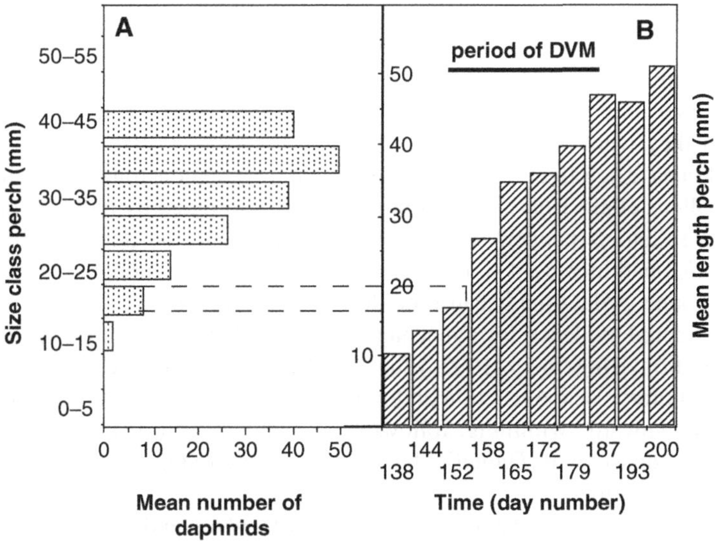

Fig. 9.8 (**a**) The mean number of daphnids in the guts of different size classes of 0^+ perch; (**b**) the mean total length of these perches in Lake Maarsseveen in 1991. Day number 152 = 1 June 1993 (slightly modified from Flik and Ringelberg, 1993; reproduced by permission of E. Scheizerbart'sche Verlagsbuchhandlung)

Table 9.2 Some dates concerning 0^+ perch (*Perca fluviatilis*), important for understanding the beginning of the migration period in two lakes

	Lake Maarsseveen	Lake Constance
Hatching	Last days of April[a]	Last days of May[b]
Tot. Length 15 mm	Day 146 (1992)[a]	Day 170 (1990)[b]
	Day 144 (1993)[a]	Day 174 (1991)[b]
	Day 144 (1998)[c]	
	Day 145 (1999)[c]	
Tot. Length 20 mm	Day 184 (1991)[d]	Day 182 (1990)
	Day 155 (1998)[c]	Day 183 (1991)
	Day 156 (1999)[c]	

[a]Unpublished data from DVM research, University of Amsterdam.
[b]Length and day numbers were calculated from a linear function fitted through the length of pelagic juvenile perch in Wang and Eckmann (1994).
[c]From van Gool and Ringelberg (2002).
[d]From Flik and Ringelberg (1993).

and the beginning of migration coincide. How can this coincidence be specified? It has been suggested that chemicals, liberated by injured conspecifics, might be a trigger for migration (Pyanowska, 1993 referring to unpublished data). Insufficient evidence is at hand, however.[4] Another suggestion is that the swimming of predating perch gives rise to a triggering cue. Large numbers of fish will certainly disturb the water, and mechanical stimuli, as proposed by Bollens and Frost (1989a), could play a role.

These authors found that the (marine) copepod *Acartia hudsonica* swam down in enclosures only when sticklebacks (*G. aculeatus*) were swimming freely, but not when they were confined to cages.[5] They therefore rejected the influence of chemical cues. The effect of mechanical stimuli was demonstrated in a subsequent paper (Bollens et al., 1994). Swimming of fish was imitated by moving rubber fish lures in a saltatory manner. After 20 min of treatment, the depth distribution of *A. hudsonica* had increased significantly by 0.4 m. Strange enough, the experiment had to be done in daytime because no descent happened in the dark. These results were considered evidence for a mechanical or visual cue, inducing daytime downward migration. Ringelberg (1995c) suggested that the downward displacement had been a temporary escape reaction and not (part of) *diel* vertical migration. Sinking or active downward swimming, following mechanical disturbances, can easily be observed in aquaria with *Daphnia*, but they are of a passing nature and, of course, without diel rhythm.

The coincidence of the beginning of DVM in *Daphnia* and the beginning of predation by perch might very well be a correlation only. Nevertheless, predation must have been a strong force in the evolutionary past for selecting DVM behaviour. Tuning to a predation-related cue must be involved. *Daphnia* is able to perceive and validate differences in kairomone concentration (Loose, 1993a; Van Gool and Ringelberg, 2002). Therefore, I suggest again that a threshold concentration or, rather, a lower concentration range is crucial to the initiation of the migration period. This *threshold range* is not the same as a sensory threshold because behaviour is involved as influenced by other factors. A physiological threshold for kairomone perception might be surpassed long before migration behaviour starts. The precise concentration within the range, at which DVM starts, must be considered variable and depending on other factors in the "set of necessary ecological conditions".

9.2.3 The Development of the Migration Amplitude

The difference between daytime and night-time depth, or amplitude, increases in the first weeks of the migration period and decreases again before the final end. This is nicely illustrated in Fig. 9.9 in which the pattern of migration for *D. galeata x hyalina* and *E. gracilis* in 1991 is depicted. We have concentrated on *Daphnia* in the previous sections because no experimental data on photobehaviour for *E. gracilis* are available. Migration behaviour was very different in the exceptional year 1991. In the first place, the start of the migration period differed. The calanoid copepod started much earlier than the cladoceran, probably already on 15 May (day 135). Even for a "normal" year, this is well in advance of *Daphnia*, which usually begins to migrate at the end of May. From Fig. 9.9 can be concluded that the day depth, and thus the amplitude (the average night depth remained constant), very gradually increased with about 10 cm per day, during about 80 days. For *Daphnia*, the day

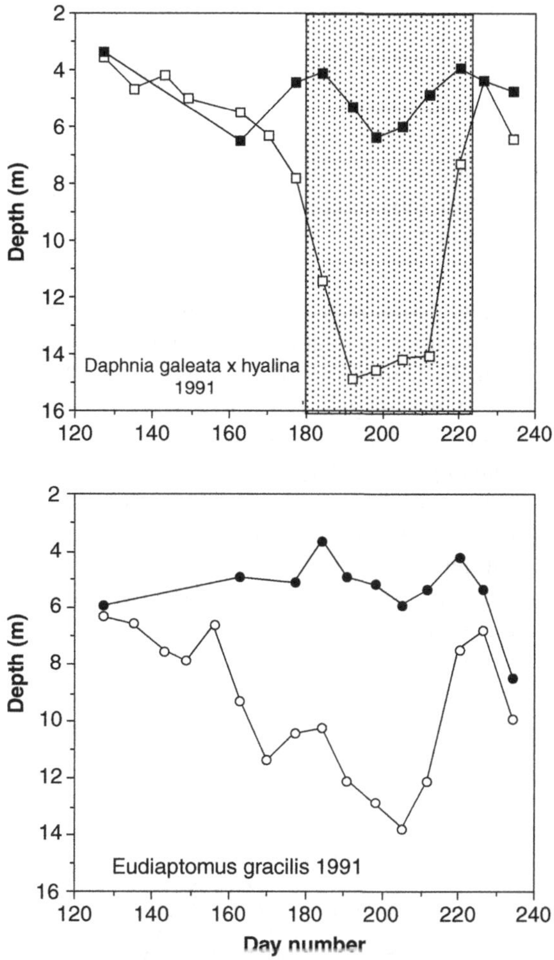

Fig. 9.9 The development of diel vertical migration of *D. galeata x hyalina* and *E. gracilis* in Lake Maarsseveen during the exceptional summer of 1991. *Black points* and *squares* represent night-time and *open squares* and point daytime mean population depths. The period of DVM in *Daphnia* is indicated by the *grey square*

depth started to increase at the same date but increased at a rate of about 5 cm per day. Probably, during this first period, for *Daphnia* also the night depth remained constant. This slow increase terminated on day 170 when a higher daily increase in day depth of about 40 cm started. Between 26 June (day 177) and 3 July (day 184), the epilimnion became free of *Daphnia* (see Fig. 9.6). Although predation must have been present because daphnids were in the guts of 0^+ perch (see Fig. 9.8), the total population did not change in that week. The period of depth increase lasted until 11 July (day 192) but the mean population depth might have continued to increase for an additional 3 weeks. Insufficient deep sampling might be responsible for the more or less constant mean depth until 31 July (212). If lines are drawn between points of day depth increase and day depth decrease, they meet at day 205, which is the same turning point as for *E. gracilis*. After this transition, day depth rapidly decreased and after 3 weeks the migration period was over for both species.

Obviously, similarities and differences in migration behaviour exist between the two species.

At least three different processes might be in operation during the first few weeks: (1) non-migrant individuals are weeded from the population by predation; (2) the number of migrating individuals in the population increases; and (3) individuals swim farther down.

1. A population of plankton animals is not genetically homogenous and the traits contributing to migration behaviour are possibly variously distributed over individuals. During the first 2 weeks of the DVM period in 1997, high numbers of a certain allozyme type of *D. galeata x hyalina* were still present in the epilimnion of Lake Maarsseveen (Ringelberg et al., 2004). They were gradually eaten by the 0^+ perch. If no distinction can be made in genotypes and all individuals figure in the calculation of a population mean depth, this depth gradually increases as a result. For Lake Constance, a differentiation in hybrids of different genetic signature was made (Weider and Stich, 1992) some years after the seasonal migration pattern of the two parent species was described (Stich and Lampert, 1981). These hybrids have a more or less intermediate vertical distribution between the parental species and, if omitted from the calculated amplitude of what was considered *D. hyalina*, the amplitude of this species would become larger. Also, the presented (Stich and Lampert, 1984) increased daytime vertical

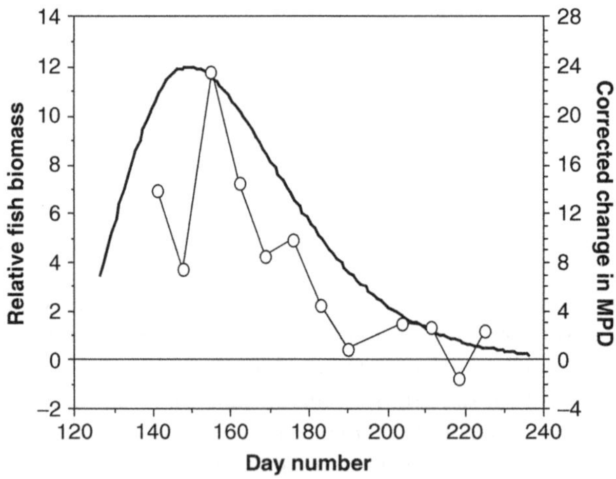

Fig. 9.10 The average vertical displacement of *D. galeata x hyalina* elicited by increases in light intensity in epilimnion water of Lake Maarsseveen, corrected by subtraction of the displacement in hypolimnion water (*open circles*). Positive values represent an increase in depth. The drawn line is a presentation of the relative biomass of 0^+ perch with time (after Van Gool and Ringelberg, 2002; reproduced by permission of Oxford University Press)

distribution of *D. hyalina* over the summer months could have been due to predation mortality of the hybrids predominantly, and not the result of a changed behaviour.

What traits are involved preventing individuals of *D. galeata* from migrating is unknown. A high threshold for light intensity changes in the morning might be thought of or a higher threshold for the perception of fish kairomones. We deal with behaviour and must also consider an absent connection between information of the light stimulus and the kairomone somewhere in the nervous system. Genetic differences in photobehaviour were described (De Meester, 1989, 1993a, b; Spaak and Ringelberg, 1997), but direct links with diel migration behaviour still have to be made.

2. If the vertical distribution of migrating and non-migrating individuals overlap and the latter also start to migrate, day depth will increase. Allozyme analysis of *D. galeata x hyalina* from Lake Maarsseveen suggests that some genotypes start earlier with DVM than other allozyme types (Ringelberg et al., 2004; see also Chapter 12). A differential sensitivity to fish kairomones might be involved. If the threshold values differ in the population, the number of individuals that smells the kairomone increases when fish biomass increases. This is again an unexplored field of behavioural physiology and ecological genetics. We are discussing intra-specific differences in migration amplitude but inter-specific differences, especially in closely related species, are also important because differences in behaviour might be large. A striking example is *D. galeata x hyalina* and *E. gracilis*, presented in Fig. 9.9. The analysis of the physiological base may help to understand differences in migration behaviour within a population or a complex of closely related species. For example, in Lake Constance, *D. galeata* hardly migrates while *D. hyalina* migrates over a large distance. Genetic differences must be responsible.
3. True changes in migration amplitude occur when all individuals of a population swim down further in the morning. This increase probably is a reaction to more fish and higher fish kairomone concentrations and thus adaptive. I have mentioned several laboratory experiments in which vertical displacements depended on the amount of fish kairomones. In all experiments water was obtained from aquaria with fish. No direct relation with field observations exists and it is unknown whether the used kairomone concentrations were comparable to those in lakes. Strictly speaking, we are not even sure whether the chemicals used in behavioural experiments were identical to the chemicals that are active in the field.

To explore the relevance of laboratory experiments with kairomones, it is necessary to combine observations from lakes and laboratory. A first step is to show that lake water enhances the phototactic swimming reaction in the same way as fish water in laboratory experiments does. A second step is to study the extent to which phototactic reactions are correlated with the size of the fish population. A third step is to demonstrate that the changing amplitude of DVM in a lake is correlated with a parameter of a changing size of the predator population.

During spring and summer of 1998 and 1999, Van Gool and Ringelberg (2002) tested epilimnion water from Lake Maarsseveen using the phototactic reaction of *D. galeata x hyalina* as caused by continuous increases in light intensity. As reference, water from the hypolimnion was used because 0^+ perch do not go as deep as that and, thus, kairomones ought to be absent. Since a phototactic reaction is nearly always invoked, in the absence of fish kairomones also, the downward swimming distance in epilimnion water was corrected by subtracting the distance in hypolimnion water. The results of 1998 are presented in Fig. 9.10. Photoresponses correlate significantly with the estimated relative biomass of 0^+ perch in the lake.[6] Between 27 May and 9 June, fish biomass and the photoresponse were both maximal. The next year, a comparative bell-shaped fish biomass curve was found but the response curve of *Daphnia* was flattened and the top was shifted for at least 2 weeks towards the end of June. As a tentative explanation for this difference, Van Gool and Ringelberg (2002) mentioned a slower development of the temperature stratification in the lake and thus a larger, unstable water column. Whether the resulting increased dilution of fish kairomone was sufficient to cause the obtained effect is hard to say. Juvenile fish biomass might have been smaller too. Nevertheless, the results of both years suggest that in lake water a substance is present that facilitates a phototactic reaction, comparable to that of water from an aquarium with fish.

The second hypothesis implies that migration amplitude is a function of kairomone concentration in the lake. To test, the 0^+ perch biomass had to be estimated and compared with the developing day depth during the DVM period (Ringelberg et al., 1997). The hypothesis is a simplification, of course. Young-of-the-year perch are confined to the upper 7–8 m in Lake Maarsseveen, which is about the depth of the epilimnion during summer stratification, but the kairomone will be mixed over a deeper part as long as a stabilised thermo-stratification is not established. For example, the concentration in the upper few metres of the water column in May might be lower than in June, even if the total fish biomass is the same. Figure 9.11a shows the mean total length of 0^+ perch in Lake Maarsseveen in 1992 together with the wet weight biomass. The arrows indicate beginning and end of the DVM period. As usual, the migration period started when perch reached a length of 15–20 mm. The figure suggests that the end of the period coincides with a biomass similar to that of the beginning. However, the end is often less clear. Of possible influence is the "memory"-effect described by Ringelberg and Van Gool (1995): once kairomones have enhanced phototaxis, the sensitisation gradually decreases in about a week in the absence of kairomones. In Fig. 9.11b, the same bell-shaped biomass curve of Fig. 9.11a is compared with the day depth of the daphnids in the lake. Obviously, the maximum migration amplitude co-occurs with the highest fish biomass. Therefore, migration amplitude possibly develops as a function of fish biomass and might be a function of the kairomone concentration in the lake. It is concluded that the proposed hypotheses have not to be refuted. However, as long as we cannot determine concentrations, this is the closest we can get at the proximate relation between fish and DVM.

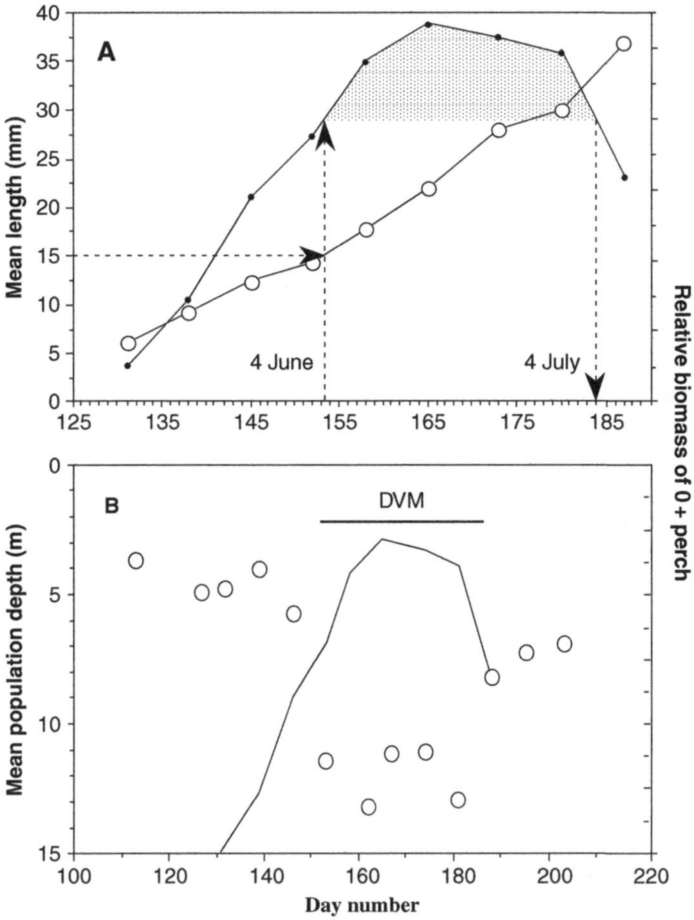

Fig. 9.11 A comparison of characteristics of 0+ perch and DVM of *D. galeata x hyalina* in Lake Maarsseveen in 1992. (**a**) Mean length of juvenile perch (*white dots*) and relative biomass (*black dots*) in the course of the season; the *shaded area* indicates the period of DVM, starting on 4 June and ending about 4 July. (**b**) The *line* indicates increasing and decreasing relative perch biomass. The *white dots* represent the daytime mean depth of the population of adult *D. galeata x hyalina*. The time axes are different in both figures (the original drawing by B.J.G. Flik was slightly modified)

The described relations between DVM and fish biomass in Lake Maarsseveen cannot be verified for other lakes because sufficiently detailed data are not available. No bioassays were done with water from other lakes.

Lake Constance differs from Lake Maarsseveen in three aspects: (1) the period of DVM starts about a month later; (2) it ends several months later; (3) amplitude is much larger. A later start can be explained by a later hatching of perch and other fish. Growth of juvenile fish is about the same in both lakes and, thus, the rapid increase in migration amplitude in June and July can perhaps be explained by the increasing biomass of the young fish. Since the eutrophication of the lake, adult perch have left the littoral zone and moved towards the open water, where they are found between 6 and 20 m depth from May to October (Wang and Eckmann, 1994). In addition, diet has changed from small fish to zooplankton (Hartmann and Loffler, 1989). The large migration amplitude of zooplankton in Lake Constance, exceeding 25 m from June until November, is therefore adaptive. The 0^+ perch stay in the open water until the end of June and then gradually move into the littoral zone. If in autumn food becomes scarce here, the juveniles move offshore again (Wang and Eckmann, 1994). The abundance of 0^+ and adult perch is not known (to me) but the vertical and horizontal distribution, as well as the predation pressure differs from Lake Maarsseveen. Adult perch are rare in Lake Maarsseveen and no threat to *Daphnia*. Fish abundance in Lake Constance might continue to be high enough to realise sufficient fish kairomones, leading to a migration period extending until November. With the drop in temperature after the overturn of the water column at the end of October, perch move into still deeper water, reaching a highest abundance at 25 m. At that time, the migration period ends.

In the Mondsee, the migration pattern of *D. hyalina* and *C. abyssorum praealpinus* is comparable to that of Lake Constance. Nauwerck (1993) does not consider fish predators important, although he observed that during spring and early summer *Chalcalburnus chalcoides* and *Coregonus spec.* efficiently suppress *D. hyalina. Daphnia* and *Cyclops* are always prominent food items for both fish species. Juvenile fish were not mentioned in the paper by Nauwerck but must have been present. If more attention is given to fish predators, their role in DVM will probably become as evident as in Lake Constance.

9.3 Diel Vertical Migration Caused by UV Radiation

Compared to fish, other factors have got little attention in the literature. If fish are present, amplitudes are large and spectacular. Small-scale amplitude migrations of a few metres are of less interest. They occur within the limits of a well-lit surface layer, where predation by visually hunting fish is still possible. If fish is absent, it is without doubt that predating fish cannot determine the vertical distribution of zooplankton. Therefore, what causes these small-amplitude migrations and what is the adaptive significance? Laboratory experiments showed that a phototactic reaction to changes in light intensity is (nearly) always possible, even without fish kairomones. Studies of alpine lakes and lakes in Arctic regions have revealed that ultraviolet radiation (UVR of wavelengths between 290 and 400 nm) might be of importance. Since the recognition of the stratospheric ozone depletion, research into the potentially harmful effects of UV-B radiation in the wavelength range of 280–320 nm has strongly increased. It had been known for a long time (Brehm, 1938) that zooplankton of high altitude ponds are often strongly pigmented and protection against UV damage was suggested. Organisms in alpine ponds and lakes are in danger because of the high UV irradiance. With every 1000 m elevation, UV-B increases by 6–10%

(Diffey, 1991 cited in Morris et al., 1995; Siebeck and Böhm, 1994). Blumthaler et al. (1997) mentioned that the wavelength of 300 nm might even increase by about 24%. The clear and dry atmosphere is responsible for a decreased absorption and the high transparency of the often oligotrophic lake water makes a relatively deep penetration possible. Many alpine lakes are without fish and the idea was born that UV-B could be a factor responsible for small-amplitude migrations and for day depths of a few metres. However, in transparent lowland lakes also, UV radiation might be a hazard. In the past, many of these lakes must have been oligotrophic and clear. Therefore, it cannot be excluded that here also radiation damage has been a selecting factor for protection mechanisms such as a downward migration out of the danger zone in daytime. Because the danger of damage was permanently present and the costs were probably low – considering the epilimnion was not left – small-distance migrations could become constitutive. If so, these small amplitude DVMs could have been retained, although many lakes have become eutrophic and UV danger is now small.

In Fig. 9.12, vertical distributions of *Daphnia* in Lake Maarsseveen are presented for 3 August 1978 and 26 June 1979. In both years, no shoals of 0^+ fish were observed, but even if they had been present, the average day depth of 5.5 and 6.3 m, respectively, would not have saved the daphnids from predation. This depth, however, would be sufficient to prevent UV-B damage because the 1 or 0.1% depth is at about 1.2 and 1.8 m, respectively.[7] For both dates, a small diel shift in depth of 1.8 and 1.6 m, respectively, occurred. In Fig. 9.13 a seasonal increase in day depth during 1985 in Lake Maarsseveen is shown. No vertical distributions for the night are available, so we cannot tell whether *D. galeata* (?) migrated or not, but the shift is evident. The seasonal difference is smaller than in the Oberer Arosa See (Winder et al., 2002) but the phenomenon looks similar. This also holds for Lake Constance, where the day depth of *D. galeata* also increased gradually in the course of the summer with a maximum attained depth in September–October (Stich and Lampert, 1981; Geller, 1986). Migration of this species is considered absent or with a small amplitude only. Fish predation is not avoided.

Small day–night vertical depth differences were also described by Müller and Seitz (1993). They observed for different genotypes of *D. galeata x cucullata* amplitudes ranging from 0.5 to 2.5 m. Distributions were confined to the upper 5–6 m and seemed to overlap with distributions of fish.

9.3.1 The Damaging Effects of Ultraviolet Radiation

Before embarking on the influence of UV radiation on diel vertical migration, damage and recovery effects must be discussed. Of the range of UV wavelengths reaching the earth (290–400 nm), the shorter ones with the highest energy per photon are the most damaging. This is UV-B radiation between 290 and 320 nm. UV-A (320–400 nm) is less dangerous and radiation between about 390 and 400 nm might even be beneficial because of it participating in the so-called photoenzymatic repair

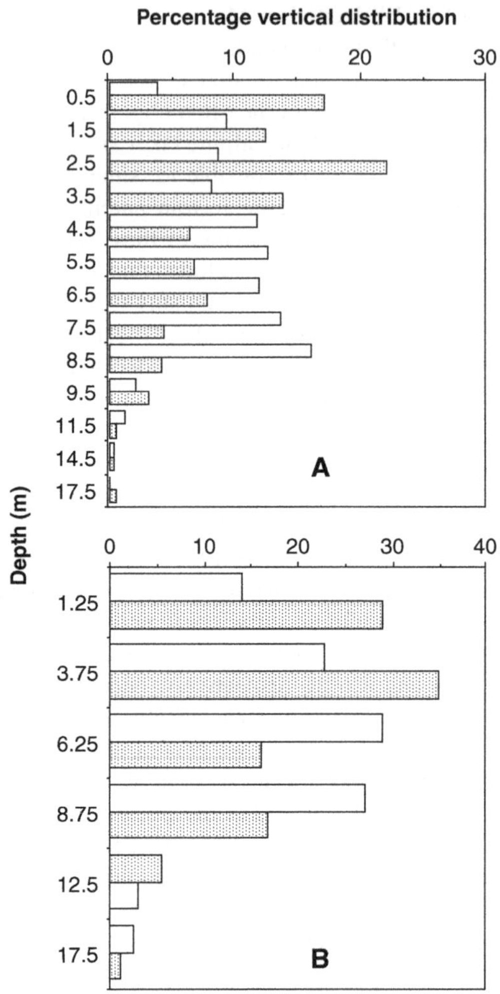

Fig. 9.12 Daytime and night-time vertical distribution (%) of (probably) D. galeata in Lake Maarsseveen on A. 3 August 1978 (day number 215) and B. 26 June 1979 (day number 177). The mean population depths were 5.47 and 3.70 m for day 215 and 6.27 and 4.66 m for day 177

(PER). The damage done by UV-B is in the first place disruption of DNA. Cell and organ functioning are affected. If the damaged DNA is part of the germ line, progeny might not be viable and birth rates will decrease. As a secondary effect of UV-B, free oxygen radicals may damage cell membranes. Cells are ruptured or osmoregulation is disordered, resulting in swellings of body tissue, especially at the joints of thorax segments in copepods (Ringelberg and Flik, 1984). Sublethal radiation can disorganise embryonic development, for example, the formation of neural connections in the growing visual system of *Daphnia* (Macagno et al., 1973; Macagno, 1977). The animals do not die but behaviour must be affected leading to a disadvantage in intra-specific competition.

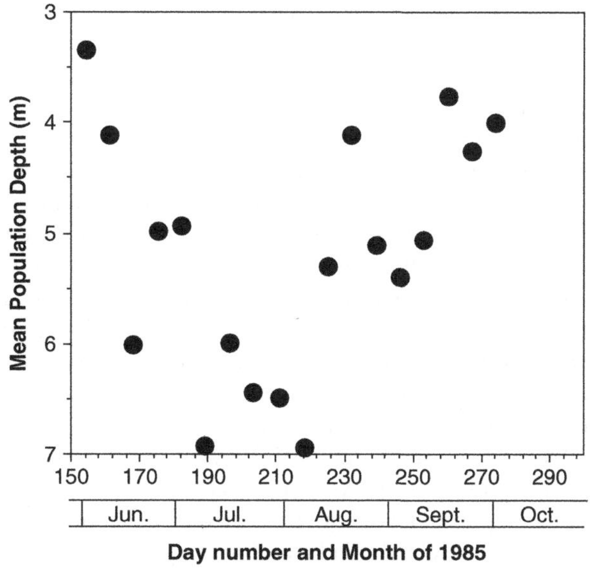

Fig. 9.13 The trend in daytime depth of Daphnia in Lake Maarsseveen for the summer season of year 1985

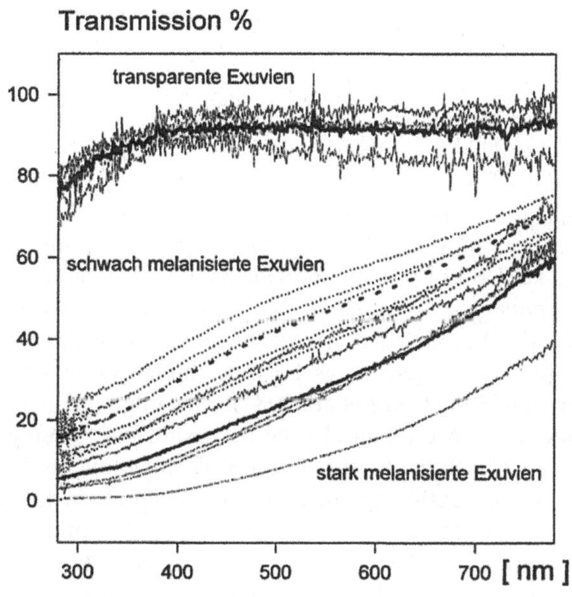

Fig. 9.14 Spectral transmission of exuvia from transparent and melanised *D. pulex*. Means are represented by *thick lines*. Compared to transparent carapaces UV radiation is highly reduced by melanin pigmentation (modified after Rhode et al., 2001 and Siebeck (personal communication, 2007); reproduced by permission of MacMillan Publishers)

Zooplankton has evolved three kinds of protection: (1) screening by pigmentation, (2) recovery by photoenzymatic repair (PER) and (3) avoidance by DVM. All three can be present at the same time although to a different extent, depending on genotype and species. A correlation with habitat is often evident.

Cladocera and Copepoda are sometimes coloured by carotenoids associated with lipids. In cladocerans, carotenoids are concentrated in ovaria, eggs and progeny, making them reddishly coloured, while the body remains relatively transparent.[8] Therefore, protection against UV-B radiation seems to be dedicated to the production of offspring (Hessen, 1994). Carotenoid content increases with the amount of lipids (Hessen, 1993) and lipid concentration is a function of food availability and used as indicator of the nutritional status (Tessier and Goulden, 1982; Tessier et al., 1983). Radiation seems to be a minor determinant of carotenoid concentration and these molecules might be of little importance in UV-B protection (Hessen, 1993). Melanin, on the other hand, is an important protector. The difference in spectral transmission through the cuticle of melanised and transparent *D. pulex* is nicely illustrated in Fig. 9.14. Especially short wavelengths are absorbed effectively. Alpine and Arctic *Daphnia* are often coloured deeply brown or black due to melanin deposits in the cuticle (Hebert and McWalter, 1983; Wolf and Hobaek, 1986; Hobaek and Wolf, 1991; Hessen, 1993). These coloured individuals are highly visible and easily preyed upon. Therefore, melanised cladocerans are often only found in ponds devoid of visual predators. A comparison of UV-B tolerance in high altitude, pigmented pond-dwelling *D. pulex obtusa* and the transparent *D. galeata* from lakes of the alpine foothill region reveals the protecting effect of pigmentation: the UV dose at which 50% of the experimental animals died (LD_{50}) was 1.6–1.7 times higher in *D. pulex* (Siebeck, 1978; Siebeck and Böhm, 1994). A melanised *D. pulex* from the Arctic had a higher tolerance for UV-B than a more transparent one from the subarctic (Hessen, 1994).

In calanoid copepods, carotenoid pigments seem to be of more importance as protectors against UV-B radiation. Distribution is more or less over the entire body, although higher concentrations occur alongside ovaria, covering nerve cell complexes, and in some segments of the antennae. Within the same species, red coloured and bluish, transparent morphs exist. For example, *Diaptomus nevadensis* in Soap Lake (USA) is red but individuals in Lake Lenore are translucent (Hairston, 1979). *Acanthodiaptomus denticornis* from Lac Pavin (France) is red but individuals in nearby Lac de Montcineyre are bluish translucent. After 48 h of UV radiation, lethal dose (LD_{50}) was 1.8 times higher in the red morph than in the translucent morph (Ringelberg et al., 1984). This might be due to carotenoid pigmentation but the difference might also be the result of a different photorepair mechanism (see Note 12). A trade-off between UV protection and predation was demonstrated by Hansson (2004). Pigmentation in copepods in lakes with a high predation pressure was lower compared to copepods from lakes with few predators. In the laboratory, under constant UV threat, and in the absence of predation, calanoid copepods responded with almost three times higher pigment levels compared to copepods in aquaria with caged fish. In the absence of UV, the level of pigmentation gradually decreased. In conclusion, Hansson (2004) suggested that pigmentation in calanoids is an inducible and adjustable defence, which depends on the prevailing danger from both predators and UV radiation. In experiments, the effect of radiation is often expressed as the lethal dose after some time, when 50% of the individuals are dead. The radiation dose ($Ws\ m^{-2} = J\ m^{-2}$) is the product of the radiation intensity or dose rate ($W\ m^{-2} = J\ s^{-1}\ m^{-2}$) with the exposure time (seconds). The interpretation and application of these LD_{50} values is not unambiguous. A short exposure at a high radiation intensity might have a stronger damaging effect than a low radiation intensity during a long time. Thus the product of dose rate and exposure time is not always constant. It is said that the "law of reciprocity" does not hold (Damkaer et al., 1981; Siebeck et al., 1994; Siebeck and Böhm, 1994). Thus LD_{50} values, as found in laboratory experiments, are not to apply unambiguously to natural situations. Throughout the year, radiation intensity changes and a certain dose will be reached after different exposure times. For example, for June, Siebeck and Böhm (1994) mention a dose of 35 kJ m^{-2} at which 50% of *D. galeata* were dead after 4 h. In September, nearly 8 h of exposure was needed to reach this dose, but then mortality was only 35%. At equal doses, a higher mortality occurred in June than in September. Weather conditions are of influence which makes LD_{50} values a first approximation of the damaging effect of UV-B.

Reciprocity and photoreactivation are not independent (Grad and Williamson, 2001; Aarseth and Schram, 2002). Of course not: reciprocity only holds when photoenzymatic repair is absent. If UV damage is the only process involved, then the product of dose rate and exposure time is constant at a certain effect. If a second process proportionally counteracts the effect of the first, the time needed to reach the same effect increases.[9] Photorepair works at wavelengths between 390 and 470 nm. To obtain a maximum effect, this largely visible light must be applied during UV-B radiation or within 2 h after the termination of the exposure. After 2 h, the recovery effect decreases linearly until after 10 h the recovery effect is zero (Siebeck and Böhm, 1994). In the lake-dwelling *D. galeata* and the pond-dwelling *D. pulex obtusa*, photorepair has a larger effect in *D. pulex obtusa* than in *D. galeata*.

The effect of UV-B and of photorepair radiation depends on wavelength. To deal with this, the notion of biological photon effectiveness has been introduced. One of the first relative effectiveness curves, based on *Daphnia* mortality, was presented by Siebeck and Böhm (1994). Mortality was maximum at about 270 nm and rapidly decreased at both sides. Wavelengths smaller than 290 nm do not reach earth and thus the curve and the solar spectrum overlap from 290 nm onwards. Siebeck and Böhm (1994) also published a biological effectiveness curve for recovery radiation. This bell-shaped curve has a maximum repair effect between 420 and 440 nm and declines rapidly at both sides.

Williamson et al. (2001) introduced the biological weighting function (BWF) . For each wavelength, the damaging result per unit dose is called a biological weighting coefficient e(λ). A plot of these unit damagings against wavelengths represents the biological weighting function. Weighting makes it possible to calculate effects at different radiation ranges and, for example, at different depths of the water column in a lake. The biological-weighted, effective exposure at a given wavelength is the product of dose and the weighting coefficient. A total radiation effect is obtained by summing the wavelength dependent effects:

$$H^* = \sum_{\lambda=290}^{500} \varepsilon(\lambda) x H(\lambda) x \Delta\lambda$$

Williamson et al. (2001) used *D. pulicaria* and mortality to estimate a biological weighting function. It can be imagined that the value of e(λ) depends on the species and the biological criterion that was used. If weighting coefficients differ much, the estimation becomes time consuming.[10]

Photoenzymatic repair is a physiological mechanism that costs energy to be maintained. This also holds for protective pigmentation, which has in addition the ecological cost of a high susceptibility to visual predation. That makes the alternative strategy of diel vertical migration superior, if possible. Pigmented *Boeckella gibbosa* lives in shallow ponds above the timberline where migration is impossible, while *B. gracilipes* migrates in deep clear lakes (Zagarese et al., 1997). The first species shows photoenzymatic repair, the second does not. Photoreactivation was not found in the calanoid copepod *A. denticornis*, living in deep Lac Pavin (Ringelberg et al., 1984).[11] Perhaps photoreactivation in migrating, lake-dwelling calanoid copepods seems less well developed or even absent. Also migrating *Daphnia* should have an underdeveloped photoenzymatic repair mechanism. This seems to be so, since *D. galeata* has a poor photorecovery compared to *D. obtusa* (Siebeck and Böhm, 1994).

9.3.2 Induced Migrations

To understand the relation between UVR and DVM, it is necessary to know to what depth UV-B radiation is damaging in lakes. This is difficult to estimate according to Siebeck et al. (1994) because several factors are of influence, for example, spectral composition, wavelength dependent attenuation, the relative spectral effectiveness of damage and recovery, dose-rate and recovery rate. It might be thought that because of a high attenuation, the shortest wavelengths are less important but if we allow for a biological weighting coefficient, depth profiles change. An illustration is given in Fig. 9.15. In A, maximum irradiance at the surface, measured on 26 June 1999, is higher for 320 nm than for 305 nm. Attenuation for the shortest wavelength is faster than for the longer one. Thus it might be thought that more damage would be experienced from 320 nm than from 305 nm. If we take effective exposure into account, as based on the biological weighting coefficient in Williamson et al. (2001), the position of the lines shifts and it becomes apparent that the shortest wavelength is more damaging up to a depth of about 3.5 m. This figure does not tell to what extent radiation is really a danger to life in a lake when exposure is over the whole daylight period. Organisms may escape to increased depths. Therefore, UV-B irradiance and the dose over the day at different depths were calculated. The interesting lake Morena East in Patagonia, with the calanoid copepod *B. gracilipes* (Alonso et al., 2004), was used as point of departure, but data from various authors were needed to arrive at the picture of UV-B isoclines, as presented in Fig. 9.16. The procedure of calculation is explained in note.[12] The dose, integrated over a day from 5 to 19 h, experienced at a depth of 1 m is 260.8 kJ m^{-2}. Survival is impossible for *B. gracilipes* at this depth. Experiments by Zagarese et al. (1997) revealed

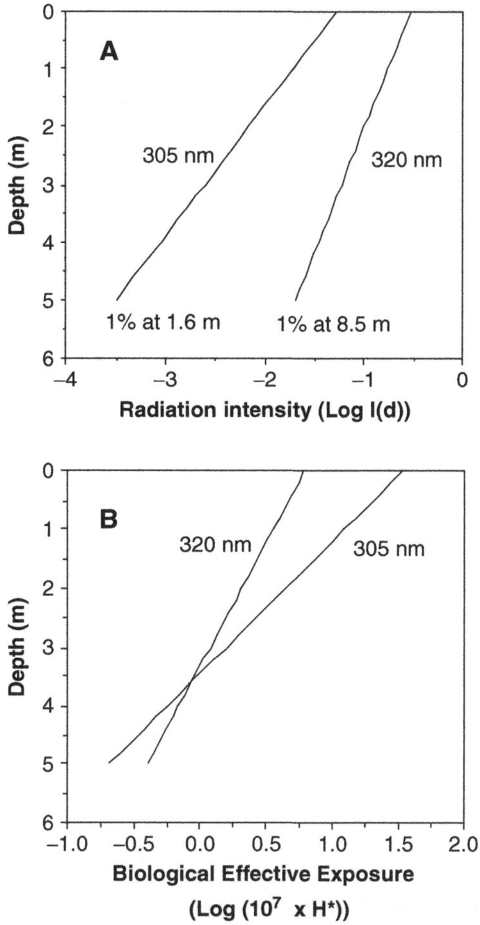

Fig. 9.15 (**a**) Radiation intensity for two wavelengths at different depth. (**b**) The biological effective exposure for these wavelengths at these depths. The attenuation coefficient for $g = 320$ nm is from Fig. 6.8, and the biological weighting coefficients are from Fig. 6 in Williamson et al. (2001). The attenuation coefficient for $g = 305$ nm is mentioned in Leech and Williamson (2000). Data hold for Lake Giles but no absolute values must be attached to the figures

an $LD_{50} = 1.3$ kJ m ^{2} for copepods from Lake Ezguerra, an $LD_{50} = 2.6$–3.8 kJ m^{-2} for *B. gracilipes* from lake Morena East, or 3.1–4.3 from Morena West (Alonso et al., 2004). Probably, these differences are not biologically significant. No photorepair mechanism seems available to this calanoid copepod. For UV-B protection, the species depends on a sufficiently deep migration in daytime. According to the modelled distribution of irradiation, this depth has to be 12 m, at least, because then the daily dose is 3.4 kJ m^{-2}. However, considering that a daily mortality of 50% is far too high for survival of the population and sublethal effects have to be prevented too, the individuals have to be deeper than 12 m in the about 50 m deep column. No precise vertical distributions for the Morena lakes were available to me, but Alonso et al. (2004) mention that the percentage of individuals in the upper 10 m in summer is very low and the majority of the population is below that depth. Avoidance of UVR was proposed to be the reason for the absence in the upper strata in this

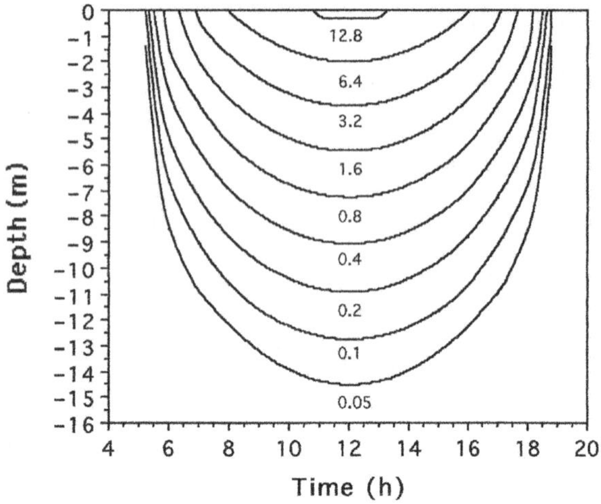

Fig. 9.16 Isoplethes of irradiance intensity or dose rate (W m^{-2}) of ultraviolet radiation between 290 and 350 nm at different depth and time in a hypothetical lake. Attenuation coefficients per 10 nm for an oligotrophic situation were used. See text and Note 12 for details

clear lake during periods of high solar radiation (Alonso et al., 2004). It is not mentioned whether *B. gracilipes* performs a diel vertical migration or that the animals stay down permanently during these summer periods. If the average distribution at night would be around the 1 m, an early descend is necessary because a dose of 4 kJ m^{-2} would have been reached well before 9 h.

We can make a simple exploration of the effect that different downward migrations have on acquiring a UV-B dose. Assuming descent starts at 7 h and occurs at a displacement velocity of 10 cm $min^{-1} = 6$ m h^{-1} (descent of *E. gracilis* took place at 4 m h^{-1} in Lake Maarsseveen (Ringelberg et al., 1991c), the dose reached after 2 h at 13 m depth would be 1.59 kJ m^{-2}. This is within the range of LD_{50} values found for *B. gracilipes* and certainly too high to maintain a population. Of course, if descent migration is faster, the received dose is smaller.

With a velocity twice as high, the dose becomes 0.89 kJ m^{-2} when reaching a depth of 13 m after 1 h 15 min. Mortality in *B. gracilipes* from Morena East is then 10%, at least.

However, another problem arises with a fast displacement velocity that late in the morning: what proximate cue will trigger the swimming response? See Chapter 11. The animals will encounter a rapid decrease in irradiance intensity that will inhibit swimming, even reverse the direction. Therefore, an earlier start of the migration down seems more obvious. If the hypothetical animals start at 5 h 30 min, which is more or less close to sunrise in early summer in the temperate zone, the dose reached after 2 h is 0.45 kJ m^{-2}. We can wonder whether this is ecologically acceptable or not, because it is still higher than the $LD_{20} = 0.29$ kJ m^{-2}, as found for *B. gracilipes*. My conclusion is that a still earlier start, before sunrise, would be better. We must then conclude that the phototactic mechanism with the resulting DVM pattern, as described in Chapters 3 and 4, offers a way out. There is one problem, though. For an amplitude that large, enhancement of the reaction mechanism by fish kairomones is necessary, if the results obtained in *Daphnia* also hold for this copepod. However, according to Alonso et al. (2004), fish do not play a role in the vertical distribution of *B. gracilipes*, although fish were present below 10 m and co-occurred during the day with the copepod. Therefore, the enhancing effect of fish kairomones in stimulation of the phototactic reaction mechanism cannot be excluded.[13] We deal with an interesting phenomenon worth of investigation because migrating into a depth with fish is kind of a problem.

If we calculate penetration of UV-B radiation for a mesotrophic situation, for example, Lake Constance or Lake Maarsseveen, the zone of mortality is less deep. No UVB attenuation coefficients are available for these lakes but concentrations of chlorophyll *a* and of DOC are and might serve as a guideline for choosing these coefficients. For a lake with comparable concentrations, Morris et al. (1995) give three K_l(d) values. If the exponential attenuation function is fitted to these values, it is possible to calculate attenuation coefficients between 300 and 350 nm, as were used in the previous oligotrophic situation. The procedure to arrive at a figure of isoplethes in time and depth was the same too. The daily doses at 1, 1.5 and 2 m were 23.2, 5.54 and 1.44 kJ m^{-2}, respectively. For *B. gracilipes*, the whole depth range would be dangerous but the LD values for *D. galeata*, 15, 35 and 53 kJ m^{-2} for 10, 50 and 100% mortality, respectively (Siebeck and Böhm, 1994), show that a daily mortality of 50% is present at depths between 1 and 1.5 m. A death rate of 50% is too high to maintain a population. Since the three calculated daily doses situate perfectly on an exponential function with depth, calculation shows that at depths of 0.7 and 2.13 m mortality is 100 and 1%, respectively. To be safe, *D. galeata* has to go down below 2 m. In both lakes, *Daphnia* is found below this depth, indeed.

In 1978 and 1979, the shoals of 0^+ perch we observed from 1988 onwards in Lake Maarsseveen, were probably still absent. In Fig. 9.13C, vertical distributions of *Daphnia* show that small amplitudes were present with mean population depths at noon of 5.47 and 6.27 m, respectively. Variation in depth is large, however, and a considerable part of the population is found higher up in the water column. If we position these mean population depths in Fig. 9.13, it becomes evident that the two anecdotal observations are probably part of a seasonal trend of increasing depth. This seems to be a general phenomenon and, for instance, found in Lake Constance where the mean population of *D. galeata* remains between 5 and 10 m throughout summer (Fig. 9.2). However, in that lake, from June onwards, 0^+ fish is present. *D. galeata* hardly migrates and is thought to have a different strategy of survival in the presence of visual predators (Stich and Lampert, 1981). Therefore, the seasonal increase in depth of *D. galeata* might be independent of the presence of fish, as it was in Lake Maarsseveen. UV-B radiation seems to be of importance. But why go deeper than 2–3 m if that should suffice according to our dose calculation? Of course, attenuation might have been less than was assumed and, if so, our calculated penetration of UV-B is an underestimation. We must also consider that sublethal radiation damage extends deeper than 2 m and selection might have led to a deeper depth distribution. Another circumstance must also be considered. Compared to oligotrophic lakes, the increased absorption in mesotrophic lakes makes that isoplethes are compressed. Upward drifts of a metre, for instance, could be lethal but of no consequence in highly transparent lakes. A reason for a deeper distribution than seemingly is necessary, given by Rhode et al. (2001), was that migration into the hypolimnion might prevent circulation to the surface due to turbulent mixing in the epilimnion. In addition, passive vertical displacements of 2–4 m of parts of a water column, with plankton included, caused by internal seiches, were described for Lake Constance by Schröder (1962). These internal seiches are also present in small stratified lakes like Lake Maarsseveen. Standing waves with an oscillation time of 6–8 h and an amplitude of 4 m were measured here (Kersting, 1981). We do not know to what extent the resulting vertical displacements of plankton are of influence on survival under UV-B radiation, so it is a speculation to suppose that they have been a factor in the development of downward migration. Whatever the ultimate aspect is, and in what way selection might have worked, the observed seasonal depth increase seems irrespective of fish presence. UV-B might be responsible.

In the Oberer Arosasee (Switzerland, 1734 m a.s.l.), *D. galeata* migrates with amplitudes of 1–3 m (Winder et al., 2002). During the ice-free period, from May to November, both the daytime and the night-time depth gradually increases with about 4 m. Therefore, two phenomena can be distinguished: (1) a small amplitude DVM and (2) a seasonal depth increase. The lake was regularly stocked with fish to accommodate anglers, but fish densities were very low and fish and fish chemicals are probably of minor importance (Winder et al., 2002). Although we deal with an oligotrophic, alpine lake, the authors also rejected a possible influence of UVR. The argumentation was that this radiation was absorbed in the upper 2 m while *Daphnia* went deeper. According to the authors, food and temperature played a crucial role: "food and temperature modulate the actual depth, at which the animals stop migrating..." (p. 126). Going deeper in the course of the season coincided with an increasing difference between a higher food concentration in the lower part than in the upper part of the water column. Also the difference in temperature over depth increased, but the direction is the other way round. "It is likely that the uncoupled food and temperature optima in the Oberer Arosasee (Fig. 1)[14] cause the continual migration behaviour of daphnids." (p. 119). It is not clear to me how this operates. It is true that the mean depth of *Daphnia* correlates significantly with the difference in chlorophyll *a* concentration between the upper and the lower layer and the same holds for temperature. However, these seasonal developing spatial differences in food concentration and temperature might be a correlation with the increasing population depth only. No causal relation needs to be involved and more evidence is needed to accept, even as a hypothesis, that "the day depths of daphnids are determined in such a way that the combination of food and temperature is optimal..." (p. 119). Optimality is a magic word in ecology, based on the maxim that selection will always realise the best mechanism. However, the physiological and behavioural mechanisms at the base of the observed phenomena are seldom studied. I think we still need an explanation for the seasonal increase in depth of *Daphnia* in the Oberer Arosasee. The paper offers a possible solution. A strong positive correlation between Secchi depth and mean population depth existed, thus, in the course of the season penetration of PAR increased. It is reasonable to assume that also UVR could penetrated deeper.[15] However, the authors found no correlation between the 1% depth of UVR and the mean population depth of the daphnids and they rejected a possible relation. UVR attenuation was not measured but calculated from the DOC concentrations in the lake, using equations in Morris et al. (1995). These authors caution that the formulae provide a rough estimate only[16] and, especially in alpine lakes, DOC is not the best predictor for UVR penetration (Laurion et al., 2000). It is generally thought that UVR is a factor of importance in alpine lakes. Therefore, I think that in the absence of a reasonable alternative explanation of the seasonal increase in depth, an increased penetration of UVR during the season offers an explanation of the increased depth distribution in the Arosasee too.

A few examples of lakes with zooplankton performing small amplitude migrations were discussed. Fish and fish kairomones seem to be of no importance. There is a seasonal trend to go gradually deeper and, at the end of summer, to come up again. In daytime as well as at night-time, mean population depth increases. Several authors have proposed that ultraviolet radiation (UVR) is a possible factor (Hessen, 1994; Leech and Williamson, 2001).

9.4 The Influence of Food

It is an open door, of course, to state that food is a most necessary factor in animals, zooplankton included. In all ultimate models on DVM, food is a key factor, as is predation. The proximate influence of predators and predation was discussed previously, and it was revealed that complicated behavioural mechanisms of prey dealt with this factor. On the other hand, the way food operates as a causal factor has far less been studied. In the impressive and extensive review of Pearre (2003), a hunger/satiation hypothesis is the central topic. Hunger is suggested to be the dominant factor at the base of ascent migration from the food-poor day depth to the food-rich near-surface water layers. When the animals are satiated at the end of the night, descent migration starts. This is a plausible hypothesis and many papers are cited in which the opinion is expressed that ascent migration at the end of the day is to feed in the upper water layers. However, it is not always clear what the authors have in mind. Do zooplankton swim upwards because food is present higher up in the water column? This would be ultimate reasoning. Or is hunger considered a causal factor that initiates ascent migration? Most papers, dealing with migrations in the field, seem to take upward swimming for granted. Problem is, however, how a tiny individual somewhere down in the ocean or in a lake, with limited sensory information and a limited capacity of its nervous system, "knows" in what direction to go and how to orient in the vast expanse of an isotropic environment? Without explanation of these aspects, the hunger/satiation hypothesis is incomplete. Researchers interested in photobehaviour (because they consider changes in light intensity at the base of DVM) have revealed how food might act as a causal factor. Clarke (1930, 1932), for example, described that *Daphnia*, in the absence of food algae, are often strongly positively phototactic, swimming in the direction of a light beam, whether it comes from above or from one side. Ringelberg (1964, 1987b) corroborated this effect. No decrease in light intensity is needed to elicit the reaction (see Chapter 2); in fact, when an animal shows a strong primary (sensu Clarke) phototaxis, increases in light intensity, otherwise causing downward swimming, have no effect. Also Haney (1993) found that prolonged exposure of *D. middendorffiana* to low food concentrations resulted in an enhancement of the positively phototactic response to food. He also showed that daphnids have no oriented response to food. An injection of *Scenedesmus* near the bottom of a 250-cm-long column caused a large part of the population of starved *Daphnia* to move rapidly towards the surface and away from the high food concentration! Therefore, *Daphnia* did not orient to food but to light for swimming upwards. This hunger effect on photobehaviour was also found for planktonic larvae of crabs (Cronin and Forward, 1980) and *Artemia* (Forward and Hettler, 1992). A mild induction of primary phototaxis might be responsible for the slow rise in the water column at the end of the afternoon, before the rapid decreases in light intensity occur and secondary phototaxis takes over, accelerating upward swimming. The absence of food or perhaps hunger or a combination of the two (we do not know what physiological trigger is of importance) must be considered a factor that affects photobehaviour mechanism 1 (PBM 1) (see Chapters 3 and 4), as do fish kairomones. Avoidance of predation and obtaining food are important ultimate reasons for DVM, but predation and food lack the properties to make adequate and directed behaviour possible.

We must consider examples of migration in which food gets more attention than in most papers.

On 21 May 1989, the population of *E. gracilis* in Lake Maarsseveen consisted of two parts (Ringelberg et al., 1991). One was found in the upper 5 m and the second part was below 10 m. This distribution remained the same during the day and the night and no DVM occurred. At the deep zone, between 9 and 11 m, chlorophyll *a* concentration was high, with a peak value of 11 $\mu g\ l^{-1}$ and in the upper part of the column, the concentration was less than 1 $\mu g\ l^{-1}$. After a week, on 7 June, the whole population of *E. gracilis* was down, where it remained also during the night. In the mean time, the deep chlorophyll *a* concentration had already decreased to 8 $\mu g\ l^{-1}$, and diminished further. One week later again, *E. gracilis* migrated in a "normal" way. It was remarkable that during this whole period *Daphnia* migrated at night exchanging the food-rich deep layer for a poor epilimnion (Ringelberg et al., 1991). Although this is an anecdotal example, it suggests that food might have played a species-specific role in modifying migration behaviour. Do these observations confirm or falsify the hunger/satiation hypothesis? Did the deep-dwelling individuals of *Eudiaptomus* not ascend because they were probably satiated? And did the deep chlorophyll *a* layer have to be exhausted before the now hungry animals started to migrate? If so, the observed phenomenon is in favour of this hypothesis. On the other hand, is the upward migration of *Daphnia*, while food is more abundant at the day depth, inconsistent with the hypothesis? Or did *Daphnia* not eat in daytime? We have a problem.

Differences in migration behaviour and diel feeding between *Daphnia* and *Eudiaptomus* were also described by Angeli et al. (1995). In Lake Geneva, the hybrid *D. galeata x hyalina* is a strong migrant, at least the larger size classes. They went down for at least 30 m and, at dusk, came up to the upper 7 m. On the other hand, *E. gracilis* migrated feebly within the 0–7 m depth layer. Differences in this behaviour were accompanied by differences in gut fullness. *Daphnia* had a full gut at dusk and dawn but no difference in gut filling was observed in *E. gracilis* during the whole day.

How must these species-specific differences be interpreted? In Lake Geneva, *Daphnia* was a strong migrant much more than *Eudiaptomus*. In this lake, *Daphnia* could only eat during the night and the diel differences in gut filling seem to be a consequence of DVM. No diel feeding rhythm is needed to explain the difference. *Eudiaptomus* could eat throughout the day and obviously did so.

The absence of a feeding rhythm in a calanoid copepod was demonstrated by Hart (1977). In Lake Sibaya (South Africa), *Pseudodiaptomus hessei* performed an extensive DVM over the whole water column. A diel difference in gut filling was observed. Adults and late copepodites stages were near or even in the mud during the day where feeding was limited or not possible. At night, they migrated upwards into the water column and the proportion of individuals with food in the gut increased strongly. Gut filling might have been a consequence of this migration or due to a diel feeding rhythm. In laboratory experiments, under different light regimes, but with food always present, no rhythm in feeding was observed. Hart (1977) concluded that the diel difference in gut filling was a consequence of migration into the water column and not because of an endogenous rhythm.

Haney and Hall (1975) studied DVM and feeding in small lakes in Wisconsin (USA). Dominant species were *D. galeata mendotae*, *D. pulex* (probably *D. pulicaria*) and *Diaptomus pallidus*. This is the only study in which the role of food was estimated as filtering rates and not as gut fullness. Therefore, the actual feeding process was object of analysis. Filtering rates were measured over short periods (less than 15 min, using the well-known Haney trap) and twice during a 24-h period. Vertical distributions and light intensities were determined frequently, especially during the twilight periods. In Little Mill Lake (maximum depth 9 m) *D. galeata* migrated through the epilimnion, whereas *D. pulex* (*D. pulicaria*?) remained below 4 m with no vertical displacement. Filtering rates for both species were higher during the night than during the day. "Hence, it would appear that an actual vertical excursion is not a necessary prerequisite for high night filtering" (Haney and Hall, 1975; p. 421). The filtering rates of *D. galeata mendotae* in Wintergreen Lake (maximum depth 6 m) and of *D. pulex* in Three Lakes (maximum 8 m) were bimodal with maxima at dawn and dusk, a depression during a few nightly hours and very low values during the day. Both species migrated over the whole length of the water column. Descent and ascent swimming and changes in filtering rate took place during the rapid changes in light intensity of dawn and dusk. Obviously, both are strongly temporally correlated but are they also coupled functionally? Haney and Hall (1975) hesitated to draw this conclusion and proposed that both phenomena are initiated by the relative changes in light intensity of dawn and dusk. Thresholds, as determined for the phototactic swimming response in *D. magna* (Ringelberg, 1964; Daan and Ringelberg, 1969), are cited to indicate that the beginning of migration and filtering were close to these thresholds.

Haney and Hall (1975) found that filtering rates of *Diaptomus pallides* remained the same throughout the day and the night. This species hardly migrated and remained predominantly in the epilimnion.

In Section 9.2.1, it was shown that a low algal concentration prevented DVM to start. Food was considered part of the "set of ecologically necessary conditions". In Section 4.5, food concentration was characterised as a modifying factor in the decision-making mechanism. On the conceptual level of cause and effect, food is not a factor of absolute necessity. In the domain of ultimate aspects, however, food is of primary importance as a principle determining factor for the number of offspring on which selection moulds the behavioural mechanisms.

On the basis of the few field observations, presented in this paragraph, some additional conclusions regarding the combination of DVM and food can be added. Noticeable in the first place is the difference in behaviour of daphnids and diaptomids. The first migrates with a larger amplitude than the latter, if the calanoid migrates at all. This might be due to a difference in reactivity to the light stimuli of dawn and dusk. The behaviour of the calanoid seems more flexible, as might be deduced from the behavioural difference between *Daphnia* and *Eudiaptomus* in Lake Maarsseveen. If the reaction to changes in light intensity is stronger in *Daphnia* than in *Eudiaptomus* (see the observation in Lake Geneva), leaving a layer with food might be the unavoidable consequence of a physiological make-up. Also Winder et al. (2004) observed in the Oberer Arosasee that *D. galeata* ascended out of a food-rich environment at night. Whatever the ultimate reason for *Daphnia* was in the Wisconsin Lakes, *Daphnia* migrated but *Diaptomus* did not. Diel differences in feeding in *Daphnia* were present in all cases. But migration is not conditional for this, as Haney and Hall (1975) found. The independent control of feeding and migration was also found for *D. longiremus* in arctic Toolik Lake (Alaska). In mid-July, Buchanan and Haney (1980) observed a change in filtering rate following upon the 0.0003 s^{-1} (= 0.02 min^{-1}) evening change in relative light intensity. Obviously, changes in light intensity suffice to trigger also the filtering process. This concluding argumentation overlooks the possible contribution of an endogenous circadian feeding rhythm.

9.5 Complications

The general pattern of DVM presented by Cushing (1951) consists of five phases: (1) an ascent from the day depth, (2) a descent over a relatively short distance during the night, (3) an ascent again at dawn, soon followed by (4) a descent to the day depth, and (5) the more or less stationary day depth. Bary (1967) divided the ascent and descent phases into a rapid and a slow-moving part, with descent swimming starting fast and then slowing down while it is the other way round for the evening ascent. These distinctions were possible because his observations were based on continuous records of migrating sound scattering layers. Bainbridge (1961) maintains that Cushing's cyclic scheme is an idealisation and only found upon the rarest of occasions. Of course, schemes are idealisations but, on the other hand, detailed knowledge is insufficient because sampling intervals are often too long to detect all phases. In the reviews of Cushing (1951), Banse (1964), Bainbridge (1961) and Segal (1970) diverse patterns were mentioned for various locations. Variation is natural in biological phenomena. Our objective is to explain them. As Pearre (1979, 2003) has pointed out, knowledge of DVM is based on differences in vertical distributions over time. The observed distributions are the results of mass movements, the sum of the swimming of many animals. We do not know what the individual animal does at a particular time. That is, however, what we want to know because without knowledge of behaviour it will be difficult to understand what causes migrations, how they are timed and what effects they have on population dynamics and communities. In the Darwinian paradigm, the individual is central, and evolutionary interpretations of DVM patterns implicitly suppose that all individuals behave in the same way. Without knowledge of individual behaviour, understanding of changes in vertical distribution is hard to obtain. Peare (1979) suggested that asynchronously migrating individuals, with a stationary time at both ends longer than the transit time in between the ends, could lead to a bimodal vertical distribution. That is true, but other explanations are possible too. In 1997, during the first 2 weeks of the migration period in Lake Maarsseveen, part of the population of *D. galeata x hyalina* remained in the upper few metres of the epilimnion, while another part migrated to the meta- and hypolimnion. Allozyme analysis showed that the genotype of the upper part differed from the rest of the population (Ringelberg et al., 2004). This observation does not exclude asynchronous swimming, in itself. However, the gradual decrease in individual numbers in the epilimnion in the course of a few weeks, while numbers in the deeper part remained the same, made plausible that we dealt with a non-migrating subpopulation and not with a population consisting of asynchronously migrating individuals.

9.5.1 Reverse Migrations

On rare occasions, zooplankton are found in the upper part of the water column around noon and deeper down during the night. Hutchinson (1967) gives examples of reverse migrations described in the older literature. One of the best documented is that of *Tropodiaptomus banforanus* in Lake Rudolf (= Lake Turkana, Africa) by Worthington and Ricardo (1936). Especially the juvenile individuals show the phenomenon. In daytime high numbers were present in the upper 5 m, but at midnight, the average number was at around 10 m and at 06 h even at 20 m.

To what extent are reverse migrations comparable with normal migrations? Do we deal with a different phenomenon, with different proximate stimuli, or are the stimuli the same but induce they an inverse pattern? Since also reverse migrations coincide with times of large changes in light intensity, a phototactic reaction to relative increases and decreases is the most obvious conclusion, as it is in normal DVM. In that case, the kinetic part of the phototactic reaction must be the same and only the orientation aspect is different. The question then is what caused the inversion of the sign of phototaxis and why. A suggestion regarding the first was given by Bayly (1963). He observed that reverse migrations of males (not females) of *B. propinqua* occurred in the alkaline lakes of New Zealand at a pH of 8 or more. Experimental data seemed to support the role of alkalinity. For example, Baylor and Smith (1957), whose paper is cited by Bayly (1963): "high pH values induce positive phototaxis while low pH values induce negative phototaxis" (p. 33). For their part, Baylor and Smith (1957) referred to Loeb (1904) who, however, found it to be the other way round. Lowering the pH with CO_2 and diverse acids made copepods and cladocerans positively phototactic and not negatively and thus swimming towards light.[17] In the mean time, reverse migrations have been found in lakes of about neutral pH with some species migrating in a reverse way, some in a normal way. In fact, reverse migrations occur in highly different circumstances and it is not always clear whether we really deal with the same phenomenon. For example, Cunningham (1972) described that under the ice of Fuller Pond (USA) *C. scutifer* and *Diaptomus minutus* aggregated at noon at a depth of 1 m but were evenly distributed from surface to bottom (14 m) at midnight. On the other hand, *D. galeata mendotae* was evenly distributed throughout most of the day with perhaps a slightly higher abundance in the first few metres during the night. Also under the ice of Silver Lake (USA), Bove and Stewart (2002) found a higher abundance in the first 2–4 m in daytime for some species, but not for other species. At night, the distribution was random over the 10-m column. Neither Cunningham (1972) nor Bové and Stewart (2002) offer an explanation for these reverse migrations under ice, or why some species are concentrated just below the ice, while others are distributed over the water column. Since we deal with differences in optical orientation, a possible explanation might be found in the specific angular light distribution in these under ice situations. In both papers, it is mentioned that a hole was made in the ice to make sampling possible. To prevent a bright spot under the hole, a box was placed over it. The box tended to cast a shadow just below the ice, as Cunningham (1972) mentioned, but from about 1 m downwards, light intensity decreased continuously as normal. It is evident from both papers that the high concentrations of individuals, argument for the conclusion of a reverse migration, occurred at the depth of this shadow of the box. It is well known that copepods and cladocerans aggregate near the edges of light intensity changes or contrasts, so my prudent suggestion is that these reverse migrations under ice were the result of optical orientation to the artificial light–dark boundaries. A repetition of observations with this possibility in mind is simple to perform.

One of the most convincing observation and explanation is made by Ohman et al. (1983). In Fig. 9.17, the day–night vertical distributions of *Pseudocalanus* and of the invertebrate predators *Sagitta elegans*, *Euchaeta elongata* and *Euphausia pacifica* in Dabob Bay (USA) are presented. The calanoid is prey for the other invertebrates.

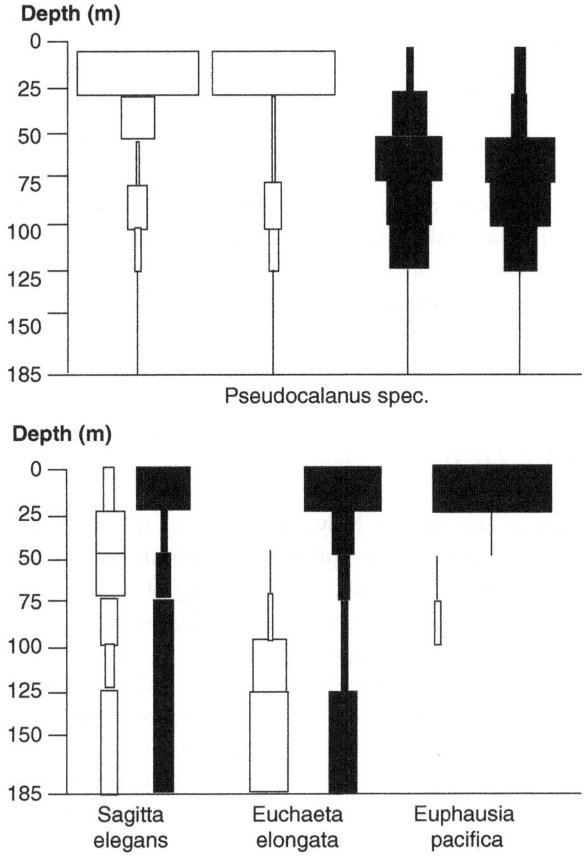

Fig. 9.17 Vertical distribution of A. females *Pseudocalanus* and B. *Sagitta elegans*, *Euchaeta elongata* and *Euphausia pacifica* in Dabob Bay (USA) on 25–30 July 1973 Open colums = day distribution; black columns = night distribution (modified from Ohman et al., 1983; reproduced by permission of The American Association for the Advancement of Science)

The difference between prey and predator distributions is striking and it is obvious why Ohman et al. (1983) concluded that predation prevention is the significance of the reverse migration of *Pseudocalanus*.

A reverse migration of a different nature but probably also induced by predators was described for scattering layers in the Northeast Atlantic by Heywood (1996). On several days, a scattering layer, at about 40 m depth in daytime, moved down to about 200 m as soon as a layer from below 300 m arrived at 40 m in the evening. The author suggested that the reverse migration was a reaction to the arrival of normally migrating predators. Although plausible, we are not sure that prey and predators were involved because the species composition is unknown.

Again, predation prevention is probably the best ultimate hypothesis to explain reverse migrations. This time non-visually hunting invertebrate predators are involved. Small species like rotifers, nauplii, *Pseudocalanus*, *Boeckella* and *Bosmina* especially migrate reverse to evade invertebrate predators such as *Chaoborus* and *Leptodora*. As predator fish do, also these invertebrate predators produce kairomones that change life history characteristics in cladocerans (Black, 1993; Lüning, 1992; Spitze, 1994; Dodson et al., 1995; Lagergren and Stenson, 2000). However, behaviour also is influenced. Water from a culture of fourth instars of *Chaoborus flavicans*, added to the first of five mutual connected experimental chambers, caused *D. pulex* to move away from that chamber (Kleiven et al., 1996). Since the invertebrate predators are in turn preyed upon by visually hunting fish they migrate downwards in the morning, thus leaving the upper water layers as a refugium for the small zooplankton. It is not far-fetched to suppose that kairomones from invertebrate predators enhance phototaxis at the base of reverse migrations. If so, only the orientation component is differently influenced. The hypothesis can be tested.

9.5.2 Drifts

Also at the end of May, *D. pulicaria* in Lake Maarsseveen begins a slow drift downwards from the epilimnion that will end deep into the hypolimnion 1 month later (Flik and Vijverberg, 2003) (Fig. 9.18). No upward migration in the evening occurred. During a 7-year-long study, this seasonal pattern was repeated annually. A relation with temperature was suggested by the authors because as soon as the temperature of the epilimnion increased above approximately 16°C, the descent started. By mid June, the water layers near the bottom at 20 m were reached and the population did not come up again before early spring of the next year. The coexisting *D. galeata x hyalina* started its DVM around the same time but stopped migration around the end of July. Whether the drift of *D. pulicaria* was initiated by the kairomone of the 0^+ perch was doubted by Flik and Vijverberg (2003) because in an experiment the addition of fish water to a cylinder with *D. pulicaria* had no effect on the vertical position. In a life history experiment, *D pulicaria* did not show any significant response to fish kairomones either, contrarily to *D. galeata x hyalina* (Vijverberg et al., 2006). On the other hand, a gradually raised temperature from 10 to 20°C made the animals drift down and Flik and Vijverberg (2003) suggested that increasing temperature might be of importance as a proximate factor initiating the behaviour leading to the downward drift. The ultimate reason for the downward drift is thought to be predation prevention. The large *D. pulicaria* is very vulnerable to fish predation and individuals are safe in the hypolimnion. The temperature increase coincides more or less with the start of feeding of the 0^+ perch. If the temperature increase is a cue to start drifting down, it is an indirect cue and thus less reliable than the direct fish kairomones used by other zooplankton. A deeper position of other *Daphnia spp.* in experimental cylinders in the presence of fish kairomones was observed by Dodson (1988), Ringelberg (1991), De Meester (1992) and others. Downward movements following upon a temperature increase were described as early as 1919 by Esterly (1919) for different zooplankton species, *Daphnia* included. Although conclusive evidence with regard to the proximate factor that makes *D. pulicaria* go down awaits further experimentation, the role of temperature

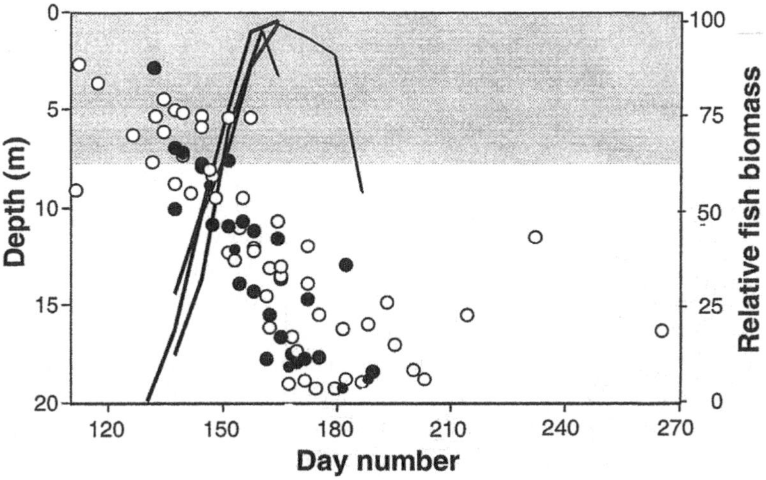

Fig. 9.18 Seasonal variation in mean population depth at noon (*open circles*) and midnight (*black dots*) of *D. pulicaria* in Lake Maarsseveen during the successive years 1992–1998. The solid lines present the relative biomass of 0^+ perch (*Perca fluviatilis*) for the years 1992, 1993 and 1994. The shaded part indicates the water layer inhabited by the perch (from Flik and Vijverberg, 2003)

increases cannot be neglected. It would be interesting to be sure that *D. pulicaria* has evolved an atypical defence strategy against fish predation.

A slow drift down upon a gradual increase of temperature of the epilimnion was also described by Johnson et al. (2007). During the arctic summer, from the end of June until the end of July, *C. scutifer* moved in Toolik Lake (Alaska) from the near surface to a depth of 10–12 m. Temperature increased from 6–8°C to 12–15°C but the copepods remained in the lower temperature range. Light intensity hardly changed during the day and no DVM occurred. Also in this case temperature increase seemed to be the proximate factor for the drift.

Tessier and Leibold (1997) studied the vertical distribution of *D. pulicaria* in six lakes. In all of them, *D. pulicaria* were down in the hypolimnion during the day, in some they stayed down also during the night, and in others they spread out over the whole column at night. Since sampling was done at the end of August and the beginning of September, the daytime distribution in the hypolimnion is comparable to that in Lake Maarsseveen. For the animals caught at night, electrophoretic analysis was performed. The water column was divided into two "habitats": the epilimnion and the hypolimnion. Clonal heterogeneity was found and the authors concluded that "habitat use behaviour" was different in different lake populations. This only holds for the night because during the day in all lakes the populations were confined to the hypolimnion. Nevertheless, Tessier and Leibold (1997) speak of a "striking habitat segregation of clones". In more traditional terms, this different "habitat use behaviour" is thus based on the observation that some clones migrate into the epilimnion at night, while some do not migrate and stay down all the time. Formulated in that way, a different vertical night distribution is at least as comprehensible as "clonal habitat partitioning" . Lake populations with a higher number of distinct allozyme types also showed a larger diffusion into the epilimnion at night; in other words, such lakes consisted of more migrating genotypes. In the sophisticated jargon of the authors this reads as "populations with a broad use of lake depth have more clonal diversity than populations displaying restricted depth use" [18] (p. 566). In Lake Maarsseveen only three allozyme types of *D. pulicaria* were present, of which one comprised more than 90% of the density (Flik and Vijverberg, 2003). This genotype did not perform a diel vertical migration.

Over the last decades, ecologists seem to be imbued with the spirit of the age to apply words and terms, etceteras, for notions that have been in use and established for a long time. For example, also Flik and Vijverberg (2003) think that "diel vertical migration can be best considered as a special case of depth selection behaviour" (p. 290), thereby following De Meester et al. (1998). Depth selection behaviour (DSB) is defined "as the behavior by which the zooplankton maintains a particular (daytime, night-time) vertical distribution in relation to the vertical stratification of the water column (light, temperature, food, predation pressure)" (De Meester et al., 1998; p. 161). Migrating animals do not comply with this definition and thus DVM is made a special case of depth selection behaviour. This does not really contribute to our understanding of a phenomenon of which the proximate as well as the ultimate aspects are reasonable well known. Of "depth selecting behaviour", if it really is a distinctive phenomenon, we do not know the underlying mechanism. No one has ever revealed the *behavioural* mechanism by which depth is *selected.* It is descriptive, phenomenological and redundant, although it suggests an explanation. Clarke's dictum (see Preface) is applicable here again. Tessier and Leibold (1997) called for that matter DVM of *D. pulicaria* "habitat use behaviour" and migration led to "habitat segregation": at night *D. pulicaria* used a different habitat than during the day. The modern use of "habitat", "depth selection" and "behaviour" erodes the meaning of these terms. The late Bob Wetzel was annoyed by this trend in ecology. In the third edition of his widely used textbook on Limnology, he devoted a paragraph to the topic called " Buzz Words: The Plague of Fuzzy Concepts and Ambiguous Terminology" (p. 464). It must be read: I cannot say it better than this eminent limnologist did.

9.6 Interpretations

Studies, extending over the better part of the summer season, revealed a seasonal character of DVM. This is consistent with the seasonal occurrence of a large number of young-of-year planktivorous fish. Small amplitude migrations, in the absence of fish, were also reported, however. For these small amplitude migrations, avoidance of UV radiation was proposed as adaptive significance.

Ecological phenomena, like DVM, are hard to explain with one-to-one relations of cause and effect. I therefore proposed that a "set of necessary ecological conditions" are responsible. This is a combination of environmental circumstances, not necessarily all of equal importance and depending on conditions. An individual animal "validates" these circumstances and the result is migration with a certain amplitude. Factors as food availability and predator presence were mentioned in the literature as important, but in the majority of cases in a qualitative sense only. The influence as a set factor is by quantity, however, and the best way to study is by experimental analysis. A first result of such analysis is the "decision-making mechanism" , presented in Chapter 4. Validation by an animal of the "set of necessary ecological conditions" is not a conscious process, of course, but a complicated process in the central nervous system, as evolved by selection. Validation leads to the best available strategy in the ambient circumstances. "Best available" is not equal to optimal. We therefore can observe at the same time and in the same lake, differences in vertical distribution, even of migration strategy. Fieldwork must describe the differences and, because the system is dynamic and always changing, must be approached sytematically and for some long time.

In Lake Maarsseveen, at the time of the experiment of nature, the algal concentration was *probably* a key factor in the set of conditions because the biomass of 0^+ perch was large enough to produce a sufficiently high kairomone concentration (although stratification was not completed and dilution of the kairomone over a large part of the water column was still possible). The combination of food[19] and kairomone concentration was for a few weeks insufficient and the validation by the "decision-making" mechanism was such that migration was not started. This conclusion is possible only because of the experimental results, presented in Chapter 4. Temperature is important as a factor determining thermal lake stratification and thus dilution of kairomones, but I hesitate to include temperature in the set. It is important to mention that predation as such is not a factor of the set. Like food concentration, kairomone concentration provides essential information for the CNS. Additional experimentation is needed to get more insight into the underlying mechanisms.

The differentiation between a small and a large amplitude is not simple. However, if zooplankton does not leave the epilimnion, it can be assumed that predation prevention is not involved. Protection against the damaging effect of UV-B radiation is plausible then. Since in very clear lakes, UVB might penetrate rather deep in the column, for example, the discussed Morena lakes, a large amplitude points not always to the presence of fish (Leech et al., 2005) (although fish were present). Since the intensity of UV is changing during the season, small amplitude migrations also might have a seasonal character. Must UV radiation be included in the "set of necessary ecological conditions"?

Photobehaviour system 2 (PBM 2) is responsible for the larger part of descent and ascent swimming. To say it bluntly, without the changes in light intensity of dawn and dusk, DVM does not occur. Therefore, the continuous change in light intensity is the primary causal factor within the set and of more importance than the secondary factors food and predation. However, although essential causally, light changes have no ultimate meaning. Predation and food directly determine mortality and birth rate and thus whether a population, a species or a genotype is maintained or not. They are important from an ultimate point of view. As causal factors initiating and steering DVM, they are unimportant because food and kairomones have no timing and no directing cue. The light intensity changes of dawn and dusk function as a clock and provide a timely, early warning signal. The angular light field makes it possible to find the direction of up and down. The perception of light changes and PBS 2 is a physiological–behavioural mechanism of the individual that was evolved in evolutionary time as a *help function* for an adequate reaction. Food and kairomone concentration are selecting forces leading to photobehaviour system 2.

The ultimate aspect of small amplitude migration is the prevention of UV damage. Is this radiation also the proximate factor, causing this DVM? *Daphnia*, at least, is able to perceive UVA. Rhode et al. (2001) suggest ultraviolet radiation acts as such. A visual pigment, extending absorption into the UVA region, is present in *Daphnia* enabling perception of near ultraviolet radiation. The wavelength region is part of the photobehaviour system and thus part of the "set of ecologically necessary conditions".

Slow drifts occur and are sometimes part of the DVM pattern. Some are disconnected from the diel cycle. Upward drifts occur, for example, at the end of the afternoon and before the large ascent movement after sunset takes place. No experimental evidence is available about the cause of this swimming but I hypothesise that an imbalance of the oscillator mechanism (see Chapter 2) might be responsible. Hunger might be the cause for an increased strength of the antennal stroke during the "hop" phase, resulting in a net upward movement in *Daphnia* and probably other planktonic animals. An example of a drift disconnected from DVM is the seasonal downward displacement of *D. pulicaria*. Over a period of about 60 days the mean population depth went down over some 17 m, thus with a displacement of 28 cm per day. Also here, an imbalance of the oscillator mechanism might be responsible. The seasonal change in depth distribution in the Oberer Arosasee is interesting because it consists of a small amplitude DVM superimposed on a slow downward drift. Fish seemed to be of no importance, thus prevention of UV damage is probably the adaptive reason. No experimental analysis is available, thus we can only guess what is going on.

A lake environment is dynamic and the direction of selection changes continuously. We see this reflected in the genotype composition of a *Daphnia* population complex (Chapter 12). The clone collection is never in equilibrium. The complex "trait of migration" is part of the pool and it may be hypothesised that each year the period of DVM starts with another combination of traits, thus contributing to the observed large variation in amplitudes.

Notes

1. The picture is more complicated since we know that the population of migrating *D. hyalina* consists of the hybrid *D. galeata x hyalina*. Behaviour of these hybrids is intermediate. For example, migration amplitude is less and they profit from higher average temperatures and food concentrations. Also life history characteristics might be intermediate. Population size of the migrating *D. hyalina*, as mentioned by Stich (1985), might be predominantly accounted for by these hybrids.

2. For the chemical perception of the kairomone molecules, a threshold concentration must be expected. This concentration will be much lower than the lowest concentration in the experiments by Loose (1993a) or Van Gool and Ringelberg (2002) and for this reason a behavioural threshold was possibly not found. On the other hand, the concentration for a maximum behavioural response can be concluded from the results. Water from the epilimnion of the plankton tower was continuously pumped into an aquarium with small fishes and the same amount flowed back into the epilimnion. In about 12 h an amount equal to the volume of the epilimnion was thus circulated. Assuming complete mixing, Loose calculated the number of fish per m^3 and thus found that a maximum day depth of the migrating *Daphnia* was reached at 3 fish (of 5 cm) per m^3. This probably is an underestimation because I think that a plume of a high kairomone concentration must have existed at the mouth of the inflow from the aquarium. For instance, with half the volume of the epilimnion mixed after 4 days when the observations started, the number of fish per m^3 is doubled. In the experiment of Van Gool and Ringelberg (2002) a maximum phototactic response was probably reached at a dilution of 12.5% of the aquarium water (10 l) with one perch of 5 cm. Complete mixing in the small experimental cylinder was realised. Thus a maximum response was realised at 12.5 fish m^{-3}. From both experiments, a maximum response of *Daphnia* can be guessed at a fish concentration of 6–12 fish of 5 cm per m^3. This is a biomass of 5–10 g m^{-3}.
   An important conclusion can be drawn from both experiments. After the sensory perception of fish kairomones in the water, daphnids do not respond with an "all-or-nothing" reaction. The response intensity depends on the concentration of kairomones. This is to be expected, of course, if these kairomones play a role in a "decision-making" mechanism.
3. Mehner and Thiel (1999) refer to fish smaller than 20 mm total length as larvae, whereas larger young-of-the-year fish were referred to as 0^+ juveniles. This was a pragmatic classification in order to deal with the variable terminology in the literature. In fact, anatomical and physiological changes accompany this critical value of about 20 mm total length.
4. Phototaxis was not enhanced if substances of injured *Daphnia* were put in the water (unpublished results by J. Ringelberg).
5. Although in two series of experiments, extending over 9 days, the day–night distributions of *Acartia* on days 8 and 9 suggested DVM; this occurred only when fish were swimming freely in the enclosures. It was remarkable that in the control and the enclosures with caged fish, the copepods were deep down during the day and the night. Why ascend when free swimming fish were present and why stay down in the absence of fish? Bollens and Frost also thought the results of these relatively simple enclosures experiments "potentially complicated" and do not exclude the possibility of an altered proportion of individuals with genetically fixed behaviour, as a result of selective predation by the free-swimming sticklebacks.
6. Juvenile perch were collected on the same days as test water from epi- and hypolimnion. Average total length increased linearly over the period. A power function was used to describe the length–weight relation. The number of fish per unit effort decreased with time according to an exponential decay function. These relationships were used to calculate the 0^+ perch biomass, resulting in the bell-shaped curve over time, as is presented in Figs. 9.9 and 9.10.
7. This value was calculated from an equation given by De Lange (1999), using a k_dPAR = 0.41 m^{-1} (see Table 9.3). The k_dUVB = 9.1 m^{-1} with a 1% value of 0.5 m was obviously measured by this author on a day with a high phytoplankton concentration because the k_dPAR = 0.83 m^{-1} is outside the range mentioned by Swain et al. (1987), based on seasonal measurements of many years.
8. The large *D. pulex* and *D. magna* of low altitude ponds may be redcoloured by haemoglobin in the body fluid Peters (1987).
9. We encountered the phenomenon when the stimulus strength–stimulus duration of the phototactic reaction was discussed (Section 3.5.2). The relation between dose rate and time for a certain damage is comparable. Without photorepair, equations (9.1) and (9.2) hold. With photorepair present equations (9.3) and (9.4) hold. With decreasing dose rate, the difference

in time at which a certain damaging effect is reached increases. At a high dose rate, it will be difficult to conclude whether reciprocity holds or not.

$$\frac{\mathrm{dEffect}}{\mathrm{d}t} = \mathrm{doserate} \quad (9.1)$$

$$\mathrm{Effect} = \mathrm{doserate} \times t \quad (9.2)$$

$$\frac{\mathrm{dEffect}}{\mathrm{d}t} = \mathrm{doserate} - \mathrm{repairrate} \times \mathrm{Effect} \quad (9.3)$$

$$\mathrm{Effect} = \frac{\mathrm{doserate}}{\mathrm{repairrate}} \times (1 - e^{-\mathrm{repairrate} \times t}) \quad (9.4)$$

10. The biological weighting function for *D. pulicaria*, as presented by Williamson et al. (2001) (see their Fig. 6), is empirically derived. The authors draw attention to the similarity of their function and those obtained for eggs of *Calanus finmarchicus* and eggs of Atlantic cod (*Gadus morhua*) as presented by Kouwenberg et al. (1999). This is not surprising because a strong relation ($R^2 = 0.999$) exists between Williamson's weighting coefficient and photon-specific energy for the UV wavelength range from 290 to 350 nm. And, of course, photon energy is a (hyperbolic) function of wavelength and thus the weighting coefficient is a function of wavelength. I think it can be concluded that the damaging effect of UVB radiation depends on the energy of the photon (and the number of quanta) with the shortest wavelength having the strongest effect. The rationale for this simplified approach is that contrarily to wavelength of the visual spectrum (thus including UVA in *Daphnia*, at least) no wavelength-specific absorption exists. Depending on the kind of organ, of physiological function, even of species, a calibration coefficient might exist but the weighting of radiation is according to photon energy. This makes it possible to weigh wavelength depending energy in a general estimate of the expected danger. In the text, examples are given for different lakes. The weighting coefficient is set at 1 for a wavelength of 300 nm ($E = 6.845 \times 10^{-19}$ J), 0.967 for 310 nm, etc.
11. Zagarese et al. (1997) illuminated *B. gibbosa* with visible light for 15 min immediately after 4 h UV-B radiation and found indeed a large reactivation effect. Perhaps the absence of reactivation in *A. denticornis* was an experimental error, as Ringelberg et al. (1984) already suspected. As Zagarese et al. (1997) remarked, neither their results nor those of Ringelberg and co-workers can be considered definitive. Care must be taken in photoreactivation experiments for unwanted exposure to longer wavelength.
12. Surface irradiance at 300, 310, 320, 330, 340 and 350 nm, as measured by Baker and Smith (1982), Hessen (1994), Rhode et al. (2001), Williamson et al. (2001), De Lange (1999) and Leech and Williamson (2000) were averaged. All measurements were made around noon on cloudless days in the summer of the temperate zone. The standard errors of the mean expressed as percentage of the averages were 25, 6, 14.5, 8.0, 5.6, 5.9 and 5.6%, respectively. The surface noon values of irradiances were multiplied by relative photon energies with that of 300 nm set to 1. The weighted dose rate over the whole range of 295–355 nm was 14.7 W m^{-9}. For Morena East attenuation coefficients for 305, 320 and 340 nm were found in Morris et al. (1995). These three values were inserted in Fig. 8 in Baker and Smith (1982). A line was drawn through them in accordance to those in the figure. The original lines hold for different DOC and chlorophyll *a* values and the position of the Morena East line fits well

for a DOC of 0.5–0.7 g m^{-3} (Alonso et al., 2004). Attenuation coefficients $K_d(l)$ were read from the inserted line and used to calculate wavelength dependent weighted dose rates at depth intervals of 0.5 m. This was done for every hour between 5 and 19 h. To this end, the noon surface irradiance was corrected. Time dependent irradiance for 320 nm, as presented by Williamson et al. (2001) were used with the highest value made equal to 1 and supposed to be the noon value in the model calculation. The data in Williamson et al. (2001) were fitted by a third power polynome ($R^2 = 0.999$) from which the surface irradiances were calculated for all wavelengths in the used UV-B range. Probably, the change with time is slightly different for the other wavelengths but this possible effect was ignored. The equal weighted dose rates for the 295–355 nm range at different times and depth were calculated by interpolation.

13. Descent swimming might be faster. For calanoid copepods *Metridia*, *Limnocalanus* and *Epischura* Wong (1988) and Wong and Sprules (1986) mentioned swimming velocities ranging from 6 to 60 cm min^{-1}. However, migrations are probably not continuous and displacement consists of periods of active, head down swimming, passive sinking and maintaining depth.
14. I think the ordinate in this figure has to be turned around.
15. Although attenuations of UVR and PAR are differently determined by factors like chlorophyll *a* concentration, particle concentration and humic substances, a correlation might be expected. Using the attenuation coefficients for PAR and 320 nm for $N = 64$ lakes mentioned in Morris et al. (1995), a highly significant regression ($P<0.001$; $R^2 = 0.855$) was calculated, going through the origin ($P = 0.906$): $K_{320} = -0.083 + 10.428 \times K_{PAR}$.
16. Morris et al. (1995) measured UV penetration in a large number of lakes and correlated UV with K_d. At the end of their paper they write "because of the uncertainty in predicting K values (due in part to variable DOC-specific absorbency) these models should not be viewed as replacements for direct measurements of UV attenuation..." "...but can be used as to provide rough estimates of UV transparency within individual lakes" (p. 1390). Laurion et al. (2000) wrote that to estimate UV penetration, DOC is less useful in alpine lakes.
17. One wonders whether Baylor and Smith (1957) have seen the brief summary paper of Loeb (1904) or not because it is clearly stated that *Daphnia* and other animals become positively phototactic on the addition of acids or CO_2. These authors also mentioned the anecdote that Loeb found the reversal of the sign of phototaxis "when he profligately poured beer into his experimental aquaria of daphnids". I could not find this in Loeb's 1904 paper so I think Baylor and Smith did not repeat the experiment but used beer in a more appropriate way.
18. I suggest to reformulate the sentence: clonal diversity is at the base of different behaviour and not the other way round.
19. At a lower level of biological integration, food must be formulated more precisely, for example, as a "concentration of blood sugar" or as a "sufficient ingestion rate". The first term informs about the "milieu interieur", the second about the "milieu exterieur".

Chapter 10
Migrations in the Marine Environment

To study the literature on DVM in the marine environment is like a "mer à boire".

Contents

10.1 Introduction

The oceans cover 71% of the Earth's surface and contain a tremendous volume of approximately 1.3×10^9 km^3. Although biodiversity of the littoral communities may be as high as in terrestrial ecosystems, the pelagic communities are relatively poor. The origin and problems of biodiversity of the pelagial were discussed by Angel (1993). Nevertheless, compared to lakes, diversity is rich with species varying tremendously in morphology and behaviour. This results in variable niches and complex food webs. Also the adaptive significance of DVM is likely to be more diverse. In lakes, predator evasion is considered the principal if not the only reason for the occurrence of DVM and this might be an important ultimate aspect in the marine biotope as well, although other adaptive reasons cannot be excluded a priori. In a very extensive, and one is inclined to think an exhaustive review, Pearre (2003) stressed the role of food as an ultimate as well as proximate factor. Food of course is

J. Ringelberg, *Diel Vertical Migration of Zooplankton in Lakes and Oceans*,
DOI 10.1007/978-90-481-3093-1_10, © Springer Science+Business Media B.V. 2010

important and food niche separation might be an adaptive reason of vertical migration too. DVM is not an isolated phenomenon, but represents a part of the composite strategies required for co-existence in the pelagic community.

Proximate causation and the physiological and behavioural basis of DVM in marine zooplankton are quite similar to that of zooplankton in lakes. This is not surprising because the environmental stimuli triggering vertical swimming are the universally present, long-duration light intensity changes of dawn and dusk. In addition, organs of light perception, varying from simple cups with a few cells to complex eyes, have evolved to be especially sensitive to changes in light intensity. All light sensitive organs have a large potential for adaptation to a logarithmic range of light intensity in common. This ensures that eyes are capable to operate during the long hours of varying daylight intensity. The consequence of adaptation, however, is that the perception of light changes is one of relative changes in light intensity and not of absolute intensity or of absolute changes (see Chapters 3 and 4). The universal presence of dawn and dusk and the comparable physiology of eyes make that proximate aspects of DVM are to a large extent independent of the species involved, whether marine or from fresh water.

The two pioneers of marine migration research, F.S. Russell and G.L. Clarke, made observations on *Calanus finmarchicus*, one of the most common calanoid copepods at both sides of the northern North Atlantic (Russell, 1934; Clarke, 1934a). Well known are the enigmatic observations made by George Clarke in the western Atlantic Ocean: in the Gulf of Maine these copepods migrated over more than 100 m, while a few days later, at the edge of George's Bank, about 100 miles from the previous station and under similar environmental conditions, the major part of the population remained in the upper strata all the time (Fig. 10.1). It has been suggested that aggregations near the surface might have to do with mating and not with specific environmental circumstances. It is not strange, therefore, why another species, *Metridia lucens*, migrated at the same time. This anecdotal example asks attention for the role of biotic factors in evoking or preventing DVM.

Russell and Clarke both recognised that light intensity was a relevant environmental factor and that light intensity measurements had to accompany sampling. However, light sensitive apparatus was difficult to obtain at the time, and for many years to come, light measurements would remain exceptional, even though most researchers were convinced that light intensity changes were crucial factors in the causation of DVM. Clarke (1934a), of course, tried to find differences in environmental conditions at the two stations in order to explain the different behaviour but he did not succeed. Even an important factor like food concentration offered no solution: the ascending animals passed right through the zone of highest abundance of phytoplankton. Therefore, he considered the physiological condition of the animals of more importance. Referring to his experiments on phototaxis with *Daphnia*, Clarke hinted at different responses to changes in light intensity as responsible for the observed difference in vertical distribution. If mating is important, the motivation of phototactic responses must be influenced. We still need experimental evidence. Also Russell published in 1934 an article on migration of *C. finmarchicus* in relation to light intensity. The observations were done in July and August

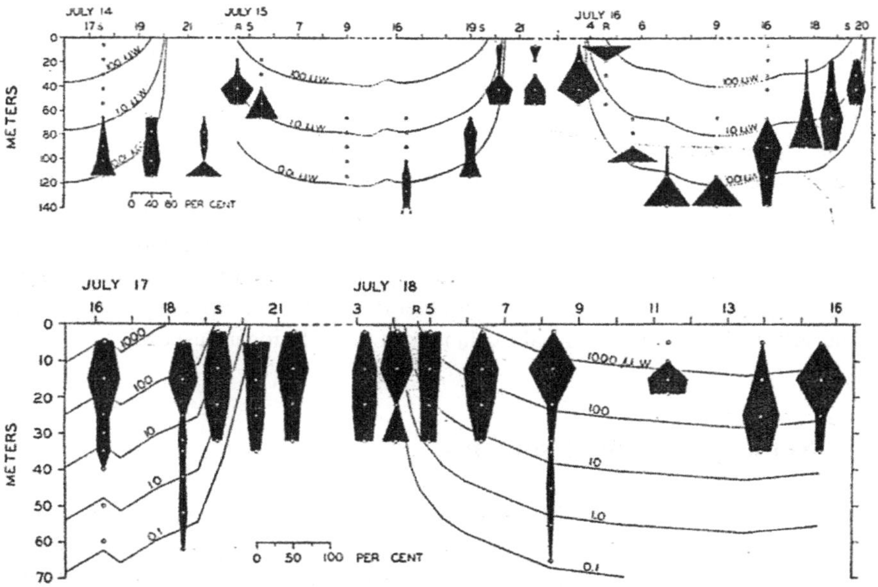

Fig. 10.1 Diel vertical migration of *Calanus finmarchicus* adult males. *Upper figure* DVM in the Gulf of Maine, *lower figure* at nearby George's Bank. Lines of equal intensity of blue light are inserted. R = sunrise; S = sunset (composed of Figs. 2 and 4 in Clarke, 1934a; reproduced by permission of the Marine Biological Laboratory, Woods Hole, MA)

1931, but the condition off Plymouth in the English Channel was quite different. His research was aimed at testing the hypothesis that migrating animals followed an optimum (absolute) light intensity. Females were found higher in the water column than males and "thus" preferred a higher light intensity. In addition, "slightly increasing optima" were found in the course of the day, "which might point to increasing adaptation throughout the day" (p. 579). In addition, the animals preferred a higher light intensity in July than in August. Although additional hypotheses had to be introduced to explain the observed differences, he did not really doubt the validity of a preferred optimum light intensity. Because he considered light as an environmental factor of a *static* nature, comparable to salinity and temperature, he thought it was appropriate to correlate daytime vertical distributions with the light intensity after descent migrations were finished. Although Clarke thought that *changes* in light intensity were essential stimuli, he nevertheless concluded that "a consistent though rough correlation between the vertical distribution of the copepods and the intensity of the submarine irradiation is apparent from the figures" (p. 437). But he also observed that early in the morning, copepods descended more rapidly than isolumes, which was difficult to reconcile with the idea of an optimum. To resolve this inconsistency he suggested that "it appears that insofar as the animals are reacting to light a higher intensity is required to initiate the migration than is required to cause it to continue" (p. 437). The dilemma had been clearly formulated in an earlier article (Clarke, 1933): "The behavior of plankton organisms may be regarded as a series of similar responses to changes in the intensity of light or as different responses to constant intensities of different magnitude" (p. 433). Clarke continued to struggle with the static or a dynamic character of the light stimulus for a long time.

I deal here with this historic controversy at some length because Clarke and Russell have had a great influence on migration research. Obviously, the idea of following an "optimum" or "preferendum" has had a greater appeal than the idea of a response to light intensity changes since it remained the paradigmatic explanation of DVM for decades to come. Although still popping up in disguised form once in a while (see further on), the preferendum theory is not a hot topic anymore because of the shift in interest from causal explanations to evolutionary aspects.

10.2 Some Interesting Problems

Although the literature on DVM in marine biotopes is large, the majority of the articles are descriptive and anecdotal, describing patterns present and features observed. Essential environmental factors are not always measured. This makes the interpretation of the data difficult, if not impossible. Nevertheless, sometimes interesting problems are evident and I will discuss three examples.

10.2.1 Ontogenetic Differences

I have always wondered about the long and complicated series of developmental stages of copepods. Daphnids emerge from the brood chamber as small replicas of their mother. In contrast nauplia stages of copepods do not look like an adult at all. It is only in the copepodite stages that the future full-grown copepod becomes recognisable. What advantage has this long road of development? Among the calanoids, species such as *Calanus* carry their eggs with them in a special sac. Others, such as *Metridia,* are broadcast spawners that liberate the eggs into the water, hoping for the best. For most the future is not bright, of course. But that is also true for eggs in a sac or a brood chamber. Females carrying eggs are more conspicuous to visual predators and should therefore migrate deeper if predator avoidance is the ultimate reason. This sounds logical and is believed to hold for freshwater cladocerans, but is belied by the available facts for copepods: generally, *Metridia* migrate earlier and with a larger amplitude than *Calanus* (see list in Cushing, 1951). Also, for the better part of the 24-h cycle, eggs carried by a migrating female develop at lower temperatures than eggs that are freely floating in the warm surface layers. This might be part of the story. Still, both life-history traits are successful. What counts, of course, is the chance of an egg making another egg, so to say. The lengthy process is the product of chances of survival at each stage, adults included. The chance of a carried egg to develop into a nauplius depends predominantly on the chance of survival of the mother and this depends largely on the chance of being eaten by a predator. Each larval stage will also have a certain chance of survival, depending on the circumstances. We deal with the product of a chain of chances until an egg of the next generation develops in a nauplius. We have no idea what this really means in terms of adaptive success. Perhaps complex environmental conditions favour broadcast spawners sometimes and carriers of egg sacs at other times.

With the development of various larval stages into adults, not only their morphology changes but also their migration behaviour and thus the pattern of DVM. Ontogenetic differences in these patterns have been described several times. Osgood and Frost (1994b) found that of all copepodite stages of *C. pacificus*, the first one (C1) was the highest in the water column of Dabob Bay and showed little indication of any migration. Also the next stage (C2) was largely distributed over the upper layers but at night the majority of individuals from below 10 m joined those already present in the 0–10 m zone, thus performing a small vertical migration. Copepodite C3 was found again deeper during daytime than the previous two stages and migrated at night from the 25- to the 10-m sample layer into the upper 10 m. Depending on the season, C4 performed a DVM from 50 to 25 m deep into the surface 10 m. In daytime, this stage was found again deeper than the previous stages, illustrating a general trend of older copepodites moving deeper. Irregularities in the pattern were also present in the last stage C5 and only the adult females underwent a large-amplitude normal DVM.

The ontogenetic development of DVM in *M. pacifica* was quite different (Osgood and Frost, 1994b). Only adult females (and not the males) performed a well-expressed migration, avoiding the surface layers during the day. On the other hand, on some dates, stages C4 and C3 were found in the upper 25 m during the day and went down during the night. Osgood and Frost (1994b) suggested that this reverse migration served the purpose of preventing predation by invertebrate carnivores such as the abundant copepod *Euchaeta elongata* and the chaetognate *Sagitta elegans*. These invertebrates performed a normal DVM in Dabob Bay. The authors supposed that the visually hunting fish and the non-visually hunting invertebrates – themselves being predated upon by the fish – induce opposite migration patterns. In freshwater systems, a similar difference in kind of predator effect is present (see Chapter 9). Obviously, this would require that the type of predator is recognised by the copepods and different infochemicals (kairomones) might be involved also in this marine environment.

A gradual development of migration behaviour has also been described for *C. sinicus* by Huang et al. (1992). In the Inland Sea of Japan this species spawns in the upper 40 m, but the actual spawning depth depends on season. The pre-feeding nauplius stages N1 and N2 were more or less similarly distributed as the eggs. From stage N3 onwards, the naupliar stages and also the first copepodites C1 aggregated in the upper part of the water column. The next stages, C3 and C4, tended to disperse over the entire water column, although C4 started to migrate. Adults and C5 performed a normal diel vertical migration.The patterns and amplitudes varied seasonally, which the authors ascribed to a varying abundance of planktivorous fish and to changes in food concentration. For example, an increasing difference in chlorophyll *a* concentration between the upper 15 m and the lower part of the column was positively correlated with migration amplitude.

The changes in behaviour of calanoid copepods in concert with the morphological development suggest that the sensory and neural systems must undergo large shifts in functioning. As yet nothing is known about the ontogenetic changes in light perception or of the potential infochemicals, nor of the evaluation and integration of the sensory inputs in the nervous system leading to different behavioural patterns regarding diel vertical migration. These fascinating developmental changes have large implications for niche differentiation and food web structure.

10.2.2 A Disguished Optimum

Buskey et al. (1989) described large-amplitude migrations for two species of copepods, *Pleuromamma gracilis* and *P. xiphias* in the Sargasso Sea. Water clarity is extremely high in this part of the Atlantic Ocean, resulting in amazingly large amplitudes of 250 and 400 m recorded for *P. gracilis* and *P. xiphias*, respectively. Day depth was indeed a significant function of the vertical attenuation coefficient K (m^{-1}) (depth = 948–3.9 × 10^4 K; $R^2 = 0.98$ and depth = 1094–3.5 × 10^4 K; $R^2 = 0.72$).[1] The largest of the two species (*P. xiphias*) migrated deepest. Aboard ship, the authors determined maximum photosensitivity at the wavelength of 480 nm. They hypothesised that light intensities at the day depths would correspond with the light-sensitivity thresholds, but downward migration proved to stop at a much higher intensity. In stead of rejecting this hypothesis, the authors tried to salvage it by suggesting that the animals could not swim fast enough to follow the descending isolumes at dawn. Photoadaptation during the descent was "leading to a reduction in their photosensitivity" (p. 213). This hypothesis is a disguised "optimum" or "preferendum" explanation of DVM and the reasoning for lagging behind the "optimum" isolume is the usual one proposed by Russell (1934) in the distant past. Again, the essential feature of the stimulus light is a static one. Since the smaller copepod also had the shallower daytime distribution, the authors state that "the relationship between photosensitivity and daytime light intensity at depth is also consistent with the predation hypothesis since *P. gracilis* is less photosensitive than *P. xiphias*" (p. 213). Now, the ultimate aspect of an adaptive value of migration is used as argument to confirm a proximate hypothesis! However, a relation between a sensitivity threshold and migration depth evokes several problems: what kind of relation must be supposed to exist between the visual threshold and the ascent migration in the afternoon? Do the animals wait until the sensitivity threshold is realised at the day depth before they start swimming? This seems unrealistic. If the ascending copepods would again lag behind the isolume, as they do during the descent migration, they would be in darkness soon and never reach the surface. Although the authors were aware of migration stimuli defined as relative changes in light intensity, they did not discuss this possible explanation. Unfortunately, light intensity was not measured continuously during the crucial times of dawn and dusk, else it would have been possible to explore the nature of the true stimulus more adequately.

10.2.3 Metabolic Advantage Revisited

Was McLaren right, after all? The suggestion of a metabolic advantage of DVM as a consequence of alternating high and low temperatures that these animals are subjected to (McLaren, 1963) has led to many comments in the past, but the issue appears to be settled now: there is no evidence of such an advantage (Stich and Lampert, 1984; Lampert, 1993). However, McLaren's idea was largely based on his experience with calanoid copepods in marine waters while disapprovals mostly originated from experiments with cladocerans in freshwater. Recently, Svetlichny et al. (2000) concluded from experiments with migrating *C. euxinus* from the Black Sea that a metabolic advantage of DVM is possible. This fascinating inland sea in south-eastern Europe and Asia Minor has a surface area of 422×10^3 km^2 and a maximum depth of more than 2000 m. It is a meromictic basin with a large variation in salinity, resulting in the stagnation of deep water and the development of anoxic layers with hydrogen sulphide. In summer, a thermocline develops at 25–30 m. Temperature in the epilimnion varies seasonally from 7 to 25°C. Oxygen concentration declines gradually until the water becomes anoxic at about 140 m, depending on the regional hydrological conditions. The main pycnocline and oxycline can be found at a depth of 70–200 m. Within a layer of 5–50 m thickness, just above the depth of high concentrations of hydrogen sulphide, oxygen concentration gradually diminishes until values of 0.3–0.2 mg O_2 l^{-1} are reached (Vinogradov et al., 1985b).

C. euxinus, a dominant species in the area, migrates in the morning from the warm, oxygenic upper layers to a depth of 120–140 m. Oxygen concentration has then declined from 1.15 to 0.35 mg O_2 l^{-1} and the temperature has dropped to 6–8°C (Vinogradov et al., 1985a). *Calanus* can easily exceed 250 ind. m^{-3}, sometimes even increasing to over 3500 ind. m^{-3} (Vinogradov et al., 1992). In daytime, aggregations occur in the vicinity of the dangerous sulphide zone. At oxygen concentrations below 0.5 mg l^{-1}, the animals become torpid and threaten to sink into the highly toxic hydrogen sulphide layer. Movements of the mouthpart appendages and an occasional thrust of the thoracic legs prevent this. This was observed in the laboratory under simulated conditions of the day depth. Svetlichny et al. (2000) determined total and basal metabolism at temperatures and oxygen conditions prevailing in the water column covered by the DVM of the animals. The energetic cost of the activity to stay below the oxycline was equal to half of the basal metabolism in the warm upper layer. Since the estimated speed of passive sinking would be insufficient to cover the migration distance of 117 m within the two available hours, the authors also determined the metabolic costs of swimming down and migrating up at different temperatures and oxygen concentrations. They estimated that 78.6% of the total energy budget was spent in the warm oxygen-rich surface layers, while only 5.4% was spent at day depth: "therefore, vertical migrations of *C. euxinus* to cold, oxygen-deficient zones (even with maximal swimming speed) provide a substantial benefit, allowing the animals to use most of the food consumed in the surface layers for the synthesis of wax esters (Yuneva et al., 1997), which are necessary for gonad development (Sargent and Henderson, 1986)" (Svetlichny et al., 2000, p. 970). Not only *Calanus* aggregated this deep but also other zooplankton, for example, the chaetognath *S. setosa* (Vinogradov et al., 1985b). In the evening, all species returned to the surface layers again.

Svetlichny et al. (2000) do not refer to McLaren's hypothesis nor to the discussion that followed in the literature. For them, hypoxia provides a substantial metabolic advantage for *C. euxinus*. Whether this is a sufficient reason to migrate is still to be answered. Egg development at low temperatures is retarded and birth rates are probably low compared to animals that would stay at the high surface temperatures. Conversely, an energy bonus might be used to make a larger clutch and thus compensate for the retardation of the egg development. Therefore, comparison of the balance between mortality and birth rates among migrating and non-migrating populations must be made before a decisive answer can be given to the question of the adaptive significance of this migration in the Black Sea. As far as I know, no mention is made of visually hunting fish in the epilimnion. Vinogradov et al. (1985b) suggest that the presence of the deep zooplankton concentrations during daytime give fishes an opportunity for intensive feeding. If so, this complicates considerations about the adaptive significance of *C. euxinus* migrations. The interesting papers on these curious migrations in a strange biotope should be followed up by further research.

10.3 Interesting Copepods in Dabob Bay

One of the many inlets in the Puget Sound area is called Dabob Bay (47° 45′ N, 122° 50′ W), a fjord-like arm of Hood Canal, 15 km long and 5 km wide. Maximum depth is 193 m. Reduced advection, realising a more or less closed system, makes Dabob Bay a fine research site for the School of Oceanography of the University of Washington in Seattle (USA). Bruce Frost with co-workers and students have described the behaviour of copepods and the role of food and predators in this bay in many papers. Copepods become interesting when interesting papers are written about them. In this case there are three main players. *C. pacificus* (2.2–2.7 mm prosome length) is the Pacific counterpart of the Atlantic *C. finmarchicus*. *M. pacifica* is closely related to *M. lucens* from the Atlantic and this name is still being used in several papers[2] (Hays et al., 2001). Since more recent papers refer to *M. pacifica*, I will do the same even if in the original paper *M. lucens* was used. The third copepod, and possibly the most interesting one, is *Pseudocalanus newmani* with a prosome length of 0.7–1.2 mm, the smallest of the three calanoids.

The extent of diel vertical migration of these species depends on season and varies annually as shown in an informative overview by Frost (1988). As illustration of this seasonal and interannual variability in Dabob Bay, some examples of daytime

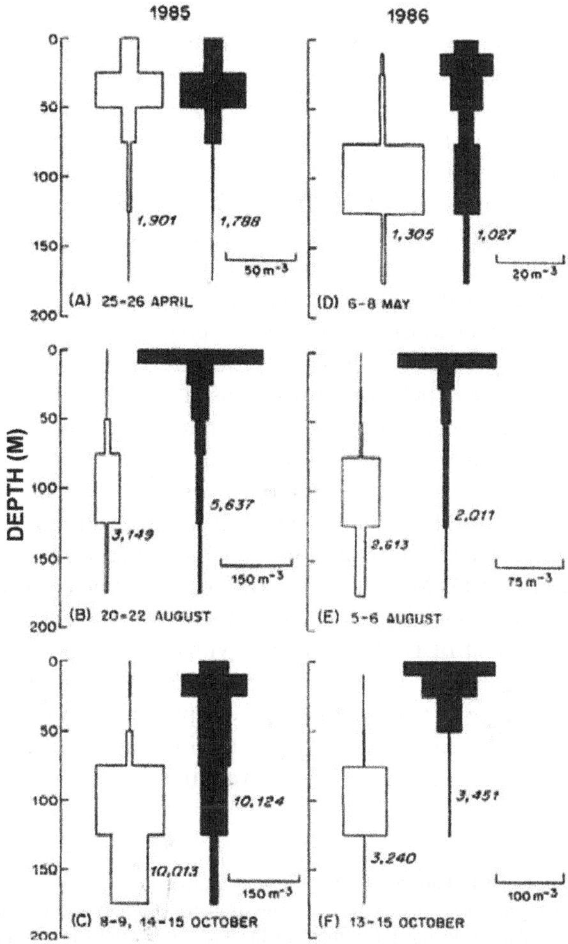

Fig. 10.2 Vertical distribution of adult females of *Calanus pacificus* in Dabob Bay in two different years and three different seasons. *White histograms* illustrate daytime and *black histograms* night-time distributions. *Italicised numbers* in the figures are numbers per m^2 in the water column (from Frost, 1988)

and night-time vertical distributions from 2 years and three seasons are presented in Fig. 10.2. A more extensive list of dates, with information on the intensity of observed migrations, are listed in Table 10.1. The seasonal character is somewhat obscured by the interannual variation. In winter-spring, March-April-May, vertical migration is absent, partial or well expressed.

In summer (August), migration is generally well expressed. In autumn (October) migration is partial or strong. The seasonal effect goes either way. For example, Dagg et al. (1989) found that about half the population of *C. pacificus* remained below 75 m depth in May 1986, while in March 1991 nearly all individuals migrated (Dagg et al., 1997).

In an attempt to explain these different patterns, arguments for and against the influence of temperature, food availability and predation were passed in review. Frost (1988) considered the absence of migration during a spring bloom of diatoms[3] as possibly related to optimal feeding conditions, but observations during summer and autumn "cast considerable doubt on the importance of food availability and temperature as controlling factors or modulators of diel vertical migrations, at least in these seasons" (p. 681). Females of *C. pacificus* aggregated at night in the 0–10 m zone (Fig. 10.2, B and E), where chlorophyll concentrations were highest, but they also migrated strongly when phytoplankton abundances were low. In a summer situation in the open Gulf of Alaska, where phytoplankton concentration was also low, *C. pacificus* was continuously near the surface. Thermal stratification of the water column in Dabob Bay was not of decisive importance. In summer, with a well-expressed thermocline, zooplankton migrated, but a strong migration could also occur in October, when the water column was nearly isothermal.

To get more grip on the effect of environmental circumstances, Frost used a model of population growth, based on life tables. Test criterion was the intrinsic rate of population growth of migratory and non-migratory populations. The results of the calculations showed that in the absence of mortality, non-migrants always had an advantage over migrants, whatever the circumstances were. With mortality included, the migrants, having a lower death rate, had a substantial higher population growth. Frost (1988) concluded that a differential mortality is the ultimate cause of DVM in *C. pacificus*, but he did not provide a proximate explanation of the role that environmental factors have. Bollens and Frost (1989b) observed that the extent of DVM correlated with an increase in abundance of 0+ planktivorous fish in the bay, especially herring (*Clupea harengis*) and stickleback (*Gasterosteus acculeatus*). Large numbers were present in August, but not in May. Analysis of stomach content revealed that both fish were feeding on the copepods. Therefore, the interannual and seasonal variation in migration was thought to depend on the

Table 10.1 Patterns of diel vertical migration of female *C. pacificus* in Dabob Bay in different years and seasons. A partial migration means that a considerable number of individuals in the population do not participate in the migration (the same migration data may have been used in different publications)

Date	Migration pattern	Author
23–24 March 1991	DVM	Dagg et al. (1997)
11–12 April 1979	DVM	Frost (1988)
25–26 April 1985	No DVM	Bollens and Frost (1989)
6–8 May 1986	Partial DVM	Dagg et al. (1989)
27–28 May 1982	No DVM	Frost (1988)
9–11 June 1986	Partial DVM	Bollens and Frost (1989b)
25–26 June 1985	Partial DVM	Bollens and Frost (1989b)
3–4 August 1973	DVM	Osgood and Frost (1994a)
5–6 August 1986	DVM	Dagg et al. (1989)
20–22 August 1985	DVM	Frost (1988)
13–15 October 1986	DVM	Frost (1988)
14–15 October 1985	Partial DVM	Frost (1988)
21–22 October 1982	DVM	Frost (1988)
23–24 October 1981	DVM	Frost (1988)

abundance of planktivorous fish. Biotic factors and not abiotic factors apparently determined the expression of migration.

Further evidence for a predator-induced DVM in the marine biotope is given by Bollens et al. (1992), not in Dabob Bay this time but in Jackles Lagoon, a small lagoon with a maximum depth of only 3.9 m, situated on San Juan Island. The extent of migration of the copepod *A. hudsonica* was correlated with the temperature difference over the water column (a measure of stratification), Secchi disk depth (a measure of water clarity) and the abundance of zooplanktivorous fish (predominantly sticklebacks). For the first two environmental factors, no correlation with migration amplitude was found. Only the presence of predators proved to be significant for the expression of DVM over the seasons. The authors suggested the existence of a plastic phenotypic response , mediated through mechanical stimuli or visual cues. This idea originated from experiments done with this copepod and *G. aculeatus* (Bollens and Frost, 1989a).

Although the concentration of algal food might not be the decisive factor determining the seasonal character of DVM, food is, of course, of primary importance for all life. The time of day when migrating copepods feed co-occurs with the period just after ascent. For example, in August 1986, Dagg et al. (1989) found that nearly the complete population of *C. pacificus* migrated from a day depth of 75–125 m into the upper 10 m, where chlorophyll concentration was highest. Most individuals arrived after sunset with empty guts. One hour later, the amount of algae in the gut had increased drastically. Ascent migration and feeding thus occurred almost simultaneously. However, Dagg et al. (1989) concluded that feeding was independent of migration because "the most intense foraging (occurred during) the hours of darkness" (p. 1069). It is not clear to me when these hours of darkness were thought to begin and end – probably with sunset and sunrise, respectively. Sunset was defined as the last interval over which a LiCor integrator (measuring radiation over 10-min periods) still registered light. Similarly sunrise began with the first interval over which radiation was registered.[4] These integrator devices are mostly used in studies of primary production and are not very sensitive compared to eyes of animals. Relative changes in light intensity are maximum after sunset and before sunrise. From the observations of Dagg et al. (1989) and also from Bollens and Frost (1989b), my conclusion is that migration into the surface layers occurred shortly before and after sunset and grazing was highest during that time. Therefore, migration and feeding might be triggered independently by these relative decreases in light intensity, implying that a causal relation between migration and feeding does not exist.

On the night of 23/24 March 1991, Dagg et al. (1997) observed a migration of *C. pacificus* out of the deep sample layer of 108–50 m into the 50–25 m and the 25–0 m layers. Most individuals remained in the middle layer.

Frequency of sampling was not high enough to pinpoint the exact time of migration, but migration took place between 1700 and 1930 hours. There was a clear coincidence between diel migration and feeding on phytoplankton by *C. pacificus* females. Feeding occurred while ascending and at 1900 hours, individuals in the upper layer had a gut filled with phytoplankton pigments. The authors assumed that

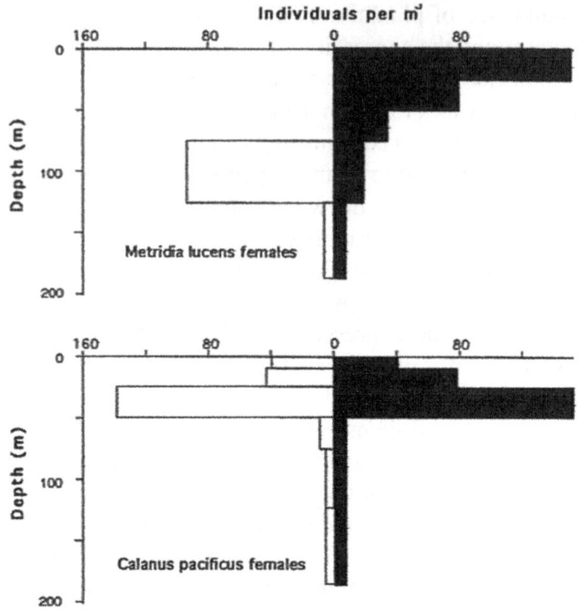

Fig. 10.3 Vertical day and night distribution of *Metridia lucens* and *Calanus pacificus* in Dabob Bay on 28–29 May 1982 (figure composed from figures in Osgood and Frost (1994))

the animals from the intermediate layer had fed actively in that layer, although the highest algal concentration was present in the upper 10 m. This might be an example of feeding bouts of individuals during excursions from deeper down into shallower food-rich layers as advocated by Pearre (1973, 2003), but Dagg et al. (1997) rejected this possibility.

Descent migration started an hour before sunrise (Dagg et al., 1989). Dagg et al. (1997) observed that just before moving down, 80% of the population ascended into the 25–0 m layer. This must have been the dawn rise described in the migration scheme proposed by Cushing (1951). Thereafter, a rapid descent took place and at 0700 hours all animals were in the deep zone again.

Additional insight into the role of food in DVM emerges from a study of Hays et al. (2001) who caught *M. pacifica* females deep down during the night, which apparently had not migrated. They had an entirely filled oil sac but an empty gut. Why had these individuals not migrated? Obviously, an empty gut had been an insufficient trigger. In vertebrates, hormones, produced by lipid cells, inform the brain about the level of hunger/satiation which, then, initiates or stops feeding. Does the oil sac of copepods play a comparable role? On the other hand, to terminate a feeding bout and rapidly leave the danger zone, an immediate working stimulus would seem more adequate. Pressure perception from a filled gut could act as such and motivate downward swimming. With photobehaviour experiments, the idea could be explored!

The difference in migration behaviour of *C. pacificus* and *M. pacifica* was discussed by Bollens et al. (1993). An example is given in Fig. 10.3. *M. pacifica* tends to be deeper in the water column during daytime. On dates that *C. pacificus* did not migrate, for example, 25–26 April 1985, or only partially migrated as on 6–8 May 1986, 25–26 June 1985 and 9–11 June 1986, *M. pacifica* made extensive vertical migrations. Bollens et al. (1993) referred to an extensive literature wherein DVM of both species was compared in different parts of the ocean. For once a clear consistent picture of migration differences emerges from all field observations: whereas the vertical distribution of *C. pacificus* is variable, *M. lucens* is always found deeper in the water column during the day and DVM is always clearly expressed. However, an explanation of the different behaviour of the two species is not easy to give. Dagg et al. (1989) suggested that *M. pacifica* might be more susceptible to visual predation even though the two species are similarly sized. Bollens et al. (1993) performed experiments with juvenile pacific herring (*C. harrengus pallasi*) to determine whether selective predation could be involved. Contrarily to expectation, the juvenile herring had a statistically significant preference for adult females of *C. pacificus*. Therefore, the hypothesis of Dagg et al. (1989) had to be refuted and the enigma remained. Visibility should be enhanced when the females carry egg sacs, as in *C. pacifica*, while *M. pacifica* liberates the eggs into the water. Considering this difference, the latter species should be less conspicuous. Gut analysis of fish, caught in the bay, pointed also to a preferential predation for *C. pacificus* over *M. pacifica*.

To reconcile the results of field observations and experiment, Bollens et al. (1993) offered several possible explanations. The most convincing one (to me, but I believe also to the authors) is that the two copepod species have different strategies to deal with predation which appear to be related to differences in escape capability. Calanoid copepods make jumps to avoid predators. Small and brief hydrodynamic disturbances, incited by approaching fish, are perceived by mechanoreceptors in the first antennae (Hartline et al., 1996). However, species-specific differences in reaction time exist. Some taxonomic groups have myelin-like sheets around the axons – as vertebrates have – which makes a more rapid spike transport along the nerves possible and results in short reaction times (Davis et al., 1999). *C. pacificus* belongs to this group. *M. pacifica* has unmyelinated nerves and longer reaction times (Lenz et al., 2000). Thus jumps should be less rapid and escape less successful in this species. Therefore, the need for predation prevention through DVM is larger for *M. pacifica* than for *C. pacificus*.

The differences in DVM of *C. pacificus* and *M. pacifica* are undoubtly based in differences in their physiology and behaviour. This proximate aspect can be studied in isolated individuals, but no results of experimental studies are available and we have to consider available indirect information. As mentioned earlier, at dawn, *M. pacifica* descend earlier. Thus, we can hypothesise that its sensitivity to relative increases in light intensity is higher, and/or its sensitivity to fish (kairomone) presence is higher. On the sampling dates in June 1985 and 1986, part of the population of *C. pacificus* remained in the upper 50 m layer of Dabob Bay, while *M. pacifica* went down. In August of the 2 years, both copepod species migrated strongly. The abundances of the most important predator fishes, *Clupea harengus pallasi, G. aculeatus* and *Merluccius productus*, was about ten times higher in August, compared to June (as deduced from Table 2 in Bollens and Frost, 1989b). Because fish, or fish kairomones, are important in facilitating the reaction to light intensity increases at dawn, we might speculate that the concentration of some kairomone was high enough for *M. pacifica* to descend in June but not for all individuals of *C. pacificus*. Bollens and Frost (1991) rejected chemical cues initiating DVM and proposed on the basis of experiments with *A. hudsonica* that visual or mechanical cues (or both) are the stimuli causing migration. Evidently, experiments are needed to underpin these speculations about proximate differences in behaviour of *C. pacificus* and *M. pacifica.*

On an ecological level, competition might also play a role, not as a proximate factor but as an ultimate reason. If these species are really as similar as they seem to be, competition must be strong. *C. pacificus* is more of a herbivore than the omnivorous *M. pacifica* (Arashkevich, 1969, cited in Bollens et al., 1993), which may explain why the former is preferentially found higher up in the water column, while the second can find food also in deeper water. Several authors (cited by Bollens et al., 1993) have suggested that potential egg production of *C. pacificus* is higher than that of *M. pacifica.* Therefore, birth rates of *C. pacificus* could be higher. If so, populations of *C. pacificus* can survive with a higher predation mortality. Some of these arguments are mutually dependent, but a picture of niche separation of these closely related calanoid copepods emerges. Vertical migration represents only one component of a complex of adaptive strategies and for a good understanding it has to be considered in relation to food web structure.

The most interesting copepod player on the scene of Dabob Bay is perhaps *P. newmani* because it makes a reverse diel migration (Ohman et al., 1983; Ohman and Wood, 1996). From mid-May through mid-September, adults were distributed deeper during the night than during the day. Feeding occurred during the day, which could effectively reduce competition with nocturnally feeding herbivores, provided that the phytoplankton composition is different during the day and the night. *P. newmani* is preyed upon by invertebrates such as the copepods *E. elongata* and *M. pacifica* (Ohman and Wood, 1996) but also by *Euphausia pacifica* and the chaetognath *S. elegans*[5] (Fulmer and Bollens, 2005). These invertebrate predators exhibit a normal DVM, coming up in the evening and going down in early morning. Ohman et al. (1983) mentioned that at times of the year when the abundance of these predators is low, moreover at a shallow station in the bay where these predators are usually rare, no reversed migration of *Pseudocalanus* occurred. Thus, the reversed migration appears to reduce the overlap with invertebrate predators, although not with predator fish. These fish prefer large-bodied zooplankton, such as the invertebrate predators. Stomach contents of juvenile chum salmon showed a negative electivity for *P. newmani*. Electivity was neutral for the common sticklebacks (Ohman et al., 1983).

The extensive and long-term study of Dabob Bay permits a close look at the biology of the three copepods of interest. Females of *M. pacifica* overwinter rather deep in the water column but are the first to reproduce: they produce eggs already in January. This first reproduction period precedes the diatom bloom, which occurs in the second half of February. Then the nauplii and copepodites reach a peak in abundance (Osgood and Frost, 1994a). Successively, the females die off but numbers increase again in May. For the rest of the year, females will remain the prominent stage. Another reproduction period occurs in April, which coincides with a new phytoplankton bloom, this time of the diatom *Chaetoceras* and the flagellate *Phaeocystis*. As omnivores, they also eat tintinnids, radiolarians and faecal pellets. In summer, the adults and C5 rise to the surface layers at night, where they predominantly consume phytoplankton.

The first and main pulse of egg production in *C. pacificus* occurs in February or March, depending on the year (Pierson et al., 2005a). This reproduction period coincides with or follows after the bloom of the *Thalassiosira* species. Contrarily to *M. pacifica*, *C. pacificus* emerges from winter as C5, which molt into females just before the first reproduction pulse. The next peak is in April, again coinciding with a phytoplankton bloom. Males and females are abundant during summer, but become scarce after having reproduced in September. For the rest of the year, C5 is the dominant stage. This stage goes into diapause and overwinters in deep water. The time-shift in reproduction between the two species should also contribute to diminishing competition.

Females of *P. newmani* are the predominant stage in February and stay so for the rest of the year. Periods of enhanced reproduction are not as evident as for the two other species. From mid-March onwards, the population declined, at least in 2002 and 2003 (Pierson et al., 2005b). Contrarily to *C. pacificus*, which is found in the 25–50 m layer, during spring *P. newmani* is concentrated in the upper 0–25 m, the layer of maximum chlorophyll. In summer, the adults perform the reversed migration.

Generally, the springtime diatom blooms are considered food pulses that initiate the early growth of marine copepods. This idea is challenged by members of the Dabob Bay research group: they think that diatoms are toxic and would hamper rather than promote the reproductive success of copepods. Based predominantly on culture experiments in the laboratory it proved difficult to convincingly demonstrate toxicity in the field. Although not directly relevant to DVM, this enigma is sufficiently important and interesting to be discussed at some length in a note.[6]

However important the role of diatoms is as a food source, the phytoplankton community of Dabob Bay does not consist only of diatoms, not even during winter-spring. During the first bloom, roughly occurring in February, small flagellates (<10 mm) were also abundant and could have been a source of food. In March, phytoplankton numbers were low, except for flagellates. Since the occurrence of migration usually starts at the end of the month (Table 10.1), the percentage composition of the most dominant algae at that time is given in Table 10.2. Data were derived from a study by Horner et al. (2005), who mentioned that about 100 species were found during the winter-spring period. The population of diatoms declined in the course of the period and their chains were often broken. A large heterotroph dinoflagellate, *Gyrodinium spirale*, heavily preyed on these diatoms.

Several ecological relations come together in a simplified version of the food web of the Dabob Bay pelagic community (Fig. 10.4). No differentiation is made

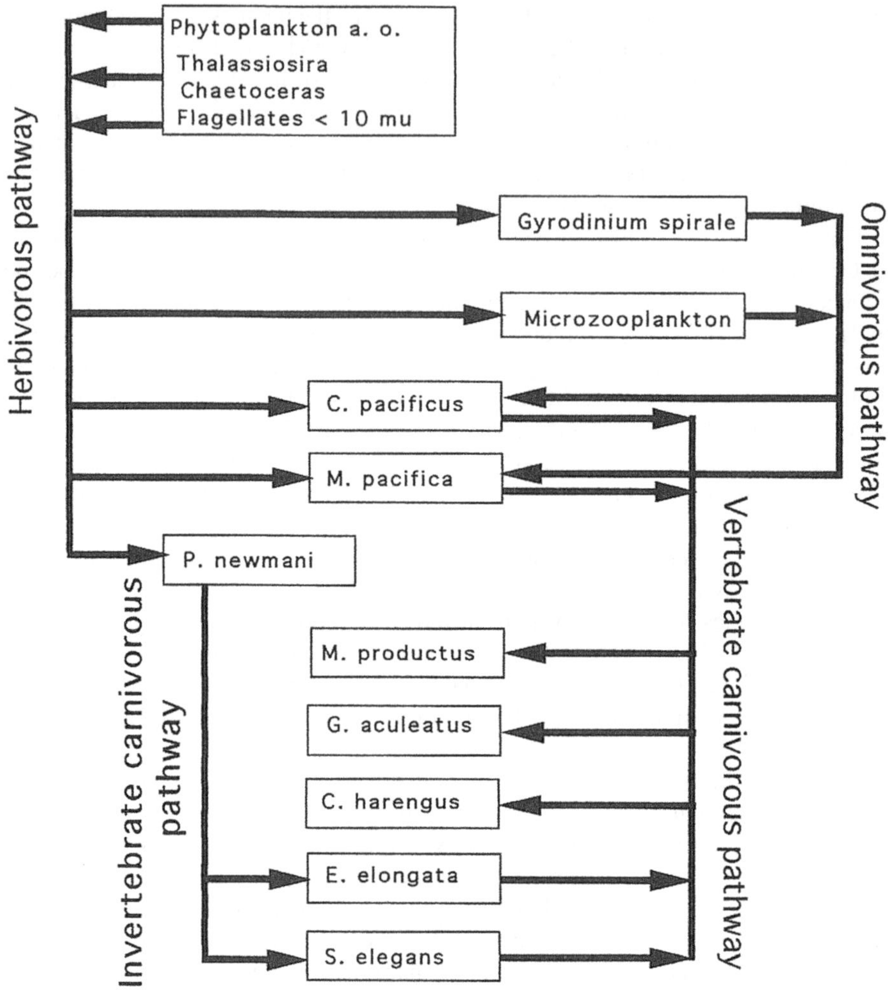

Fig. 10.4 Simplified food web of the pelagic community of Dabob Bay

between adult animals and larval stages. This makes the web surveyable, but also less realistic. Four pathways are presented. Because *S. elegans* and other invertebrate predators are consumed too, two carnivorous pathways are distinguished. No attempt has been made to quantify the web relations. If possible at all in view of the seasonal succession, that would be a further complication although the relative importance of the individual relations would become more evident. The intermediaterelation of the microzooplankton and *Gyrodinium* between the diatoms and the copepods is according to Fulmer and Bollens (2005). The position of *C. pacificus* and *M. pacifica* is made central in the figure. Their importance in the web can be read from the number of first-order relations with other species. When the population of *S. elegans* increases in summer (Fulmer and Bollens, 2005), this invertebrate

Table 10.2 Percentage species composition of the seven most dominant algal species or groups in the upper 30 m of Dabob Bay between 27 March (day 86) and 24 April (day 114) 2002. Percentages based on a total of 1,751,598 cell l^{-1}

Species or species group	Percentage
Other species	5.0
Phaeocystis	12.8
Flagellates < 10 mm	3.4
Heterotroph dinoflagellates	3.1
Thalassiosira species[a]	11.8
Chaetoceras species[a]	60.0
Other diatoms	3.9

[a]The two diatoms *Thalassiosira* and *Chaetoceras* were declining towards the end of the period. Data derived from Horner et al. (2005).

becomes a main predator for *P. newmani*. A timely retreat of the copepod to greater depth at dusk should reduce the threat of predation. This small copepod might be less visible to the fish and therefore its strategy could be to remain in the food-rich surface water during the long daytime hours, whereas *Sagitta* is more of a problem to the small calanoid and has to be avoided by descending to deeper water at night.

In a food web, species are linked by eating and being eaten and thus by flows of matter. From the discussion above, it is evident that the quantities involved in these flows are affected substantially by behaviour of both predators and prey, and thus involve mutual flows of information. This information web is superimposed on the traditional food web. In Chapter 13, a start is made to combine these two webs for a tritrophic pelagic lake system.

10.4 Krill

Norwegians have always been, and still are, whale hunters. So it is not surprising that the now internationally used name "krill" is originally the Norwegian word for "whale food". This generic name for Euphausiacea (Crustacea) comprises a large number of marine bioluminescent shrimp-like species that live free of the bottom, ranging in size from 8 to 140 mm (Plate 3). They are an important link in the pelagic food web because they serve as food for fish, squid and, of course, as bulk food for baleen whales, which eat tonnes of krill per day. To be able to exercise that function, numbers of euphausiids have to be very large. Indeed, swarms of kilometres across with densities of thousands per m^3 have been reported in Antarctica but also elsewhere in the oceans they are abundant. Most species are omnivorous and opportunistic feeders, filtering algae in the upper water layers at night or sucking copepods into their feeding basket at a deep day depth. Many, but not all, perform extensive diel vertical migrations. As in other zooplankton, migration behaviour in euphausiids may be variable. For example, Balino and Aksnes (1993) (cited in Onsrud and Kaartvedt, 1998) mention that *Meganyctiphanes norvegica* ascended to the surface at one location but remained below 40 m at another location within the same Norwegian fjord. Behaviour is apparently flexible but hardly anything is known about the factors underlying the expression of a particular type of behaviour. Various euphausiid species have been found to be differentially distributed over the water column, up to a depth of thousand metres. Their rich variety in behaviour and ecology makes these animals an interesting object to study, but it is difficult to arrive at general conclusions with regard to their vertical distribution. It is clear, however, that we deal with quite different and more complicated phenomena than we have come across in lakes. Traditional sampling with nets disclose details about qualitative and quantitative species compositions but the vertical distributions obtained are static and do not record the continuous movements like echo sounders do by following sound scattering layers (SSL). Although krill generally represents the major component of SSL, I discuss euphausiids in two separate sections because both methods reveal different aspects of behaviour and ecology.

Vertical distributions are more clearly expressed in oceans than in lakes, which suggests a stronger habitat differentiation. The closely-related species of euphausiids appear to have a large niche overlap in terms of food which should lead to strong competition. To avoid the threat of competitive exclusion, spatial separation might be one important ecological way to prevent this. In examples derived from the rich community of Euphausiacea near the Canary Islands in the Atlantic Ocean (Baker, 1970), this separation was apparently successful, although we do not know to what extent the vertical distributions are really stable. Before we go into details about this interesting case, a few other observations are given as an introduction.

Hu (1978) sampled four species off the west coast of Ohahu, Hawaii (Fig. 10.5). The percentages (inserted in the figures) indicate the overlap in range occupied during day and night for each species[7] as an indication of the extent of DVM. *Nematobrachion sexspinosus* migrated least, while *Thysanopoda aequalis* migrated strongest. Why do these euphausiids migrate? Avoidance of visual predators, considered to be the most important reason for DVM in lakes, is not mentioned by Hu (1978). He suggested that the different vertical distributions of the four species may serve the reduction of a potential interspecific competition for food. They all eat phytoplankton but there is a difference in their feeding chronology. *N. sexspinosus* eat during the day while the strongly migrating species *T. monacantha* and *T. aequalis* feed at night. *T. pectinata* has no daily feeding rhythm. As true omnivores, all eat zooplankton too and the difference in body size suggests that they can handle different prey sizes. The largest species, *T. pectinata* (31 mm) and *T. monacantha* (28 mm) prey on copepods, chaetognaths and other euphausiids. The two smallest species, *T aequalis* (16 mm) and *N. sexspinosus* (23 mm) feed on copepods, but at different times of the diel cycle. Differences in time of feeding among ecologically similar animals can effectively reduce competition, provided that the resources differ at those different times. This was realised in this case because the ratio of the copepod species, identified in the stomachs, was different: the ratio of *Pleuromamma* to *Oithona* in *N. sexspinosus* was 37 to 1 and in *T. aequalis* 2–14.

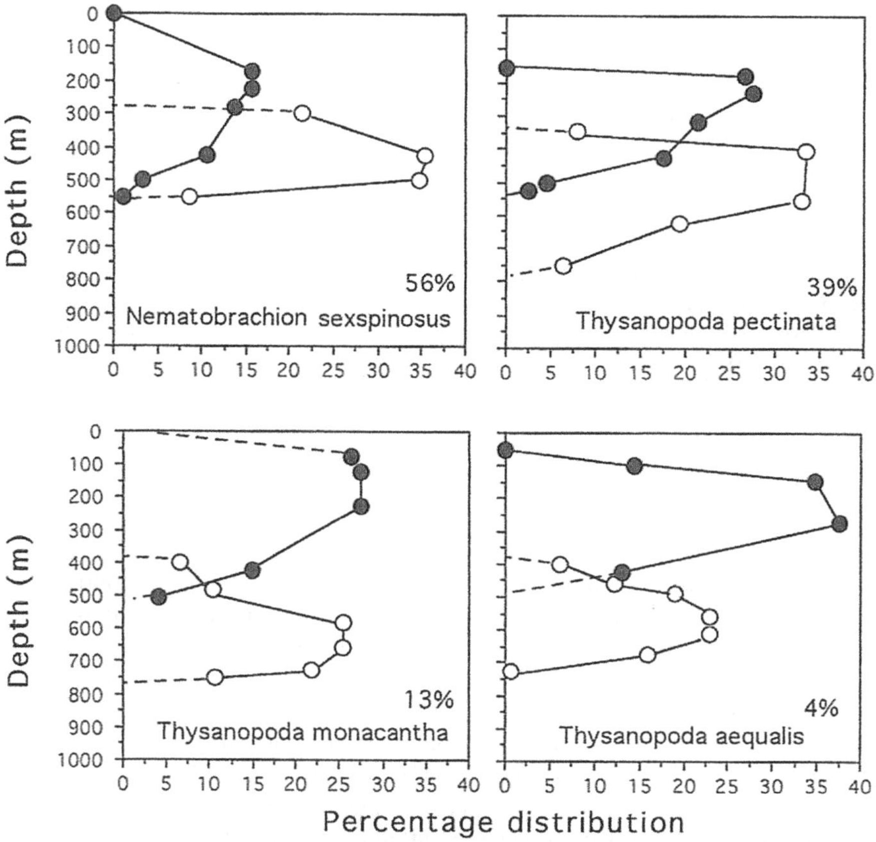

Fig. 10.5 Example of the vertical distribution of four euphausiid species off the west coast of Ohahu, Hawaii. *Black points* represent the night-time and *open points* the daytime distributions. The *percentages* indicate the overlap between day and night distributions (Modified after Hu, 1978; Reproduced by permission of the American Society of Limnology and Oceanography, Inc.)

This suggests that the copepod composition in the upper water layers may have differed during the day and the night. DVM of *Oithona* could be responsible, but size-specific prey handling may also be at the base of the different stomach content, since *Pleuromamma* is larger than *Oithona.* These interesting observations indicate that food niches in co-existing euphausiid species are different. Together with other ecological aspects, such as vertical habitat-partitioning and DVM (of both euphausiids and prey), a complex of factors emerges that is responsible for sufficiently different niches to make coexistence possible. We have to look for more examples, but DVM of euphausiid species serves as an illustration of vertical space partitioning. The adaptive significance of DVM in the marine biotope might be more complicated and different from the simple predation-prevention issue, considered to be the reason for migration in lake zooplankton (Chapter 9).

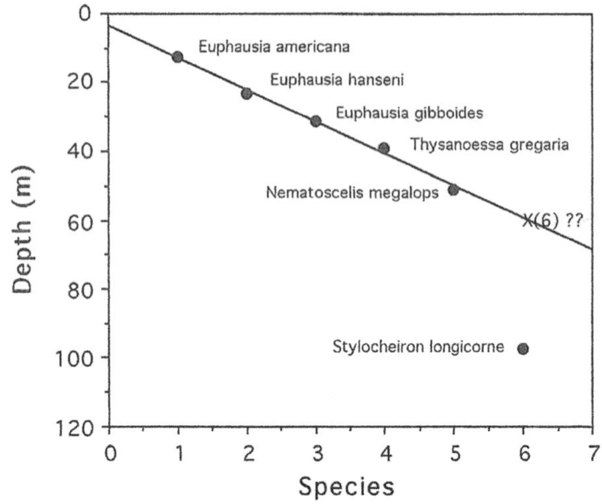

Fig. 10.6 Mean population night depth of euphausiids in the northern Benguela upwelling system (18° S, 10° 30′ E) arranged according to depth. Depth segregation conforming to the function Depth = 3.52 + 9.3 (species number). If used, *Stylocheiron longicorne* would have been number 10. The space in between number 5 and 10 could have been used by four species of which one, X(6), is indicated as an illustration (Figure based on data in Barange, 1990)

Vertical habitat-partitioning has also been suggested by Barange (1990) for six euphausiid species from the northern Benguela upwelling system off the coast of Namibia (Fig. 10.6). Four species migrated, viz. *E. america*, *E. hanseni*, *E. gibboides*, *Nematoscelis megalop* while *Thysanoessa gregaria* and *Stylocheiron longicorne* did not migrate. For all species, the mean population depth at night proved to be significantly different, although the vertical distributions overlapped. According to the author, space segregation was independent of the distribution of the major food items, consisting of copepods and phytoplankton. At night, the highest abundance of the copepods was present above 40 m and the mean population depth of three euphausiids was also above this depth. Why (and how) vertical space segregation occurred is not satisfactorily answered in the paper.

Extensive sampling of daytime and night-time vertical distributions of plankton was made during the 1965 SOND cruise by R.R.S. "Discovery" off Fuerteventura (Canary Islands). Baker (1970) described the distributions of 28 species of euphausiids. In addition, Angel and Fasham (1973, 1974) carried out factor and cluster analysis of the zooplankton distribution, including euphausiids. Not all species occurred in sufficiently high numbers in the catches to reach reliable conclusions. For 15 of the most abundant ones, the mean depth of the adults during the day or the night is presented in Fig. 10.7. As Baker (1970) remarked, there tends to be a vertical separation of the depths of the taxa that do not migrate. Overlap of adjoining species is often considerable though and the extent of this will be more closely inspected. The daytime vertical positions of the migrating species lie more closely together. Remarkably, these migrant species seem to interrupt the sequence of the non-migrating species between 400 and 600 m depth (Baker, 1970). This was also concluded by Angel and Fasham (1974) based on the result of their statistical analysis. At night, these species migrated into the 250–350 m zone (Fig. 10.7). The line drawn by me in these figures may suggest more than is really present in the

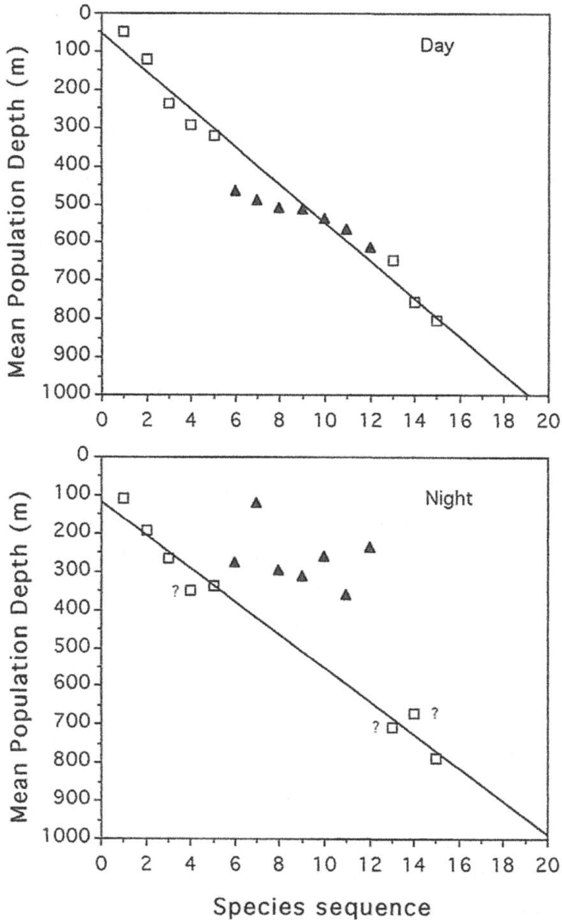

Fig. 10.7 Depth distribution during the day and the night of euphausiids near the Canary Islands. Migrating species are indicated by *triangles*, non-migrating by *squares*. If the total catch was less than 20 individuals, the species is provided with a *question mark*. The species are as follows: 1. *Stylocheiron suhmii*, 2. *S. affine*, 3. *S. longicorne*, 4. *Euphausia brevis*, 5. *Stylocheiron elongatum*, 6. *Euphausia gibboides*, 7. *E. krohnii*, 8. *Nematoscelis tenella*, 9. *Euphausia hemigibba*, 10. *Nematoscelis megalops*, 11. *N. microps/atlanticum*, 12. *Thysanopoda obtusifrons*, 13. *Nematobrachion boopis*, 14. *Thysanoessa microphthalma*, 15. *Thysanoessa parva*. The drawn line was fitted by least squares but has no meaning other than to accent the distribution over depth (data from Baker, 1970)

data, although the mutual distance between the mean depths of the non-migrants is comparable. Of course, everything with properties differing in magnitude can be arranged in a series: if 15 random numbers between 0 and 1000 are ranked, a more or less straight line would be obtained. The deviation from the mutual average distance among successive random numbers would be much larger. A parameter such as mean population depth must be interpreted carefully because sample size and the distance between samples are in many cases an approximation of what is needed for a precise description of vertical distributions. It can not be concluded that " habitat differentiation behaviour" or active selection for a certain depth is present because such behaviour is unknown and in the rather homogenous water column, few cues[8] are present to direct the animal. This does not imply that the observed distributions are random. Vertical habitat separations might be maintained by competition since this is heaviest at the overlapping borders of the vertical distributions present.

It is a strange phenomenon that in daytime, the migrating species are confined to the relatively small vertical distance between 400 and 700 m because their abundance is not low. In terms of weight, migrantspredominated over the non-migrants. Angel and Fasham (1974) remarked that no upward migration from below 700 m occurs. It must be borne in mind that juveniles are not included in this discussion, although some of them also migrate. For a further discussion of the community of euphausiid species, a quantification of the extent of mutual space occupation can serve as a guideline. Overlap between pairs of species *i* and *j* was calculated using the equation:

$$O_{ij} = O_{ji} = \frac{\sum p(iH_m)^* p(jH_m)}{\sqrt{(\sum p(iH_m)^2 * \sum p(jH_m)^2)}} \qquad 10.1$$

with $p(iH_m)$ being the frequency of species *i* in a certain part of the vertical habitat H_m. (Pianka, 1974). The value of this overlap ranges from 0 (no overlap) to 1 (complete overlap).

The results for the daytime and night-time distributions are presented in Table 10.3. The larger the value of vertical habitat overlap is, the higher the potential threat of competition between the two species. Of course, habitat overlap need not result in competition. This will only be the case if resources are in short supply and if the ecological niches of the species are similar to some extent. Underlined numbers in the table refer to migrating species. One glance at the daytime table makes clear that the daytime distributions of the migrating species overlap considerably. Generally speaking, the kind of food and feeding period seems to be the most important niche-dividing character. Therefore, the feeding habits of some of the species are given in Table 10.4. The migrating euphausiids are marked as omnivores that forage predominantly at night. The day depth seems to be largely a "parking space", where competition does not play an important role. At night, they invade the upper 300 m, which is the habitat of the *Stylocheiron* species. However, these are predators, eating during the day. The migrating euphasiids graze on algae and if they also prey on copepods and other zooplankton, this may involve in part vertical migrants from deeper down and, therefore, other prey items than the *Stylocheiron* species eat. Being omnivore is an elastic notion. The ratio between algal food and animal food consumption not only depends on species but also on the kind of food that is available in abundance.

Nevertheless, the feeding types according to Roger (1975) indicate a food niche difference between species. For example, *S. elongatum* (5), with a ratio of 41, is an omnivore feeding more on algae than *N. tenella* (8), having the ratio 60. *E. gibboides* (6) is more of a herbivore than *E. krohnii* (7) with which it nearly completely inhabits the same column part during the day. If indeed, *E. gibboides* (6) is active during the day and night but *E. krohnii* (7) predominantly during the night, the ecological niches are sufficiently different to prevent competitive exclusion. At night *E. krohnii* (7) has a high overlap with *S. affine* (2). However, *S. affine* (2) is a day-active predator and thus has a different ecological niche than the migrating *E. krohnii* (7).

Table 10.3 Vertical (column) habitat overlap between pairs of euphausiid species

1	2	3	4	5	6	7	8	9	10	11	12	13	14	15
1	0.17	0.02	0.00	0.01	0.00	0.00	0.00	0.00	0.00	0.00	0.00	0.00	0.00	0.00
	2	0.65	0.00	0.01	0.01	0.00	0.00	0.00	0.00	0.00	0.00	0.00	0.00	0.00
		3	0.28	0.47	0.09	0.07	0.17	0.13	0.02	0.05	0.01	0.11	0.00	0.00
			4	0.78	0.03	0.03	0.10	0.04	0.00	0.02	0.00	0.06	0.00	0.00
				5	0.31	0.19	0.37	0.30	0.10	0.10	0.00	0.17	0.00	0.00
					6	0.92	0.59	0.60	0.75	0.67	0.00	0.08	0.00	0.00
						7	0.83	0.63	0.84	0.83	0.09	0.20	0.17	0.01
							8	0.84	0.66	0.48	0.45	0.71	0.15	0.12
								9	0.74	0.71	0.57	0.54	0.01	0.16
									10	0.93	0.44	0.29	0.00	0.11
										11	0.52	0.57	0.17	0.22
											12	0.56	0.41	0.29
												13	0.41	0.32
													14	0.50
1	2	3	4	5	6	7	8	9	10	11	12	13	14	15
1	0.17	0.20	0.69	0.02	0.04	0.19	0.15	0.29	0.02	0.13	0.08	0.07	0.01	0.00
	2	0.28	0.38	0.04	0.17	0.87	0.25	0.55	0.00	0.46	0.26	0.16	0.00	0.00
		3	0.36	0.03	0.68	0.17	0.93	0.67	0.65	0.66	0.66	0.01	0.01	0.00
			4	0.09	0.07	0.16	0.30	0.27	0.01	0.57	0.27	0.14	0.13	0.24
				5	0.68	0.01	0.82	0.53	0.71	0.59	0.77	0.02	0.02	0.00
					6	0.14	0.83	0.72	0.98	0.36	0.09	0.08	0.01	0.02
						7	0.16	0.65	0.00	0.14	0.07	0.02	0.20	0.00
							8	0.74	0.82	0.66	0.86	0.08	0.06	0.01
								9	0.65	0.37	0.57	0.16	0.06	0.20
									10	0.29	0.79	0.01	0.01	0.00
										11	0.65	0.00	0.23	0.14
											12	0.00	0.00	0.00
												13	0.31	0.43
													14	0.41

Complete overlap = 1, no overlap = 0. Upper table represents daytime, lower table night-time distribution. Underlined numbers are migrating species. Species as in legend of Fig. 10.2. Data from Baker (1970).

The three non-migrating species within the upper 300 m are day-active predators. *S. affine* (2) and *S. longicorne* (3) overlap and this especially holds for the juveniles. For example, the juveniles of *S. affine* (2) and adults of *S. suhmi* (1) inhabit the same column part (overlap 0.99) and both predators must be strong competitors. Even size is comparable, 6–9 mm and 7–8 mm, respectively (Roger, 1975). Interspecific competition must be severe and co-existence a problem. However, the problem of interspecific competition changes in intraspecific competition if Mikkelsen (1981) is right that the three belong to one species-group. Are we witness of evolution and are different genotypes of one species already separated over vertical space? The enigma becomes even more interesting!

In Table 10.3, vertical habitat overlap of pairs of euphausiids was presented and for some of the species with a high overlap, we looked for circumstances that could release a possible interspecific competition. Niche differentiations might be subtle but in combination sufficient for co-existence. The "solutions" will be found in physiological and ecological factors. The available information is not sufficiently detailed to permit definite conclusions. Although the study of Baker (1970) was extensive, the number of individuals in the catches was often small. Therefore, the species distribution over the large column cannot be as definite as one would like to have in order to draw conclusions. However, I think it was possible to recognise phenomena and formulate problems.

Since DVM is our main interest, one problem is why some species migrate and others do not. For what reason are the migrating species packed within a specific column part? One possible explanation might be that the *Stylocheiron* species, as visually hunting predators, have to be avoided in daytime, and even so at night, since the migrants swim into the lower realm of the predators only. To graze, they make perhaps occasional feeding excursions higher up in the water column. Pearre (2003) called attention for asynchronous feeding excursions of individuals into the food-rich upper water layers. As soon as stomachs are filled, these individuals move down again. Several examples were given by him. Some euphausiids with algae-filled guts were found deep down in the water column long before sunrise.

The non-migrants seem to have realised prevention of interspecific competition by spacing out the migrants by exploiting the phytoplankton resources during the night. Resource partitioning was suggested by Angel and Fasham (1974) and Mauchline (1980) concluded that "these adaptations, including the evolution of diurnal vertical migration, allow the pelagic community to exploit, to the full, the three-dimensional environment of the oceans..." (p. 442). I think this is true. For understanding the community, essential relations have to be considered, including the role of juvenile stages, because they have their own strategies of survival. Taxonomic names are convenient indications of different species but not sufficiently refined since the functional groups divide space. The community is a collection of individuals in space–time with subtle differences in anatomical, physiological and behavioural properties. For example, Brinton (1967) suggested that migrating genera are usually regarded as least specialised morphologically with rather uniform, coarsely setose, comb-like thoracic legs that apparently sweep or strain the water to gather algae. Several authors have found migrating euphausiids to have phytoplankton in the stomachs (Roger, 1975; Hu, 1978; Gibbons et al., 1991) but also copepods. Nevertheless, even if copepods are present in high concentrations at greater depth, DVM remains compelling (Onsrud and Kaartvedt, 1998; Sameoto, 1980). This indicates a strong drive to migrate and a behaviour insufficiently flexible to suppress it when – to us – it is obvious to do so. Available food is perhaps insufficient at the "vacant" depth range between 400 and 600 m to sustain resident populations such as seen in the upper 400 m. So, this space is only available to herbivores or omnivores that migrate.

If a large number of migrants make nightly, asynchronous feeding excursions into the food-rich upper water layers, the night-time distribution of these species

Table 10.4 Some characteristics of the 15 euphausiids discussed in the text

	Species	Length (mm)		Feeding type[a]	Feeding type[b]	Feeding time
1	*S. suhmi*	6–9			100	D
2	*S. affine*	7–8			90	D
3	*S. longicorne*	7–13		P	73	D
4	*E. brevis*			F-O		
5	*S. elongatum*				41	D
6	*E. gibboides*	22–27	M	F	7	D-N
7	*E. krohnii*	14–18	M	F-O		N
8	*N. tenella*	15–21	M	O	60	D-N
9	*E. hemigibba*	12–16	M	F-O		N
10	*N. megalops*	21.8	M	O-P		N
11	*N. microps/atlanticum*	17.8	M	O-P/P	60/78	N
12	*T. obtusifrons*	20	M			D-N
13	*N boopis*	19–21		P	96	D-N
14	*T. microphthalma*					
15	*T. parva*	9–11				

[a]Feeding type according to Mauchline (1980): P = carnivore, O = omnivore and F = herbivore.
[b]Feeding type trophic level according to Roger (1975) ranging from 100% = animals in stomachs to 0% = purely algae in the stomachs.
The indication of the time of feeding is D = diurnal, N = nocturnal and D-N = feeding during the whole day. M = migrating species.

ought to be more spread out over the column than the daytime distribution. This was examined by comparing daytime and night-time vertical distributions by means of Simpson's measure of concentration,[9] but no differences were found. This does not imply that asynchronous feeding excursions did not occur, but simply that it could not be demonstrated in this way.

We might expect that predators play an important role in euphausiid migration because krill represent important links in the pelagic food web and form prey for many higher trophic-level consumers (Greene et al., 1992; Onsrud et al., 2004). How *M. norvegica* is at the mercy of predating fish is revealed by an extensive study in the relatively shallow(120 m) Oslofjord (Norway) by Onsrud and Kaartvedt (1998) and Onsrud et al. (2004).

A layer of these krill of 20–40 m thickness was sandwiched between deep-living Norway pout (*Trisopterus esmarkii*) and epipelagic whiting (*Merlangius merlangus*) and herring (*Clupea harengus*). *M. norvegica* were deeper than 70 m during the day and ascended at dusk to a depth of 10–30 m. The diel feeding rhythm was associated with these vertical migrations. However, Onsrud and Kaartvedt (1998) suggested that individuals must make feeding excursions from the upper dense layer of krill to the surface by night to obtain the observed gut fullness. At the day depth, *M. norvegica* also preyed upon overwintering *Calanus* spp. but nevertheless carried out migrations. DVM occurred irrespective of fluctuations in phytoplankton or copepod concentrations, as Sameoto (1980) had noted also. Ascent and descent coincided with dawn and dusk and, obviously, light changes were compelling stimuli. The benthopelagic pout moved also upwards in the evening, forming a layer in midwater and then fed on the lower part of the krill layer. Although having large eyes with presumably a high visual acuity, the authors thought it unlikely that visual predation would be possible at night. Onsrud et al. (2004) suggested that the lateral-line system enables these pout to find euphausiids by the perception of tactile stimuli. The whiting and herring schools situated above the krill in daytime dispersed at sunset and invaded the ascending krill layer. Both species fed on *M. norvegicus* during the day and the night (Onsrud et al., 2004). Descent migration of the fish took place before sunrise.

The emerging picture of feeding and migration by *M. norvegica*, herring, whiting and pout is complicated. The euphausiids move upwards to feed but do so when light intensity is low and feeding by herring is hampered, albeit not impossible. The descent migration started presumably before light intensity was high enough for the clupeids to start feeding efficiently by sight (Onsrud and Kaartvedt, 1998), but stopped before they entered the depth of the large-eyed, deep-living Norway pout. DVM for *M. norvegica* is, therefore, not a straightforward strategy to evade predators. As usual, the interspecific relations in the Oslofjord community are complicated and require behavioural compromises. Migration alone might be insufficient to maintain the population. Onsrud et al. (2004) suggest that in daytime, in the proximity of deep fish assemblages, the euphausiids performed instantaneous escape reactions.

10.5 Sound Scattering Layers (SSL)

Continuous recordings of SSL in the ocean can provide excellent temporal and spatial coverage of diel vertical migrations, but a disadvantage is that the species composition within these layers is not always known. Small fish and especially krill (euphausiids) have often been mentioned as the possible constituents. Later studies were accompanied by catches of animals to identify the components. In the Gulf of St. Lawrence estuary, Sameoto (1976) found his samples taken within SSL to consist of the euphausiids *T. raschii*, *T. inermis* and *M. norvegica*. Dominance varied throughout the year and among different regions of the Gulf, but euphausiids were important ingredients everywhere and all the time. Onsrud and Kaartvedt (1998) found *M. norvegica* to be the major acoustic target in Oslofjord and Kaartvedt et al. (1996) observed another euphaussiid, *T. inermis*, to be the major constituent of an SSL off the coast of northern Norway. If continuous recordings are accompanied by light-intensity measurements, the data become important in the discussion about the proximate stimuli responsible for DVM. When more than one frequency is applied, the technique allows a distinction between specimens of different size, for instance, between krill and small fish. However, the technique is inadequate for understanding questions concerning the adaptive significance of migration unless additional data on species-specific distributions, feeding and growth are collected.

Different species may migrate together for some time, but often the layers split and each part then continues at a different vertical velocity (Clarke and Backus, 1956, 1964). Bary (1967) classified the ascent and descent migrations of SSL in

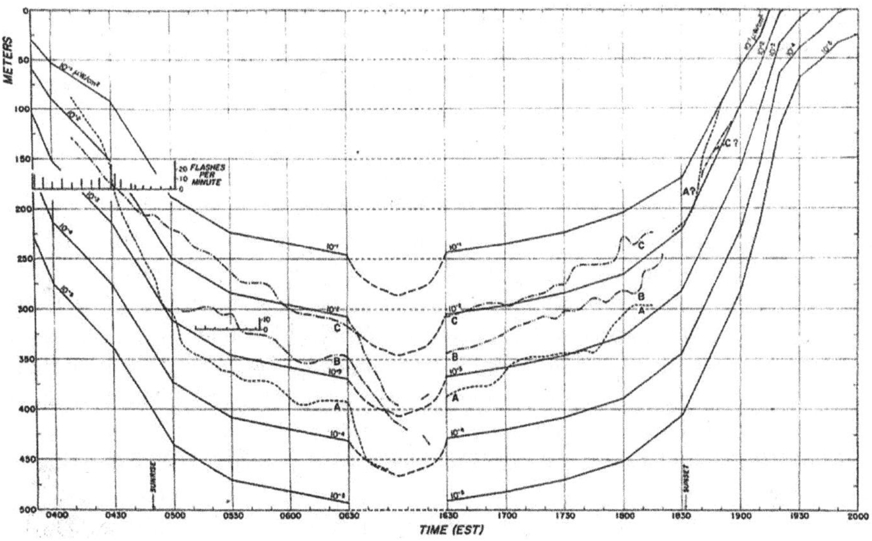

Fig. 10.8 Vertical migration of sonic scattering layers (SSL) off the northeast coast of the United States during 15 August 1959. *Layer A* migrated downwards at a rate faster than the descent of isolumes, *Layer C* more slowly. In late afternoon, the layers migrated towards the surface at a rate somewhat faster than the isolumes (from Clarke and Backus, 1964)

Saanich Inlet into four phases, but these phases can not always be recognised in echograms. However, a fast and a slow phase can easily be distinguished, as in Fig. 10.8, one of the earliest recordings made by Clarke and Backus (1964). Morning descent of line A appears to start with a fast displacement, but changed into a slow drift shortly after sunrise after splitting into two layers (A and B). Between 4 h 10 min and 4 h 50 min, displacement velocity of A was 4.4 m min^{-1}, thereafter, between 5 h and 6 h 30 min, it reduced to 1.0 m min^{-1}. Afternoon migration began with a slow upward drift, which rather suddenly changed into rapid swimming just before sunrise. Near the surface, or rather at the end of dusk, vertical movement came to a stand still. In general, fast displacements make up for more than 60% of the total migration amplitude. Gibbons et al. (1991) thought that this pattern of "slow-fast (-slow)" ascent was a behavioural strategy to remain for longer periods within 'preferred" food concentrations if the environment is vertically stratified. There is no evidence for this suggestion. As I will discuss (see also Chapter 11), the animals in the layers react to changing light intensities: "migrating" light intensities change depth simultaneously with the animals.

This has been shown, for example, by Boden and Kampa (1958), Kampa and Boden (1954) and Clarke and Backus (1956, 1964). Some SSL moved as fast as the isolumes and Boden and Kampa (1958) proposed that the constituting species followed their "optimal" light intensity, in line with what had been the generally accepted explanation for decades. Clarke and Backus (1956, 1964) also observed a close relation with isolumes. However, they also saw that during ascent the SSL could move into light intensities that were 100× higher than the one they had experienced at their day depth. In the morning, the SSL sometimes moved faster down than an isolume, implying that the animals were swimming into gradually lower light intensities. This is what happened to the SSL animals represented by line A in Fig. 10.8. However, many displacements were not fast enough to keep pace with the isolumes. Lagging behind, Clarke (1933, 1934b) had already observed such in his study of *C. finmarchicus* in the Gulf of Maine. This pioneer of quantitative phototaxis experiments obviously found the "causal" explanation of an "optimum" light intensity difficult to accept, and realised, that some physiological mechanism was responsible. As a possible explanation of the differences in migration speed, Clarke and Backus (1964) suggested that once a change in light intensity had triggered upward or downward swimming, migration continued independently of light. In the discussion of the causal factors and mechanisms at the base of DVM (Chapter 11), I will return to this idea of Clarke again.

Table 10.5 compiles vertical displacement velocities of the SSL from the literature in comparison with estimated values of "velocities" of isolumes.[10] All three possible differences in displacement velocity appear to be present. Moving at about the same velocity with isolumes is predominant, especially after sunrise and before sunset, when light intensity changes slowly and consequently also isolumes change depth slowly. One might think that only then animals are capable to go as fast as their preferred isolumes, but this cannot be so because the same layers (thus animals) move much faster when light intensity changes rapidly before sunrise or after sunset. Oceanic SSL, coming from a deep day depth, move faster than SSL migrating above the shallower continental slopes (Bary, 1967; Heywood, 1996). SSL within the upper 100 m tend to go as fast as isolumes. This is illustrated in Fig. 10.9: *M. norvegica,* the main component of this SSL, ascents in Oslofjord from about 70 m depth at approximately the same speed as an isolume (although not at the same light intensity) at different times of the year. Heywood (1996) thought that the shallow migrations are initiated by relative changes in light intensity, but she had no suggestions for the stimulus that triggered migration of animals coming from deep down in the water column.

From her extensive study of migrations of SSL in the northern part of the Atlantic Ocean, I have chosen to describe the migrations of a shallow and a deep layer. On 4 August, at 52° N, 17° W, a layer, ascending from about 80 m depth, reached the near surface between 1930 hours and sunset at 2053 hours, moving at a velocity of about 0.6 m h^{-1}. On the same evening, at 2030 hours, an SSL appeared on the echogram at a depth of about 333 m. Before sunset, this layer moved upwards at a velocity of about 1.4 m min^{-1}. Thereafter, it accelerated to 4 m min^{-1} until the ascent stopped at 2153 hours, at 50 m from the surface.[11] In Chapter 5, I showed that photobehaviour, triggered by relative changes in light intensity, is able to produce all three migration patterns, including the "moving faster than isolumes pattern".

It is not possible to determine precisely the beginning of the fast downward migrations in the morning because the scattering signal is often lost in the noise of the near water surface. From figures in Clarke and Backus (1956, 1964), it can be guessed that descent migrations start at least 50 min before and end 20–30 min

Table 10.5 A collection of vertical displacement velocities (m min^{-1}) of scattering layers (second number) and of isolumes (first number) in the oceans

Author	Specification of date and layer	Before sunrise	After sunrise	Before sunset	After sunset
(1)				0.9; **0.9**	2.3; **2.3**
(2)	20-7; deep				3.8; **4.6**
(2)	21-7; deep			0.7; **0.9**	3.5; **4.4**
(2)	21-7; middle			0.7; **1.8**	
(2)	21-7; upper			0.7; **0.7**	
(3)				0.6; **0.6**	3.0; **5.1**
(4)	14-8; layer A	4.0; **4.3**	0.8; **1.1**	2.3; **0.9**	
(4)	14-8; layer B			2.3; **1.7**	
(4)	14-8; layer C	3.3; **1.1**	1.2; **1.3**	2.1; **1.5**	
(4)	15-8; layer A	2.7; **4.4**	0.6; **1.3**	0.7; **0.9**	
(4)	15-8; layer B		0.6; **0.6**	0.7; **0.7**	
(4)	15-8; layer C	2.7; **1.9**	0.7; **1.4**	0.7; **0.8**	
(5)	4-8; upper			**0.6**	
(5)	4-8; lower			**0.7**	**3.7**
(5)	8-8	**1.6**	**1.3**		
(6)	November			0.1; **0.1**	0.5; **0.5**
(6)	March			0.1; **0.1**	0.4; **0.7**
(6)	June			0.1; **0.1**	0.4; **0.3**

(1) Kampa and Boden (1954); (2) Clarke and Backus (1956); (3) Boden and Kampa (1958); (4) Clarke and Backus (1964); (5) Heywood (1996); (6) Onsrud and Kaartvedt (1998).

after sunrise, thus lasting about 80 min. From Table 1 in Bary (1967), the beginning of downward migrations, observed at different times of the year, can be averaged to 53 min before sunrise while they last for about an hour. According to Heywood (1996), the deep SSL started to move down about 1 h before sunrise. Although the time course of the change in light intensity differs with latitude and time of year, there is a reasonable conformity in the beginning of descent. We can conclude that the fast phase of downward migration coincides with the period of highest relative increases in light intensity. From the two echograms in Clarke and Backus (1956), the beginning of the fast phase of upward swimming is about 15–20 min before sunset and the whole period lasted for at least 1 h. The same appears to be true for *M. norvegica* in Onsrud and Kaartvedt (1998).

Heywood (1996) mentioned that normal DVM were seen throughout her cruise over the Atlantic Ocean from west of Ireland to Greenland and vice versa, mostly going down to about 400 m, and both ascent and descent being more rapid than corresponding isolumes. However, reverse migrations appeared at dusk: as soon as a deep layer reached the surface, a near-surface layer would go down. I drew attention to a similar reversed migration of *P. newmani* in Dabob Bay. Heywood (1996) remarked that the descent migration in the evening seemed to be correlated with the arrival of predators or competitors. Since the species involved were unknown, it is hard to conclude what initiated the early descents. Gibbons et al. (1991) observed that the omnivorous *E. lucens* rapidly fed at the surface and went down early to reach

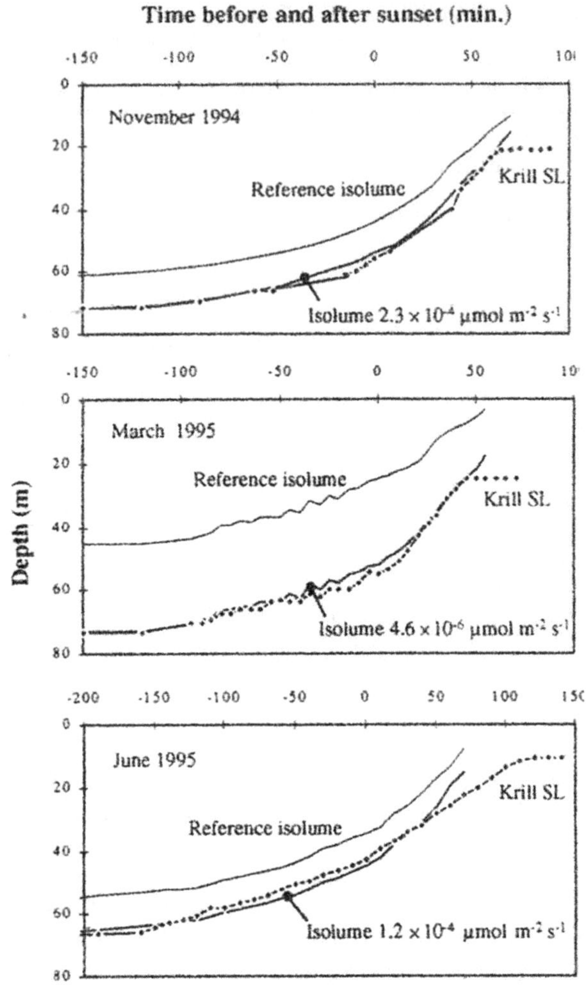

Fig. 10.9 The *upper* boundary of upward migrating krill in Oslofjord (Norway), as represented by sonic scattering layers on three different dates. The position of the 10^{-3} mmol m^{-2} s^{-1} isolume is given as reference (from Onsrud and Kaartvedt, 1998)

the day depth of 100–200 m at about midnight. He suggested that this occurred as soon as individuals were satiated.

I briefly introduced the principal aspects of migrating sound scattering layers. The factors causing the migrations are treated in the next chapter.

Notes

1. Attenuation coefficients were calculated from data in Table 1 in Buskey et al. (1989).
2. Dagg et al. (1989) wrote that the systematics of the coastal Pacific population are inconclusive. These authors follow the opinions of Fleminger (1967) and Park (1968) (see their paper for reference) in assigning the name *M. lucens*.

3. Recent papers have strongly suggested that high concentrations of diatoms are "toxic". If true, this rather controversial idea has implications for the development of spring populations. See Note 6.
4. Dagg et al. (1989) mentioned that "times of sunrise and sunset so defined generally agreed to within 30 min with predictions (U.S. Dep. Commerce 1985), but on the mornings of 6 and 7 May heavy cloud cover caused a 45–55 min delay between predicted and observed sunrise" (p. 1063). This explains why "night" in Fig. 3 lasted for 10 h 30 min while the period between sunset and sunrise was 9 h 13 min according to information of the U.S. Naval Observatory; Astronomical Applications; aa.usno.navy.mil/data/docs/AltAz.html.
5. Although Pearre (1973) found a negative selectivity for a deep-living *Pseudocalanus* species in Bedford Basin (Canada).
6. Cultures seem to have demonstrated that springtime diatoms of the genus *Thalassiosira* (see Pierson et al., 2005a for references) have a deleterious effect on the development of marine copepods. Confirmations from nature are, however, scarce and contradictory (Frost, 2005). The Dabob Bay research group devoted an extensive study to the enigma. An account can be found in Progress in Oceanography 67 (3–4), 2005.

The toxic chemicals were identified as polyunsaturated aldehydes or PUAs (Wichard et al., 2005). Ianora et al. (2004) added different concentrations to cultures of the dinoflagellate *Prorocentrum minimum*. It was assumed that the chemical would adhere to the non-toxic dinoflagellate cells to make ingestion by the copepods possible. Females of *C. helgolandicus* were fed with the dinoflagellate and at the highest concentration the born juveniles were dying at the copepodite stage C1 (p. 405). At lower concentrations of PUA some newborns reached the adult stage but mortality was still extremely high. When fed *P. minimum* only, survival was better but adulthood was reached in a few cultures only. There is no doubt that monocultures of diatoms like *Synedra costatum* are bad food nor that aldehydes derived from these diatoms have a toxic effect. However, survival and larval growth on the non-toxic dinoflagellate were not impressive either. That pure diatom diets present sufficient food quality, as the authors claim, seems to be a bold assertion.

This is indeed the critique that Jones and Flynn (2005) have on the application of monocultures. These authors fed *A. tonsa* with two *Thalassiosira* species. Using a monodiet, all copepods died. The rapid effection nauplii was attributed to aldehyde toxins. *A. tonsa* continued to eat the diatoms, but egg survival rates were poor. If dinoflagellates were added, population growth was even better than if fed with dinoflagellates alone. Compared to monocultures, mixed food gave always better results. The authors think this indicates that diatoms are not toxic but of a lower nutritional value than dinoflagellates: "any evaluation of toxicity not only requires the use of a mixed diet, but also that an assessment of the nutritional status (e.g. N:C) of the prey is critically important" (p. 1459). This has been traditional wisdom for a long time. To quote Friedman and Strickler (1975): "attempts to culture copepods have shown that for long term maintenance of reproducing populations, multi algal diets and/or bacteria, vitamins, or other 'critical substances' must be supplied" (p. 4187). The physiological status of a culture of food algae is important. Freshwater zooplankton grow better on exponentially growing algae than algae in the stationary phase, let alone those from the declining phase. The presence of long-chain polyunsaturated fatty acids (PUFAs) in the diet is important for the development of freshwater zooplankton (Gulati and DeMott, 1997; Weers and Gulati, 1997) as well as of marine organisms (Levine and Sulkin, 1984; Roger, 1975). The ecological relevance of laboratory cultures is, therefore, not always clear (Frost, 2005).

A study of diatom toxicity in the field was performed in Dabob Bay (Halsband-Lenk et al., 2005). During February and March, abundance of *P. newmani* was estimated and egg hatching success and nauplier survival to stage N3 were determined. The essential ratio of aldehyde to non-aldehyde producers must indicate the possible toxicity of the algal community. Egg production rates did not vary much, except on day 100 (20 April) in 2002. Hatching success was high during the whole period. In 2003, hatching success was lower during the period of a high R, as compared to the periods before and after, although differences between dates were not statistical significant. R being a ration, a decrease might indicate an increase of non-aldehyde producers or a decrease in aldehyde producers, but I find it difficult to arrive at a definite conclusion. The authors are very positive about the toxic effect of winter–spring diatoms on the development of *P. newmani*. Calanoid copepods may be highly selective with regard to ingestion of prey and selectivity may be variable. A chemical sense is involved (Friedman and Strickler, 1975). Leising et al. (2005) mentioned that *C. pacificus* may one week reject a diatom species while consuming it some time later. Highly abundant species may be avoided often after encounter. The *Thalassiosira* species need not to be toxic all the time, depending on the growth phase of the population and selective feeding may reduce a possible harmful effect.

7. The following equation was used:

$$P = \frac{D_n - S_d}{D_d - S_n} \times 100$$

D_n and D_d are the deepest depths at which a specimen was caught during the night and the day, respectively. The shallowest depths are indicated by S_n and S_d.

8. Obvious stimuli are pressure and light intensity. However, pressure perception seems to be absent in pelagic invertebrates and due to a large capacity for adaptation, the perception of a particular light intensity is difficult too.
9. Simpson's measure of concentration is also used as a measure of niche breadth by Levins (1968):

$$B_i = \frac{1}{\sum P_{ih}{}^2}$$

where p is the chance of species i in habitat part h.

10. The estimation of the displacement velocities from the figures in the literature could not be done with high precision. Precision might be about 0.2 m min^{-1}.
11. Estimates of time and depth were made from Fig. 4 in Heywood (1996). They could not be done with high precision. Sunset was calculated from Astronomical Applications; aa.usno.navy.mil/data/docs/AltAz.html.

Plate 3 *Euphausia superba,* a species from the Antarctic Ocean. Many species of Euphausiacea are found everywhere in the oceans where they are important constituents of sound scattering layers (SSL). Sometimes they occur in very large swarms, as the species portrayed. Then they are bulk food ("krill") for baleen whales. (Photograph courtesy by Jan Andries van Franeker, Imares, Wageningen University, the Netherlands.)

Chapter 11
The Confrontation of Experimental and Field Studies

Contents

11.1 Introduction

Evolutionary thinking has made DVM research undeniably richer, but the historical perspective also led to an alienation from science. For disciplines like physiology, molecular genetics or developmental biology, with a strong base in rigorous experiments, this has not happened. In DVM research, speculations about the evolutionary development and adaptive significances adorn discussions, but remain often to a large extent free of obligations to present factual evidence. To serve the idea of "optimal" performances, too many assumptions about animal behaviour have to be made, not hampered by knowledge. New terms and notions have been introduced, suggesting progress. At a fine special ASLO symposium in 1978, I heard for the first time about "proximate" and "ultimate" causes. At the time, I considered them old wine in new sacs, just replacing terms like causality and finality, which had been in use for a long time. In Newtonian physics, only causality exists and in most of biology too, cause and effect are unambiguously connected. In ecology, a contributing factor is not often called a cause anymore, but is replaced by less well-defined phrases like "trigger" and "cue". For example, a generally accepted opinion is "that light plays an important role, but DVM seems to be triggered by a

J. Ringelberg, *Diel Vertical Migration of Zooplankton in Lakes and Oceans*,
DOI 10.1007/978-90-481-3093-1_11, © Springer Science+Business Media B.V. 2010

chemical signal". Is triggered the same as caused and thus, is DVM caused by the presence of a chemical in the water? What then is that important role of light? Has Loose (1993b) not convincingly demonstrated in the Plöner Plankton Towers that the presence of fish kairomones did not evoke vertical migration if changes in light intensity were absent? There are no experiments demonstrating unambiguously that fish kairomones alone cause or trigger – whatever you please to call it – periodical upward and downward swimming. Therefore, either a statement like the above one is false, or "triggered" is not equal to "caused". If the latter is the case, what does trigger mean? Before we can continue with a discussion about the explanation of DVM, a closer look at causality in ecology is required.

Part of the confusion arises from the fact that the meaning of causality depends on, and is restricted to, the level of biological integration: I will use the whole lake experiments of Dini and Carpenter (1988) with planktivorous fish to illustrate the difference (Ringelberg and Van Gool, 2003). After Dini and Carpenter (1988) removed minnows from Tuesday Lake (USA), *Daphnia* stopped migration. They resumed DVM next year, 6 weeks after the re-introduction of minnows. At this scale of the ecosystem, the role of fish must certainly be called the cause of DVM. This conclusion holds for an artificial fish introduction, as in Tuesday Lake, but also for "natural experiments", when for instance in spring 0^+ fish start to populate the open water. Causality is less simple at the next integration level, that of the population. Dini et al. (1993) thought of the potential presence of migrating and non-migrating genotypes in a lake. In the presence of predating fish, the migrating genotype would be dominant and sampling would reveal a migrating population. In the absence of predatory fish, and because of the costs inherent to migration, the non-migrating genotypes would soon take over. Consequently, sampling would not show migration. In such a convenient arrangement, predation by planktivorous fish could be held responsible for a shift in genotype abundance, but these fish are not the proximate cause of DVM: they represent the boundary condition.

Let us turn to the level of individuals. *Daphnia* individuals swim down upon increases in light intensity and swim upwards when light intensity decreases. This has been shown ample times in experiments. It is not necessary for fish or for their kairomones to be present. The light signal is strong enough to evoke a response even though the notion of fish being present through kairomones may evoke a stronger reaction. In the absence of light-intensity changes, *Daphnia* has never shown diel vertical swimming. Therefore, our conclusion is that at the level of the individual, fish and their kairomones are not the cause and they do not "trigger" migration behaviour. They only influence the responsiveness. This reasoning seems to be against the current. In the recent literature, fish are considered to play a more important role as the cause of DVM and light intensity changes may often even not be mentioned at all. How is this possible?

I must admit that a term like proximate factor has become handy now. I define a proximate factor as a property of the environment that plays a role in the expression of a certain phenomenon, for instance a particular behaviour or a life-history character. A causal factor is also a proximate factor, but not all proximate factors are causal factors. Some are modifiers, without the power of eliciting the response by themselves. A fish kairomone is undoubtedly a proximate factor affecting DVM. Within the "set of necessary ecological factors", fish kairomones play an important role. As the complex food/satiation/hunger does (Pearre, 2003). However, in DVM, kairomones or food do not cause the *diel* vertical migration and their role is secondary to the role of light-intensity changes. At the ultimate sphere of interpretation, these two secondary factors become of primary importance. In evolutionary times, during which migration behaviour (thus adaptive photobehaviour, by the way) has been selected, with the ratio between mortality rate and birth rate as the main *criterion* of selection,[1] these two factors must have been crucial. The difference between the two entirely valid ways of reasoning is whether we analyse a present-day situation – the result of evolution thus far – or look into the past to imagine what led to the present situation. Causality in ecology is not as simple as in Newtonian physics. The realisation of this and the shortcomings and discontent with the old term has perhaps led to the introduction of new terms like "trigger" and "cue". However, these new ones are vague and have confounded the issue. Thinking about DVM has become imprecise.

If a correlation between DVM and some environmental factor is used as part of an explanation of the phenomenon, it is necessary to have at least some idea about a relevant mechanism at the base. If no idea exists about the causal mechanism by which the correlation has come into existence, the power of explanation, even when the correlation is statistically significant remains questionable. Again, I refer to Clarke's dictum (see Preface). If, for example, the start of a seasonal period of DVM is correlated with the appearance of a certain difference in temperature between epilimnion and hypolimnion, it is implicitly supposed that the temperature difference can be perceived by the planktonic animal. As far as I know, there is no indication that animals are able to obtain this knowledge, or even how to use it. Therefore, the correlation has no biological meaning or relevance. A correlation between the beginning of DVM and of tourists arriving at the shore could be significant as well, but nobody would believe in a significance of this relationship. Because I consider knowledge of mechanisms as very important, I started this book with physiological–behavioural mechanisms before turning to migration in the field. In this chapter, I discuss the validity of these mechanisms in interpreting and causally explaining the observations of DVM made in lakes and oceans. This is necessarily limited to descriptions of migrations that have been accompanied by measurements of the relevant proximate factors. Regretfully such factors have seldom been included. Although light has been mentioned as an important factor for more than a century, observations of DVM that have been coupled with light measurements remain rare.

A fundamental problem exists. In experiments, the performance of individuals is analysed, while descriptions of migrations in the field are based on performance of groups (Pearre, 1979). Only if all individuals in a population act and react synchronously, experimental and field data can be matched. Realising this, one must make the best of the information available!

11.2 Three Examples from Very Different Species

11.2.1 Phantom Larvae

The fourth-instar larvae of many *Chaoborus* species are often burrowed in the mud of lakes during the day, emerging from the sediment in the evening and being limnetic during the night (Parma, 1971). Before sunrise a downward migration sets in and they disappear in the bottom again. Haney et al. (1990) made an extensive study of the evening migration of *Chaoborus punctipennis* at different times of the year and in several lakes. The authors correlated the start of the rapid ascent migration with the moment at which the rheobase stimulus value – as determined for *Daphnia* (Chapter 3) – was reached. The correlation was highly significant(Fig. 11.1).

The authors concluded that the particular threshold (in casu $17 \times 10^{-4}\ s^{-1} = 0.10\ min^{-1}$) was a good predictor of the time at which *Chaoborus* larvae begin their rapid upward migration in the water column. Haney et al. (1990) used advanced sonar equipment to make precise and continuous recordings and could also estimate the rate of upward displacement of the leading edge of the population (Fig. 11.2) Displacement velocity and the rate of the relative decrease in light intensity were

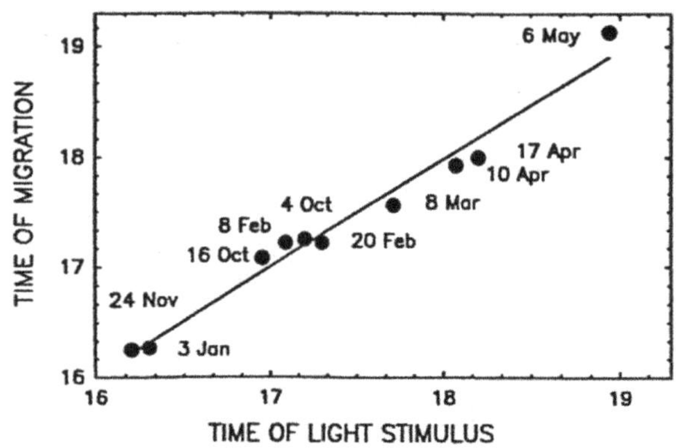

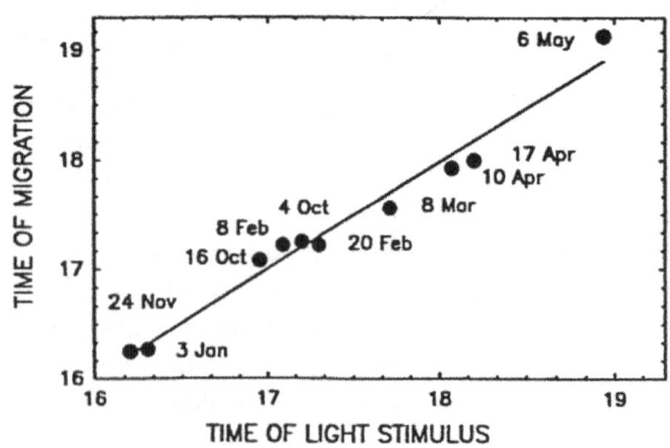

Fig. 11.1 Relation between the time of rapid ascent migration of *Chaoborus punctipennis* in the evening and the time at which the rheobase threshold of the relative decrease in light intensity occurred. The date of the observations is indicated beside each data point. $R^2 = 0.98$; $P < 0.001$ (from Haney et al., 1990)

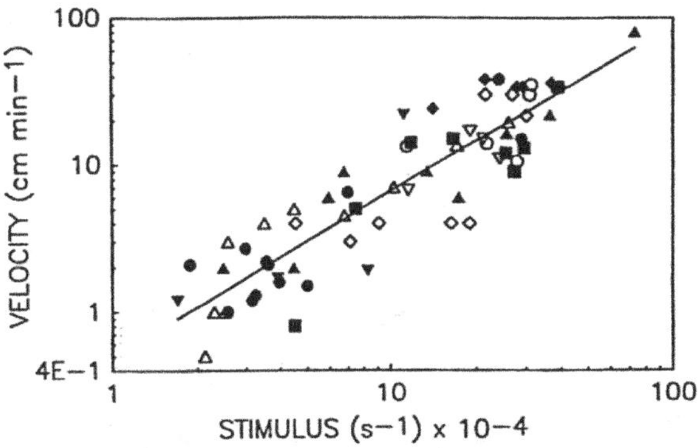

Fig. 11.2 The relation between the displacement velocity of *Chaoborus punctipennis* larvae and the relative change in light intensity during the upward migration in the evening. The symbols represent different dates: *open circles*, 3 January; *black dots*, 8 February; *open triangle*, 20 February; *black triangle*, 8 March; *open square*, 10 April; *black square*, 17 April; *open reversed triangle*, 6 May; *black reversed triangle*, 4 October; *open diamond*, 16 October; *black diamond*, 24 November (from Haney et al., 1990)

highly correlated over a wide range of swimming velocities (1–80 cm min^{-1}). When the light-intensity decrease reached the threshold stimulus, a break in the function line was expected but this is not visible in the data. The authors suggested that no threshold existed for the upward swimming and that the velocity was a continuous function of the rate in relative-light change.[2]

The larvae migrated at light intensities varying from 10^{-4} to 10^{-6} W m^{-2}, with one extreme value between 5.3 and 26×10^{-10} W m^{-2}, observed under thick ice with a heavy snow cover. The authors concluded that such a wide light-intensity range does not support the preferendum hypothesis: no particular light intensity is followed. A similar conclusion was reached by other investigators of *Chaoborus* (Teraguchi and Northcote, 1966; Swift and Forward, 1988).

An interesting illustration of the timing of DVM in *C. flavicans* in three lakes and throughout the year is presented in Fig. 11.3 (Lorke et al., 2004). The moment at which the temporal change in backscatter signal of an Acoustic Doppler Current Profiler was strongest was chosen as criterion for the onset of migration. Therefore, migration must have started somewhat earlier. Evidently, the strongest temporal change in depth occurred about an hour before sunrise and 1 h after sunset. The threshold value of the relative change in light intensity at which *Chaoborus* started migration, according to Haney et al. (1990), occurred about 70 min before sunrise. Time to sunrise and sunset of migration in Fig. 11.3 can be roughly estimated at 50–60 min, which is, considering migration had started earlier, close to the rheobase threshold of the relative change in light intensity. Lorke et al. (2004) found that cloud cover (or the lunar cycle (!)) was of no influence on timing and concluded that "migration is not solely triggered by light intensity, but also by a coherent 'internal clock' of the larvae" (p. 1289). Obviously, the authors had overlooked the pertinent articles on the effect of relative rather than absolute changes in the literature.

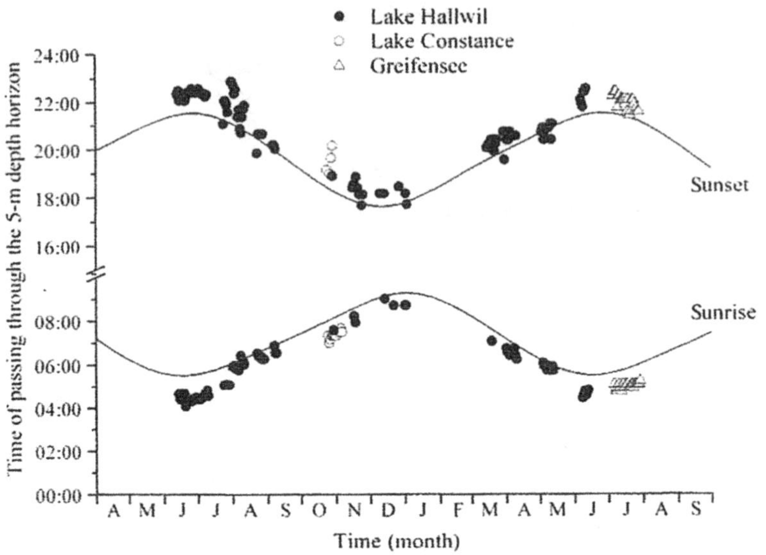

Fig. 11.3 Seasonal variation in the timing of DVM predominantly by *Chaoborus flavicans,* as estimated by the Acoustic Doppler Current Profiler. As migration time was chosen the time at which the strongest temporal change in backscatter signal occurred at a depth of 5 m. After Lorke et al. (2004)

Not all authors arrived at the same conclusion as Haney et al. (1990). Swift and Forward (1988)[3] observed that only the leading edge began to migrate at a threshold of $17 \times 10^{-4}\ s^{-1} = 0.102\ min^{-1}$ but not the peak and the trailing edge of the population. Therefore, they thought the relative change in light intensity difficult to accept as a stimulus. There was also no relationship between the rate of migration and the rate of relative change in light intensity. The authors suggest that the difference in method might be one source to explain the different result: Haney et al. (1990) monitored migration continuously, while their measurements were limited to samples at three depths with 15 min intervals. It is a common finding that not all individuals in a population start migration (or any other behaviour) at the same time and thus have reacted to the same threshold value but this is insufficient reason to reject relative changes in light intensity as a stimulus. The emphasis on a threshold value might overlook that some individuals also respond to suprathreshold stimuli. An endogenous rhythm makes *Chaoborus* larvae leave the mud. Their timing is, of course, never precisely the same for all individuals and some might have "missed" the threshold decrease. Nevertheless, like those already swimming upwards in the water column, they respond to the continuing light intensity change. DVM is a continuous response to continuously changing light intensities as long as these are above the physiological threshold value.

On the basis of laboratory experiments, Swift and Forward (1988) proposed an alternative explanation for the migration observed, viz. that ascent migration starts when light intensity falls below a certain absolute value in the evening and descent migration starts, when ambient light intensity reaches a (different) absolute light threshold in the morning. This would not explain why not all individuals have reacted, to the same absolute intensity. More importantly, if migration is started in this way, its continuation has to be an "all-or-nothing reaction" with an unexplained end.

A third study of *C. flavicans* migrations and light changes has been performed by Wagner-Döbler (1988, 1990). Her interpretation of the observed migrations in three small lakes was that the onset of upward migration occurred at a relative decrease in light intensity of $1.5 \times 10^{-3}\ s^{-1} = 0.09\ min^{-1}$. This value is quite close to the value from Haney et al. (1990) given above. However, the data also made her draw the conclusion that light intensity had to drop below an absolute value of about 10^{-4} W m^{-2}. Wagner-Döbler (1990) hesitated between an absolute light intensity as trigger and a relative change because of the "intimate relationship between both aspects of the light regime in nature" (p. 301). She suggested that both could act together in initiating the onset of migration. Based on field observations only, this hypothesis can neither be accepted nor refuted. Indeed, absolute light intensity determines the sensitivity to relative changes in light intensity (see Fig. 11.4) and the experiments of Swift and Forward (1988) point in that direction too. However, although the value of the stimulus threshold depends on absolute intensity, the stimulus-response mechanism remains the same.

It may look surprising that such a completely different species reacts to the same light stimulus and even to the same threshold value as the *Daphnia* species investigated (Chapter 3). However, dawn and dusk are important anchors for many animals to change their activity and behaviour and the light changes during dawn and dusk apply universally. Therefore, these can only be expected to be used universally in explanations.

11.2.2 *Daphnia*

On 11 June 1992, our research activities in Lake Maarsseveen started early in the morning. It was still dark when the first samples were taken 3 h before sunrise. Each half an hour, we would go back and forth ten times to the deepest part of the lake until one hour and a half after sunrise, taking samples at eight depths simultaneously. Light intensity was measured about 30 m from the shore at a depth of 0.30 m and registered continuously in a laboratory van at the shore. I was sitting outside the van, staring over the feebly reflecting surface of the lake. Suddenly, at the far shore, a pair of oyster catchers awakened and called out. Within a minute, I heard but could not see them flying over the lake. Just in time, I went inside the van to see the horizontal line on the flatbed recorder starting to rise: dawn had come. It was 2 h before sunrise.

One result of this early-morning activity is presented in Fig. 11.4. The relative change in light intensity rose quickly and within about half an hour the rheobase threshold of phototaxis was reached. Now, *Daphnia* ought to start its downward migration. Indeed, after all the daphnids in the samples were counted, weeks

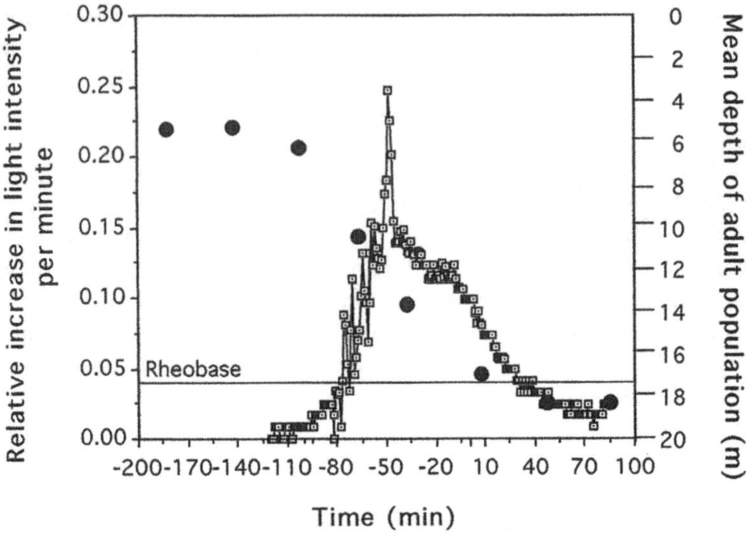

Fig. 11.4 Downward migration of the adults of *D. galeata x hyalina* in Lake Maarsseveen on 11 June 1992 and the relative increase in light intensity on the same morning. The rheobase or absolute threshold for phototaxis is surpassed at –77 min before sunrise and again at 30 min after sunrise. The descent migration over 14–15 m occurs in between these 107 min

after that serene morning at the lake shore, the position of the black dots showed that migration started somewhere between 102 and 97 min before sunrise. Within 35 min, the adults had descended over a distance of 4.26 m. The position of the dots shows that descent migration must have stopped between 8 and 48 min after sunrise. During this period, the relative increase in light intensity had sunk below the rheobase threshold again. DVM of the adult *D. galeata x hyalina* took place within about 2 h when the relative increase in light intensity was above the experimentally determined threshold of phototaxis.

Three similar dawn programmes and two sunset programmes have been performed: sunrise, 27 June 1989; sunrise, 30 May 1990; sunrise, 11 June 1992; sunset, 29 May 1992; sunset, 22 June 1992. We will now look at the data in a more formal way. The vertical displacement velocities (dv) of the adult *D. galeata x hyalina* were calculated as the difference in average depth over one half hour. These values were compared with the logarithm of the absolute light intensity ($\overline{LnI(d)}$) and with relative changes in light intensity ($\bar{S}$), averaged over the same periods (Ringelberg and Flik, 1994). Linear and multiple regressions were calculated with $\bar{S}$ and $\overline{LnI(d)}$ as independent variables and dv as the dependent variable:

$$dv = a + b \times \bar{S} \tag{11.1}$$

$$dv = a + (b_1 \times \bar{S}) + b_2 \times \overline{LnI(d)} \tag{11.2}$$

Table 11.1 Results of linear regression (11.1) and multiple regression (11.2)

Date	*N*	Regression	Significance	Equation
Sunrise,	8	Linear	NS	dv = 2.05 + 1.64 dv = 2.5 m
27 June 1989		Multiple	NS	h^{-1} (SE = 0.35)
Sunrise,	8	Linear	NS	dv = 2.75–1.61 $\bar{S}$
30 May 1990		Multiple	NS	dv = 2.6 m h^{-1} (SE = 0.44)
Sunrise,	8	Linear	$P(b) = 0.001$	dv = 0.87 + 35.51 $\bar{S}$
11 June 1992		Multiple	$P(b_1) = 0.007$	dv = 2.98 + 32.33 $\bar{S}$– 0.37
			$P(b_2)$ NS	$\overline{LnI(d)}$
Sunset,	10	Linear	$P(b) = 0.033$	dv = –0.22 + 31.47 $\bar{S}$
29 May 1990		Multiple	$P(b_1) < 0.0001$	dv = –5.50 + 53.96 $\bar{S}$– 0.46
			$P(b_2) = 0.003$	$\overline{LnI(d)}$
Sunset,	8	Linear	$P(b) < 0.0001$	dv = –3.32 + 78.81 $\bar{S}$
22 June 1992		Multiple	$P(b_1) < 0.0001$	dv = –3.58 + 74.70 $\bar{S}$– 0.05
			$P(b_2) =$ NS	$\overline{LnI(d)}$

For the first two sunrises, none were significant. Displacement velocities were virtually the same in the 2 years, being 2.5 and 2.6 m h^{-1}, respectively (Table 11.1). Surprisingly, this result suggests that light is not a factor of importance. However, if the animals were not swimming actively downwards, but sinking passively in response to the light increase, displacement velocity would be constant. In 1992 at sunrise, the downward displacement velocity was significantly correlated with the relative change in light intensity, indeed, and higher than that of passive sinking in the previous years.

P denotes the probability that (particle) regression coefficients b, b_1 and b_2 are equal to zero; *N* is number of data used in the calculations. Means of dv are given if the linear regression is not significant, otherwise 95% confidence limits of the linear regression coefficients are given (slightly modified from Ringelberg and Flik, 1994).

Ascent migration cannot be passive and must involve active swimming. The sunset of 29 May 1990 revealed a significant linear regression indeed, although the probability of the regression coefficient being different from zero is not large. In 1992 at sunset, the linear regression coefficient was highly significant.

Before and after the maximum, relative changes in light intensity have the same value, although at different absolute light intensities.

Displacement velocity was lower prior to the maximum than at similar relative decreases after the maximum. After introducing absolute light intensity as an independent variable in the multiple regression analysis, the significance of $\bar{S}$ as a factor increased.

There is another explanation for the difference in displacement velocity at both sides of the maximum rate in relative change in light intensity. Before this maximum at dawn and after the maximum at dusk, rates increase, thereafter decrease (see Figs. 6.9 and 6.10). The experiments in Chapter 4 showed that swimming velocities to relative changes in light intensity are larger when the rates of these changes are part of a sequence of increases, thus accelerate. Both explanations imply a high correlation with absolute light intensity. Therefore, it is not surprising that the significance of $\bar{S}$ increased after the introduction of absolute light intensity in the multiple regression but this does not necessarily mean that absolute intensity is an important factor in the physiological response mechanism. As recognised by Wagner-Döbler (1990), the interdependence of the different modalities of the light factor in nature makes it difficult to estimate the significance for behaviour of each modality separately. These two examples for *Chaoborus* and *Daphnia* show that the correlation could have been interpreted wrongly, if experimental knowledge of the photobehaviour mechanism had been absent.

In Fig. 11.5, the linear regressions of dv and $\bar{S}$ are illustrated for 1990 and 1992. In 1992, the downward displacement velocity at sunrise is significant all right and the migration amplitude was much larger than in 1990 (Fig. 9.4), suggesting active

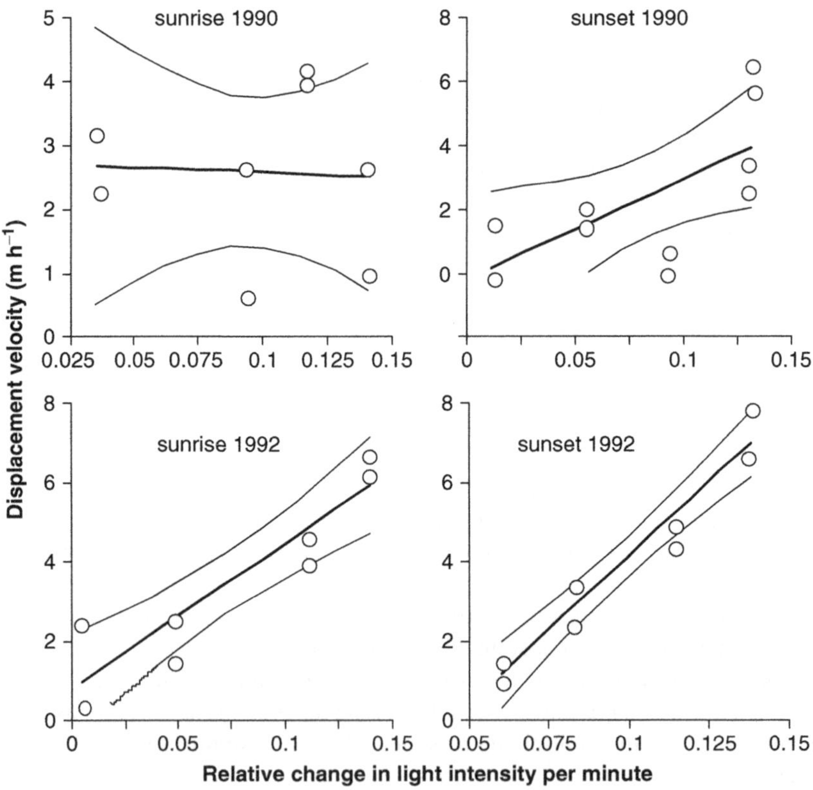

Fig. 11.5 The vertical displacement velocity as a function of the relative increase or decrease in light intensity during morning descent and evening ascent of adult *D. galeata x hyalina* in Lake Maarsseveen. The dates are as in Table 11.1. Regression lines and 95% confidence limits are presented. Regression functions are given in Table 11.1 (Figure slightly modified after Ringelberg and Flik, 1994)

swimming. The explanation might be that the fish kairomone concentration in 1992 was relatively high and thus the acceleration effect of the light stimulus was more important than in 1990. A large part of the downward migration then occurred already before the beginning of the deceleration of $\bar{S}$. The regression parameter for the absolute light intensity was neither significant at sunrise nor at sunset.

A comprehensive overview of the influence of the relative change and absolute light intensity on swimming velocity for sunrise and sunset in 3 years is presented in Fig. 11.6 by plotting the percentage contribution of each factor including the unexplained part – residuals to total variance, expressed as the sum of squares in an ANOVA. In 1989 and 1990, the amplitude of DVM was small and the stimulus $\bar{S}$ did not explain part of the variance of downward displacement. This must not be misunderstood. Without the relative changes in light intensity of dawn and dusk, DVM would not have occurred! As explained above, $\bar{S}$ is not a determinant of dv because displacement is by passive sinking. Nevertheless, the stimulus is responsible for initiating this sinking all right. For sunrise and sunset 1992, the picture changed drastically because the relative change in light intensity explained nearly all the variance.

These field observations of *Daphnia* endorse the importance of the relative increase and decrease in light intensity as a factor initiating and stimulating descent and ascent swimming during DVM. Although the light changes must have been about the same in all years, DVM clearly was not. Additional factors must have been involved and the concentration of fish kairomones represents an obvious candidate.

11.2.3 Sound Scattering Layers (SSL)

In 1964, Clarke and Backus still wondered by what stimulus and mechanism a particular sound scattering layer (SSL) could move as fast, faster or slower than

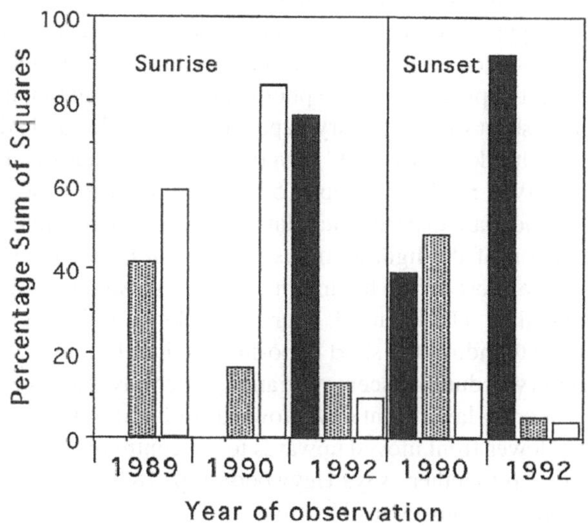

Fig. 11.6 The percentage contribution to the total sum of squares of the relative change in light intensity (*black columns*), the absolute light intensity (*grey columns*) and the percentage of the residuals (*open columns*) in an ANOVA. The vertical displacement velocity during DVM was a dependent variable. See also Table 11.1 for dates of the migrations. Figure slightly modified from Ringelberg and Flik (1994)

the isolumes on the same day, although they had concluded that migrating animals tend to follow optimal light intensities. Not satisfied, however, they added that the behaviour by which an isolume is followed is modified by "the nature of the layer (that is species composition, J.R.), by light adaptation and by the response to the rate of change of the light" (p. 32). A suggestion made in 1956 (Clarke and Backus, 1956) was repeated, viz. that once a change in light intensity had caused upward or downward swimming, migration continued independently of light. The reinterpretation of their data from 1956 by Ringelberg (1964) was rejected because this author "apparently used the change of light intensity at a constant depth, instead of at the changing depths where the migrating animals were found. His calculations thus are not applicable to our results" (p. 10). Also Aarseth and Schram (1999) used this argumentation to reject the role of relative changes in light intensity. However, the relative light change is independent of depth although perception might be modified by upward or downward swimming as was discussed previously (Chapters 3 and 5).

Of course, swimming independent of light can be at velocities lower, equal and higher than that of an isolume. However, that connection does not make sense because the continuous light intensity change is no factor anymore according to Clarke and Backus (1964). The consequence is rather far-reaching. No experimental evidence exists for such "all-or-nothing" response. An erroneous assumption behind the thinking of Clarke and Backus (1956, 1964) is that migrating animals swim continuously. Vertical displacements of scattering layers look continuous, but the participating individual animals probably swim discontinuously. If swimming is interrupted by periods of no swimming as in the photobehavioural model (Chapter 5), full changes in light intensity can be perceived again during these periods. In Chapter 3 was shown that continuous changes in light intensity evoke "discrete response packages" resulting in stepwise vertical displacements. Swimming of animals in an SSL must be faster than *Daphnia* is able to do in order to make an SSL move faster than an isolume but this is no problem because euphausiids, for example, easily do that. Some modification of reactive swimming must be added: for instance an initial burst once the threshold is surpassed. After a short, variable time, the processes of the photobehaviour model that terminate swimming take over. This short time may vary dependent on species and stimulus strength as swimming velocity does. All kinds of vertical displacements can be realised in this manner as the DVM model in Chapter 5 has made plausible (see Fig. 5.15).

There are experimental data in favour of interrupted swimming and renewed perception of the light stimulus. In *Daphnia*, the swimming response not only stops after a short period but might even be followed by a short swimming in the opposite direction. The feedback change in relative light intensity must then be sufficiently large to induce the short opposite reaction. Possibly, this was what Heywood (1996) observed during ascent migration, when fast moving animals in the upper part of a scattering layer went down towards the centre. During descent migration animals in the lower front moved upwards to the centre of the layer. As if "the zooplankton like to keep together" says Heywood on p. 183. The photobehaviour model provides a more mechanistic explanation.

Migrations can be compared with an ascending or descending absolute light intensity but the two have no direct relation. For decennia, authors assumed a correlation, even a causal relation. That erstwhile paradigmatic explanation of DVM seems difficult to forget and can be found in some modified form in the literature occasionally. Some investigators think that even light intensity changes are not proximate stimuli governing migrations of SSL. For example, Gibbons et al. (1991) concluded that "euphausiids seem to regulate vertical speed as a strategy to remain for longer periods in 'preferred' food regimes". Nevertheless, "the influence of small-scale changes in light intensity and cloud cover cannot be ignored, and the present conclusions should be viewed with caution" (p. 483). Indeed, no evidence for a correlation between food concentration and migration speed was presented nor were small-scale changes in light intensity measured. Why not refer to the literature and mention the role of the large-scale changes in light intensity of dawn and dusk?

11.3 Special Circumstances

Exceptional environmental conditions may be used as natural experiments if one or more factors are completely different from those encountered usually, and thus might provide clues regarding the causation of phenomena. Light conditions around dawn and dusk are aberrant in polar regions because the sun may not rise or sink below the horizon. During a solar eclipse an additional dawn and dusk is added. Both situations may shed light on the diel pattern of initiation of photobehaviour and DVM.

11.3.1 Polar Regions

The Arctic and Antarctic summers are characterised by the specific daily changes in light intensity. During midsummer, the sun remains above the horizon all the time and no large and rapid increases and decreases during dawn and dusk occur, as is the case in temperate and tropical zones (Fig. 11.8). Illumination still varies in a diel cycle, but the rate of change could be too small to initiate DVM. Investigators differ in opinion. Some found marine zooplankton not to migrate under the midnight sun (Bogorov, 1946; Longhurst et al., 1984; Digby, 1961; Sameoto, 1984). Also freshwater copepods and cladocerans in Toolik Lake (Alaska, 74° N) did not migrate or only feebly migrated during summer (Buchanan and Haney, 1980; Johnson et al., 2007). Fischer and Visbeck (1993) found no migration of a sound scattering layer under the ice during the period May–July. DVM is absent too during the continuous darkness of the polar night (Kosobokova, 1978). As the final conclusion of their literature study, Conover and Huntley (1991) wrote as follows: "we conclude that some copepods do migrate in the polar seas near mid-summers days, but we also conclude that few, if any, studies of zooplankton distribution to date have been designed well enough to make this statement unequivocal supportable" (p. 28). Based a.o. on a literature study, Blachowiak-Samolyk et al. (2006)

concluded: "despite extensive research, the question of whether Arctic zooplankton migrate diurnally during periods of constant illumination has not been resolved" (p.102). Based on observations in the Barents Sea (76–77° 30′ N), these authors concluded that Arctic zooplankton do not perform DVM during periods of midnight sun. Although the light climate makes polar regions special, and is considered by all authors an important factor, few light-intensity measurements have accompanied zooplankton sampling programme. Seasonal differences in the light regime do exist, however, and in spring and autumn, DVM has been observed (Buchanan and Haney, 1980; Conover et al., 1988) (Fig. 11.7).

I shall approach the problem by applying our knowledge of photobehaviour and the experimental evidence that relative changes in light intensity are initiating stimuli that have to surpass a threshold value in order to evoke vertical swimming. Although photobehaviour in Arctic copepods has not been studied, it is plausible to suppose that also these animals react to changes in light intensity. However, relative changes in light intensity in the Arctic summer are far smaller compared to temperate regions (absolute intensities is not necessarily lower). Threshold values, as determined for animals from lower latitudes, are not reached. Therefore, the absence of DVM in polar regions would not be surprising and still be in accordance with the photobehaviour model developed for animals from the temperate regions. However, this explanation is at least incomplete. Migration may be absent in the pelagic area of the open sea but still be present under the sea ice (Conover et al., 1988). This remarkable phenomenon has also been described in a fine paper by Fortier et al. (2001). They mention that especially the medium- to large-calanoid copepods, such as *Calanus hyperboreus*, *C. glacialis, Metridia longa* and *Pseudocalanus acuspes,* performed extensive migrations under ice. Ascents coincided with periods of maximum rates of the relative decrease in light intensity but descent migration started before light increase again, which is contrarily to the idea that descent migration is evoked by relative increases in light intensity. In the ice-covered Resolute Passage (Canadian Arctic), Hattori and Saito (1997) found *Pseudocalanus* to ascend between 20 and 23 h and to descend into deeper water around midnight. A third surprise was that even in spring and autumn, the light changes are about a factor 10 smaller than the thresholds found for *Daphnia* and other animals.

Can these phenomena be interpreted in accordance with the initiation of DVM, as discussed thus far? Fortier et al. (2001) suggested that the absence of migration in the open water "is not linked to a low $\Delta I/I$ (which was sufficient to trigger migrations under the ice), but rather reflects a lack of refuge from visual predators in the continuously illuminated surface layer" (p. 1276). This absence of a pelagic refuge in ice-free waters is due to the relatively high light intensity, also during the "night". Consequently, the death toll would be too high if the copepods ascended into the surface layers. Under ice, absolute intensity is about 100 times less and the potential for visual predation is therefore limited. This ultimate explanation might be true, but offers no mechanistic, proximate explanation. The relative change in light intensity is the same under ice as in open water, but absolute intensity is not. An explanation for the difference in migration behaviour might be found in this difference. The threshold stimulus for phototaxis depends on the adaptation light intensity

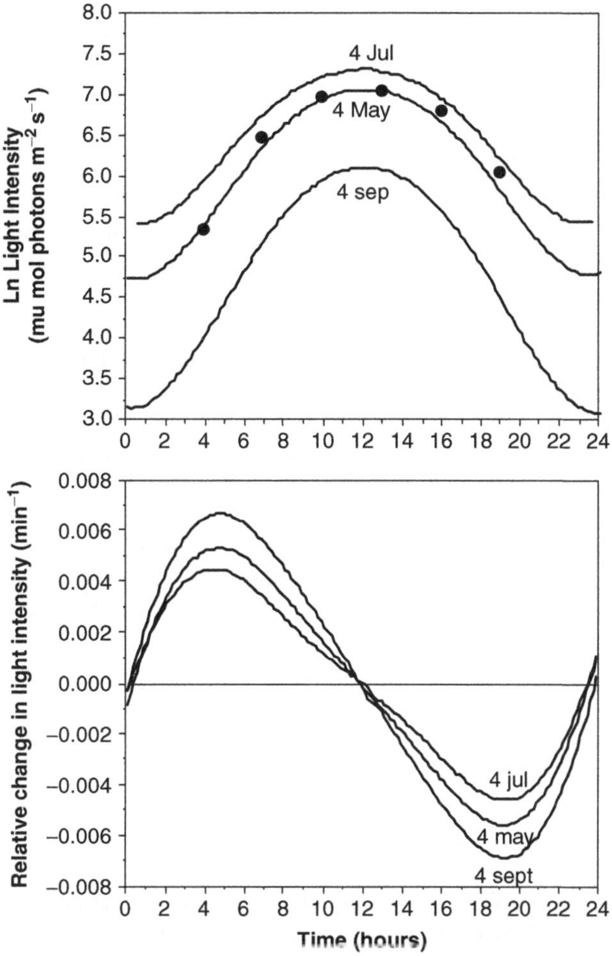

Fig. 11.7 Diel changes in absolute light intensity (*upper figure*) and relative changes in light intensity (*lower figure*) on 4 May, 4 July and 4 September in Barrow Strate (Canada, W 96° 00′, N 74° 30′). Calculations[4] were based on light intensity measurements of Fortier et al. (2001) and altitude of sun data from US Naval Observatory, Washington, DC. Time was shifted such that sun transit was at 12.00 h. The inserted dots are PAR intensities presented by Blachowiak-Samolyk et al. (2006) for 3 May 1999, measured at Bjornoya (74° N and 20° E). At the two dates and both places the sky was clear

(Chapter 3). Although this dependency was determined for instantaneous light decreases only, the data suggest that the effect can be large. In the marine calanoid copepod *Acartia tonsa*, a 100-fold increase above the adaptation light intensity at which sensitivity is highest would result in no reaction at all (see Fig. 3.4). A tentative hypothesis, therefore, might be that the absence of migration in the open water is due to a higher stimulus threshold as compared to the one under ice. Determining the mutual relation between threshold stimuli and adaptation intensity in polar copepods might test this hypothesis and could perhaps offer an explanation for the difference in what seems to be a tenfold higher sensitivity for light changes in these copepods.[5] If true, this would point to selection for a high sensitivity of the light organ. *P. acuspes*, a species also found at more temperate latitudes, seems to be a good experimental animal to study this sensitivity.

When the sun sinks below the horizon again, relative changes during dawn and dusk increase (Fig. 11.8), but maximum values are still small compared to those of temperate regions. The highest absolute intensity becomes considerably smaller than in mid-summer. A hypothetical high sensitivity of the copepods and a lower threshold associated with the lower light-adaptation intensity might be sufficient for some species to react and resume DVM, also in the open waters.

Our third problem is the early descent migration under ice, long before light intensity increased. Fortier et al. (2001) suggested that the cue lies in the presence of visual predators feeding actively under the midnight sun. Hattori and Saito (1997) mentioned that the amphipod *Thimisto libellula*, which is generally considered a carnivore, ascended to the ice–water interface around 2230 hours in high concentrations. However, they found no evidence for predation because the guts contained no *Pseudocalanus* but diatoms. Whether the copepods are able to perceive the presence of these potential predators directly is unknown. Anyhow, such a reaction to the "knowledge of predator presence" should be inhibited until sufficiently food has been filtered from the water. According to Conover et al. (1988), filling of the gut in *P. acuspes* takes about 5–10 min. Just below the ice–water interface algae are abundantly present (Fortier et al., 2001) and satiation might be rapid. Also Hattori and Saito (1997) initially thought that the midnight descent of *Pseudocalanus* corresponded to a satisfied situation. However, gut contents in terms of pigment were higher during the day than at night. Their explanation was that temperature at the ice–water interface drops during the night sufficiently to make the ice-algae inaccessible to *Pseudocalanus*, which in turn could be the reason for a midnight descent. Obviously, this explanation requires a lot of assumptions and is not really satisfactory.

Early descents are not restricted to polar regions. In his extensive review, Pearre (2003) made it plausible that satiation is at the base of these early descent migrations of individuals. Even when light intensity is still decreasing, downward swimming seems to occur occasionally. This would be in disagreement with the model of phototactic swimming. Must we therefore reject this model? No alternative mechanism is available and we only need a modification or expansion. If a filled gut, or some other indicator of satiation (such as for instance the hormone leptine, produced by fat cells in our body, which is responsible for a feeling of satiation), motivates downward swimming, photobehaviour might still be responsible. As discussed in Chapter 3, phototaxis consists of a kinetic and an orientation component. In reverse DVM, the only difference from normal DVM is in the orientation of the animal. This might also apply to predawn downward swimming. Satiation could provide for the different orientation, resulting in a negative phototactic response. Experiments with fish kairomones (Chapter 7) have shown how orientation of the body axis of pinned

Daphnia changes nearly 180° when kairomones were added. Thus orientation is flexible and subject to other cues than just light condition. True enough, hypotheses can be "salvaged" by superimposing new ones, which may lead to constructs like the cycles upon cycles of Ptolemy's explanation of planetary movements. But in the present case I think it is justified. In fact, I do not introduce a new hypothesis. The orientation component of photobehaviour is only made plastic and responding to changes in the physiological state of the animal. Satiation is just considered the motivating factor, as the fish kairomone turned out to be in the orientation experiment with *Daphnia.* Importantly, the light field still serves as the anchor directing the animal where to go, away from the light in this case.

Although attention has been paid here predominately to the few positive examples of migrating copepods in high latitudes, most papers stress the absence of migration. This was to be expected from the photobehaviour model because thresholds of relative changes at "dusk" and "dawn" are too small to cause a reaction. That is also why the positive examples are especially interesting for our discussion. Of course, more field observations and especially experimental work is required to test the hypothesis but any other explanation should first address the question what physiological–behavioural mechanism is responsible for guiding the animals up and down.

11.3.2 Solar Eclipses

The rapid and large changes in light intensity during solar eclipses represent natural experiments in the study of photobehaviour. We measured the total eclipse shown in Fig. 11.8 on 11 July 1991 in Managua (Nicaragua). After setting up the equipment we were afraid that the sun would be obscured by Cumuli clouds – as seen initially in the spikes – but some 20 min before the precious moment of totality the clouds disappeared and the sky became clear as glass. With the sun completely covered by the shadow of the moon, darkness was deep enough to see stars faintly in the sky. Soon, light intensity started to increase again and life continued as usual.

On several occasions, especially in the marine environment, vertical distribution of zooplankton during solar eclipses have been studied. What can be expected during these strange short events evoking a rapid sequence of dusk and dawn? No reaction at all, according to the conviction that predation and search for food are the all important forces initiating and driving migration?

Zooplankton is at the day depth, out of reach of predators, so why would they suddenly start to rise? Or is the zooplankton misled by these aberrant light stimuli and respond with up- and downward swimming as in DVM? What can be expected from the photobehaviour model?

In Table 11.2, information on duration and relative changes in light intensity is presented. Firstly, the maximum rates of relative change are much faster than those usually around sunrise or sunset in the temperate zone (compare with Figs. 6.9 and 6.10). Secondly, the periods during which the relative changes exceed the threshold

Table 11.2 A comparison of some characteristic properties of a normal day dawn (30 May 1990) and dusk (29 May 1990) change in light intensity and these changes of a solar eclipse (11 July 1991). Measurements of normal day at Lake Maarsseveen, the eclipse in Managua

	Normal day		Solar eclipse	
	Dawn	Dusk	Decrease	Increase
Above threshold period	130 min	120 min	30 min	30 min
Maximum relative change	0.16 min^{-1}	0.14 min^{-1}	3 min^{-1}	2 min^{-1}

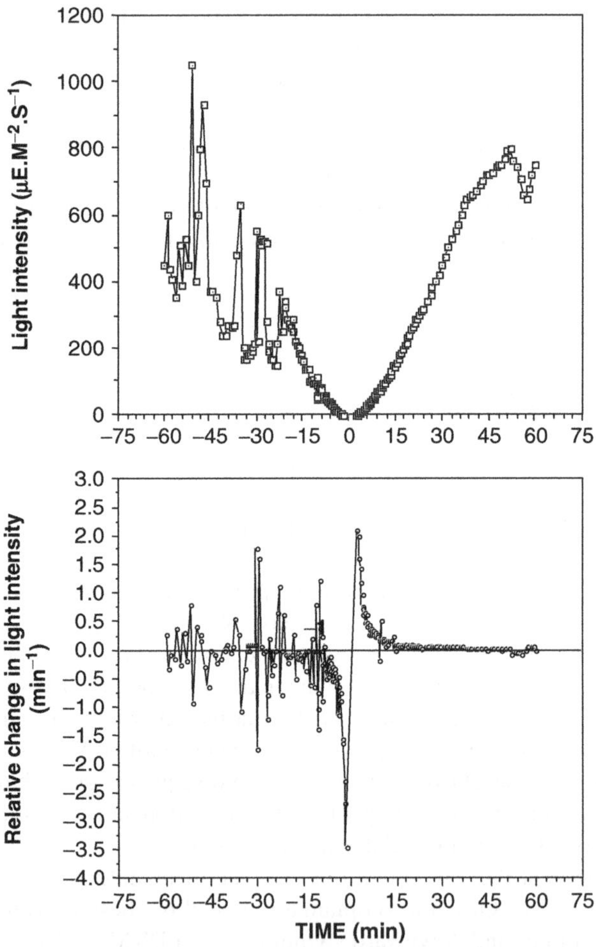

Fig. 11.8 Light intensity changes during the solar eclipse, observed in Managua (Nicaragua) on 11 July 1991. Clouds, passing for the sun prevented a smooth decrease before totality. These changes can be large and relative changes are larger than the threshold for photobehaviour. However, duration might be too short to cause a photoresponse (see the stimulus strength–duration curve of Fig. 3.6)

of 0.04 min^{-1} (see Chapter 3) are much shorter during each phase. Although displacement velocity depends on stimulus strength, there is of course an upper limit to swimming velocity and higher rates of relative change will have no further effect. Therefore, we must expect that during the short period of about half an hour, the vertical displacements can only be relatively small. In addition, upward swimming will soon be followed by downward swimming, when light intensity increases again. Thus, sampling with traditional net hauls would have to be carried out at a high frequency to detect any vertical displacement. Although the light changes are spectacular during an eclipse, their short duration makes that vertical movements will not be easy to detect, unless we can measure individual animals in situ.

Nevertheless, several examples of migration behaviour have been observed. During the total eclipse of 20 June 1963, Backus et al. (1965) observed the ascent of three scattering layers in the North Atlantic Ocean, 322 km south of Woods Hole. Before and after the eclipse of 7 March 1970, several sound scattering layers were observed to perform normal dawn and dusk migrations by Franceschini et al. (1970) in the Gulf of Mexico. Caruthers et al. (1972) suspected that these scattering layers had responded but that vertical displacements had been small and undetectable, partly because of a heavily overcast sky. For the same place and eclipse, Bright et al. (1972) reported the results of net catches taken in 10-min tows at 25 m and the surface. The copepods *Nannocalanus minor*, *Scolecitrix danae* and *Undinula vulgaris* migrated all to the surface and numbers even exceeded those present there at night. The euphausiid *Stilocheiron carinatum,* with a daytime distribution between 90 and 270 m, showed a strong response with again high numbers near the surface. On the other hand, the deep-living species *Euphausia tenera*, *E. brevis*, *E. mutica* and *E. americana* did not appear in the surface catches during the eclipse, although they do migrate to the surface water layers at night. The authors suggest that time was too short to ascend over such a large distance.

Deep SSL have also been studied during the total eclipse of 10 July 1972 in the North Atlantic. Tont and Wick (1973) observed a definite upward reaction, but "they did not follow isolumes".

Also freshwater zooplankton in Lake Geneva reacted to the eclipse of 11 August 1999 (Giroud and Balvay, 1999). *Cyclops praealpinus* and *D. hyalina* swam upwards during the decrease in light intensity and moved down again after the moment of totality. *Eudiaptomus gracilis* showed a less well-expressed reaction.

It is evident from these examples that reactions to changes in light intensity caused by solar eclipses do occur, and apparently do so independently of whether predators or food are present. The photobehaviour mechanism provides the only explanation of the events observed during solar eclipses.

11.4 Nymphs on Drift

The validity of the relative change in light intensity as a stimulus initiating the changing behaviour at dawn and dusk would get even more general weight if non-planktonic species also react.

The study of Haney and co-workers (Haney et al., 1983; Schloss and Haney, 2002; Schloss and Haney, 2006) on mayfly nymphs (Heptageniidae) provides an excellent example. In daytime, these insect larvae are hidden below stones in small streams, but at dusk they crawl to the upper side in order to graze algae. In doing so, they can be washed from the stones, although they may also actively enter the stream and drift over considerable distances sometimes. Maximum drifts occur in the evening and Haney et al. (1983) determined the beginning of the drift in two woodland rivers while at the same time measuring light intensity. The drifts started at about the time when the relative decrease in light intensity had surpassed the threshold value found for *D. magna.* However, absolute intensity had an effect as well and neither the relative stimulus nor the absolute light intensity alone seemed to regulate the start. The authors proposed a model of light control consisting of a photokinetic activity, regulated by the relative decrease, and a phototactic component, controlled by the absolute light intensity.

These hypotheses were experimentally tested by Schloss and Haney (2002) in laboratory streams. Nymphs could be observed with video recorders and light intensity and light intensity changes could be controlled with sophisticated instruments. The timing of increased locomotor activity underneath the stones and changes in the vertical location on the substrate was strongly correlated with the strength of the relative decrease in light intensity. Experimental decreases in light intensity were applied at two adaptation intensities. The results suggest that the light changes appeared to be the stronger controlling factor, although also absolute intensity was of influence. In fact, the response characteristics are highly comparable with those of *Daphnia.* For example, the configuration of the stimulus strength–duration curve, presented in Fig. 11.9, looks much like the one for *Daphnia* in Fig. 3.6. The minimum stimulus evoking a reaction in mayfly nymphs is small compared to *D. magna.* Obviously, these nymphs are highly sensitive to light changes (Schloss and Haney, 2002). Therefore, the problem arises how these animals deal with the effects of fluctuations and low light intensities outside the periods of twilight. Clouds will pass before the sun, and shadows of branches of trees in forest streams vary depending on the changing position of the sun. Compared to zooplankton in open water, ambient light intensity varies considerably, both spatially and temporally. Because these fluctuations do not appear to lead to responses in natural streams, the animals must somehow recognise the difference between transient light changes and sustained twilight changes. They must move to exposed locations to feed while at the same time avoid multiple predators. Schloss and Haney (2006) dealt with this problem. Although the shape of the stimulus strength–duration curve is highly comparable in *Daphnia* and *Stenonema modestum*, the response parameters are different. For example, the latent period, reflected in the rheobase value, is much more extended in the mayfly nymphs (R = 0.0001–0.0006 min^{-1}) than in *D. magna* (R = 0. 055 min^{-1}). Consequently, also the disintegration constant of the accumulated excitatory state is different, viz. $3.3 \times 10^{-5}\ s^{-1}$ and. $2–7 \times 10^{-3}\ s^{-1}$, respectively (Schloss and Haney, 2006). Therefore, daphnids react to shorter continuous changes in light intensity than mayfly nymphs. Nymphs are not inclined to leave hiding places underneath stones easily!

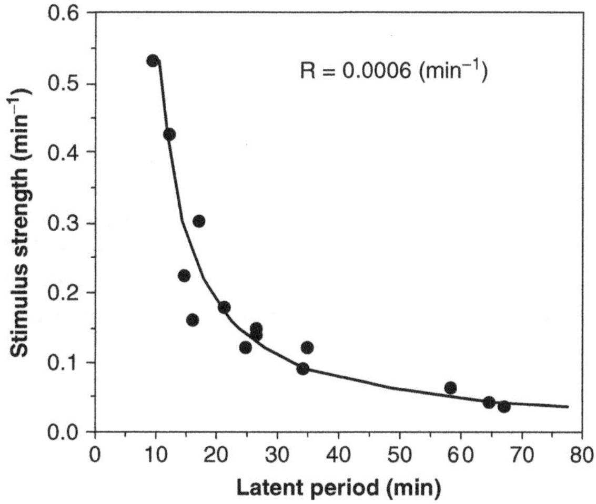

Fig. 11.9 The stimulus strength–duration curve for enhanced locomotor activity of the mayfly nymph *Stenonema modestum*. Data points are averages. The curve is the best fit of the function Ln S/(S-R)= $-1.1278 \times 10^{-3} + 2.1706 \times 10^{-4}$ latent period. See also Section 3.5.2, especially Fig. 3.6 for comparison. The figure is a modification of Fig. 4 in Schloss and Haney (2002)

Haney et al. (1983) suspected that absolute intensity also had an effect on the timing of drift in the woodland streams. Light intensity as a proximate cue, initiating locomotor activity or crawling on the substrate, was not supported by the experiments of Schloss and Haney (2002). Nevertheless, they found that nymphs adapted to reduced light intensity responded earlier than those adapted to bright light.

This indicates a lower response threshold at lower light intensity. One is reminded of the typical dual relation between absolute intensity and thresholds of relative changes, found for *D. magna* and *A. tonsa*, as presented in Fig. 3.4. Although insufficient data points are available to really construct a comparable figure for

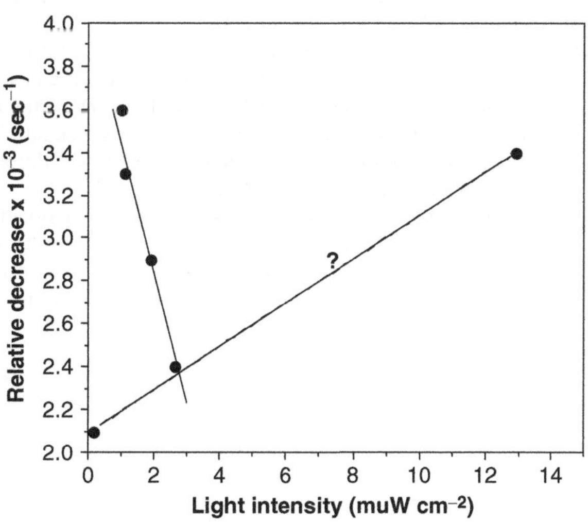

Fig. 11.10 Light intensity (I, mW cm^{-1}) and relative decrease in light intensity (s^{-1}) at the estimated time of the initiation of evening drift of mayfly nymphs in two woodland streams. Composed from data in Table 1 in Haney et al. (1983)

S. modestum, the data given by Haney et al. (1983) were sufficiently tantalising to present a preliminary Fig. 11.10.

The study of Haney and co-workers is a scholarly example of how to do ecological research: field work, measurements of relevant environmental factors, followed up by sophisticated laboratory experiments in order to test hypotheses derived from the field observations.

The study has shown that a species taxonomically very different from *Daphnia* but more importantly a species with very different behaviour and ecology show comparable responses to changes in light intensity of dawn and dusk. This not suprising because dawn and dusk are universal periods of transition in the mode of living of all animals.

Notes

1. The ratio between birth rate and mortality rate is a criterion for selection of migration behaviour. Both are calculated values of the population level but composed from the chance of having viable progeny and the chance of death at the individual level. Both are a function of food availability and a.o. predation. These are often called ultimate factors but are not different from the proximate factors that are in operation in present time. If thus a continuum in time exists and selection in the past is not different from present day selection, no distinction is necessary between proximate and ultimate factors because ultimate factors lose their meaning if we not only look into evolutionary past but also go back in time. However, proximate and ultimate factors need not be identical. The sometime used "driving forces" seem to be superfluous.
2. Although this conclusion is certainly acceptable, there seems to be a conflict between the Figs. 11.1 and 11.2. One expects a bend in the function line of Fig. 11.2 because in the echograms a change in displacement velocity could be recognised, upon which Fig. 11.1 was based. A slow phase of upward migration, changing into a fast phase, is encountered often and will be discussed.
3. It might be found strange that Swift and Forward (1988) refer to Haney et al. (1990) and thus to a paper that had not appeared yet. However, they refer to these authors as "in press" and obviously had seen it as a manuscript.
4. Light intensity (PAR) for 4 June 1995 as presented by (Fortier et al., 2001) for 74° 30" N and 97° W was correlated with sun altitudes for the same date and place as presented by the Astronomical Applications Department of the US Naval Observatory (http//aa.usno.navy.mil/cgi-bin/ aa_altazw.pl). A quadratic function was fitted to the data. This function was used to calculate Ln light intensity for different dates. From the light data relative changes per minute were calculated.
5. Baldwin (1993) studied stream drift in mayfly and stonefly nymphs in a stream in Subartic Alaska (62° 17′ N). He found in June that drift started at relative light intensity decreases of 0.034–0.039 min^{-1} or at least a factor 3 lower than those found at lower latitudes (see Section 11.3.1). He suggested that with increasing latitude sensitivity increases.

Chapter 12
From the Individual to the Population and Beyond

Contents

12.1 Introduction

Zooplankton are important links in pelagic food webs of oceans and lakes. This holds for herbivores like most cladocerans, or omnivores like cyclopoid and calanoid copepods, but also for predators as many species of *Euphausia* are, for example. Ontogenetic changes between herbivores, omnivores and carnivores are common and intertwined, which makes that the food web cannot be linearly arranged. Diel vertical migration has considerable consequences for the dynamics of pelagic populations and the composition of the relational web. Migrations change intra- and interspecific competition, and thus community composition. These are all species-specific effects and the previously discussed uniformity of responses to light stimuli, which made generalisations about the causation of DVM possible, has disappeared.

The ultimate consequence of DVM for population dynamics boils down to costs–benefits balances. Ecological costs of migration are large. For example, increased egg development times due to lower temperatures at the day depth or decreased fecundity as a result of poor food quality deep down in the water column. Both determine birth rates, an essential parameter of population growth. Mortality is another one. In the presence of visually hunting predators, leaving well-lit surface water is often more profitable than staying near the surface. The ratio of birth and mortality

J. Ringelberg, *Diel Vertical Migration of Zooplankton in Lakes and Oceans*,
DOI 10.1007/978-90-481-3093-1_12,

rate determines whether DVM is worthwhile or not. Migration as a strategy for prevention of predation is phenotypically induced and not constitutive[1] or genotypically fixed. Migration must be plastic with a possibility to "choose" (Chapter 4) from a set of alternatives. Populations consist of several genotypes with different "reaction norms" of migration behaviour and life histories which result in a mosaic of adaptive traits and leads to variable migration amplitudes. Consequently, not all genotypes in a lake behave "optimally" at a certain moment but rapid temporal changes of the lake environment make coexistence nevertheless possible. At the ecological level, the species concept is pushed into the background and for *Daphnia*, for instance, has to be replaced by a *clone collective*.

The relative fitness of a *Daphnia* genotype is not exclusively determined by the kind of migration strategy but also by life-history traits. For example, in the presence of fish kairomones, daphnids mature earlier at a smaller size and females have larger clutches of small eggs (Machacek, 1993; Reede and Ringelberg, 1995; Stibor, 1992). This is also considered a defence against visually hunting fish, although from small eggs small-sized neonates are born which are less resistant to starvation (Cowgill et al., 1984; Tessier and Consolatti, 1989). We deal again with a trade-off between two profitable traits having antagonistic effects of which the adaptive outcome depends on environmental circumstances.

The variety in life-history responses, or the reaction norm of the phenotype, might be different for different genotypes and overlap is often considerable. In the ever-shifting configuration of environmental circumstances combinations of induced life history and migration properties realise the highest fitness for a particular genotype at a certain time. Selecting forces slide continuously over the community landscape and parthenogenesis in *Daphnia* makes short-term shifts in the species and clonal community possible. The system is highly dynamic.

Genotypes with different migration amplitudes create different "Merkwelts" and the changed relative fitnesses result in altered dominances in the clone collective as became apparent after allozyme analysis in Lake Maarsseveen (Section 12.6). Interaction between life history and DVM was also described for migrating *D. hyalina* and non-migrating *D. galeata* in Lake Constance. Culture experiments revealed that both species do equally well under favourable food conditions, whereas *D. hyalina* is superior under the poor conditions as present in the hypolimnion (Stich and Lampert, 1981, 1984). In the self-created circumstances of seasonal migration, *D. hyalina* even has a higher abundance than *D. galeata.*

In the absence of DVM competition would be severe and one can speculate whether DVM is not only responsible for a decreased predation but also for less competition. DVM is intricate and influences the total pelagic community.

12.2 Changes in Population Size During Periods of Migration

A sufficient large and coherent body of information about population development of zooplankton is rare, and I have to turn again to *Daphnia* in Lake Maarsseveen.

In Fig. 12.1, the changes in number of juveniles, adult females without and with eggs are presented for the first years after 0^+ perch (*Perca fluviatilis*) and DVM

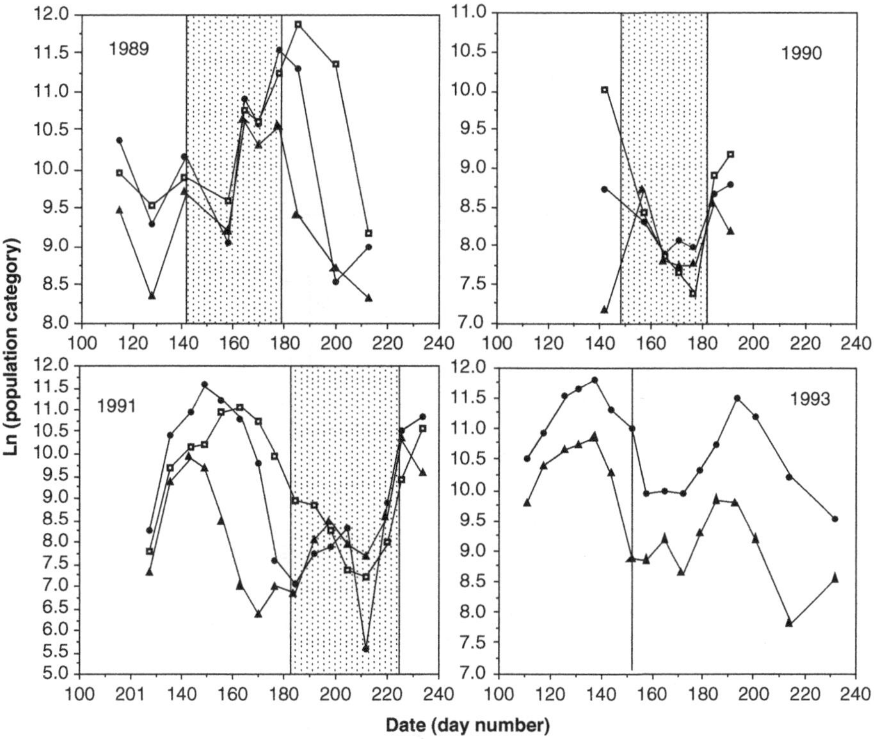

Fig. 12.1 Number of individuals of different population categories in the water column of Lake Maarsseveen under 1 m^2 in different years. *Black dots* indicate juveniles, *open squares* are adults without eggs and *black triangles* are adults carrying eggs in the brood chamber. The *grey area* indicates the period of diel vertical migration. In 1993, the end of the migration period was unclear. Day number 100 is 10 April, 140 is 20 May, 180 is 29 June and day number 240 is 28 August

were encountered in the lake. During the first 2 years, the amplitude of migration was small but increased in 1991 and 1992 to decrease again in the following years (Fig. 12.2). We had to wait until 1998 before a large amplitude was encountered again.

Population size varied from year to year but during spring and summer a pattern of change can be recognised. Before the period of DVM starts[2] (around day 151, 31 May except 1991, see Chapter 9) the three population categories decreased and this tendency continued during the two succeeding weeks. Thereafter, population size started to increase until migration was over and size decreased again.

During the first 3 weeks of the seasonal migration period, the number of adult *Daphnia* in the upper 5 m decreased rapidly (Table 12.1). This cannot be attributed to a downward migration because in the lower strata no increase occurred (unless the daphnids went down below 17.5 m, which is improbable in 1990 considering the mean population depth in Fig. 12.2). Predation by 0^+ perch seems to have been responsible and thus not all *Daphnia* had migrated. A comparable disappearance from the epilimnion was observed in nearly all years which raises the question

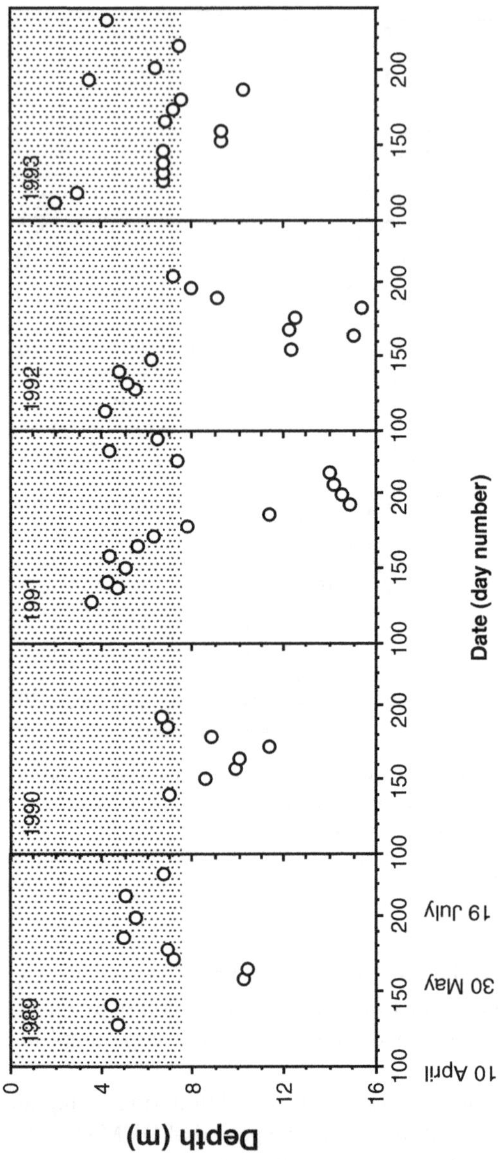

Fig. 12.2 Mean depth of adult *Daphnia* at noon in Lake Maarsseveen around and during the seasonal period of DVM. *Grey* indicates the epilimnion (when stratification is present), *white* the hypolimnion

"why not?" and also how does the photobehavioural mechanism differ in these individuals. I return to this question again later on in this chapter.

During the migration period, population size became very small. This is not readily visible in Fig. 12.1 because of the logarithmic scale. The strong increase in population size during the final weeks of a migration period seems strange and the most obvious conclusion is that predation pressure became gradually less, which is true (Fig. 12.18). However, more circumstances changes during migration periods. For example, phytoplankton growth increased at the end of the period (Fig. 12.22).

It must be stressed again that more often than not a single factor cannot be held responsible for an ecological phenomenon to occur. The realisation of a "set of necessary ecological conditions" is necessary.

Table 12.1 A comparison of the numbers of adult *D. galeata* and *D. galeata x hyalina* in column parts under 1 m^2 in Lake Maarsseveen in 1990. The percentage difference in numbers with one sample date earlier is given between brackets. Note that a percentage change can provide a false impression of an absolute number change which is what is important in population dynamics

	Number of adult *Daphnia*			
Column part	Day 142	Day 149	Day 157	Day 165
0–5 m	23,056	8726 (–62%)	98 (–99%)	195 (+99%)
5–17.5 m	21,896	20,585 (–6%)	17,199 (–16%)	12,623 (–27%)

12.2.1 Two Examples of Population Dynamics

To facilitate a description of population parameters that determine changes in population size, mutual relations are presented in Fig. 12.3. In Fig. 12.4, information pertaining to the changes in population size of *Daphnia* in Lake Maarsseveen for the year 1989 is presented in the format of this figure.

During the pre-migration period, birth rates were low, became maximum during the period of DVM and thereafter suddenly decreased. The high birth rates during migration is surprising because egg development times, figuring in the denominator of the birth rate equation, increased due to low temperatures at the day depth, but that negative effect was obviously counteracted by an increased egg ratio. The pre-migration decrease in fecundity also came to a halt after *Daphnia* started to migrate, but fecundity plummeted again as soon as DVM was over resulting in a large decrease in population size (Fig. 12.1). The number of adults with eggs was low before and after the migration period. Clutch size decreased continuously during the whole observation period except during migration. It seems that in 1989 the short period of migration interrupted an ongoing decline of the *Daphnia* population. Although predation does not figure in this short narrative, it was present because guts of 0^+ perch were filled with daphnids (Ringelberg et al., 1991b); therefore, predation must have played a role in the dynamics of the population also.

The influence of fish predation on population dynamics in early summer is subject to controversy. For the shallow eutrophic lake Tjeukemeer, Boersma et al. (1995) were of the opinion that predation by 0^+ fish could not be responsible for the early summer decline in zooplankton. Only in the second half of the summer

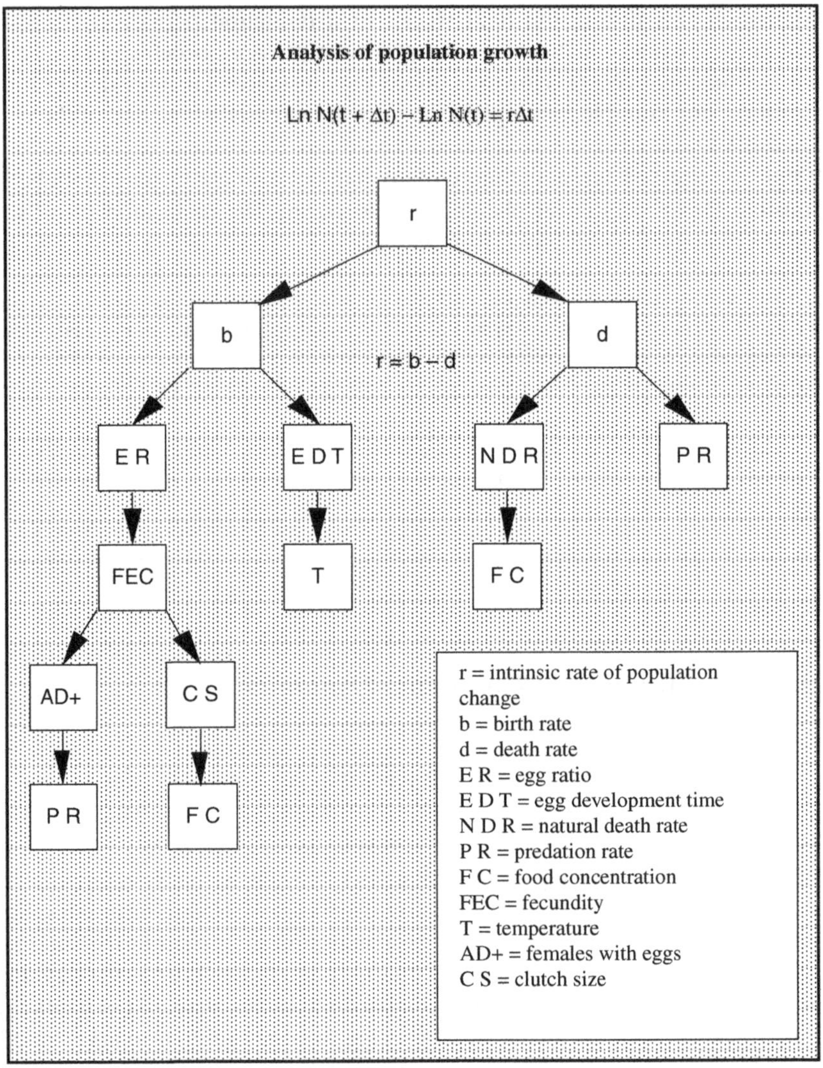

Fig. 12.3 The route of population analysis

fish predation had a large impact on *Daphnia,* but during the early summer months, daphnids died of starvation. Also Mehner et al. (1998a) found no evidence for a strong predation by 0^+ perch and zander (*Stizostedion lucioperca*) in Bautzen reservoir during early summer. The final conclusion of Mehner and Thiel (1999) at the end of a review paper was that, generally, predation by 0^+ fish could not be held responsible for the summer decline of large cladocerans in lakes. Only during late summer and autumn, zooplankton density was thought to be controlled by juvenile fish predation.

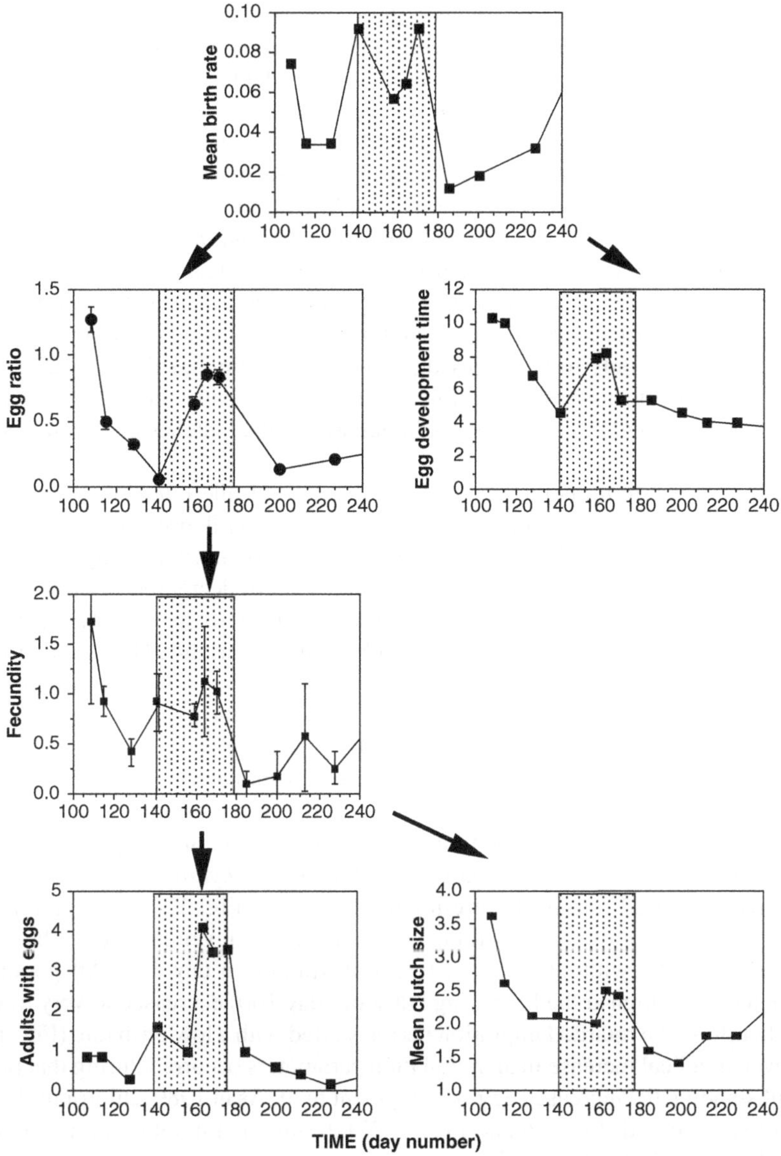

Fig. 12.4 Birth rates for *Daphnia galeata x hyaline* and the component population parameters necessary to calculatate the birth rate for the year 1989 in Lake Maarsseveen with emphasis on the period of diel vertical migration (*grey area*). Egg development time in days; fecundity as number of eggs per adult female; adults with eggs as number under a column of 1 m^2. *Arrows* indicate the route of analysis; compare with scheme in Fig. 12.3. *Vertical bars* indicate 95% confidence limits

If fish predation is of minor influence, the surprising increase of the *Daphnia* population during the migration period needs an explanation. During the pre-migration period chlorophyll *a* concentrations were low at all depths and that remained so in the epilimnion up to about day 200. However, around day 141 chlorophyll *a* concentration had strongly increased in the metalimnion between 7 and 11 m. In general, chlorophyll *a* concentration increases in the course of summer (Sommer et al., 1995), also in Lake Maarsseveen (Fig. 12.22). We must reject the idea that *Daphnia* had found the high metalimnetic algal concentration and, *therefore*, had started to migrate downwards because they returned to the surface again at night. Why would they have done that? To benefit from the high temperature? Anthropomorphic reasoning like that must be strongly rejected. DVM was facilitated by kairomones of 0^+ perch, leading not only to a decreased predation mortality but DVM made grazing on the metalimnetic algae also possible. That certainly contributed to an increase in the number of gravid females and in a larger size of the clutches and, in the end, to a larger population size.

Soon after the termination of the migration period, the number of adults with eggs decreased steeply and the population of *Daphnia* collapsed (Fig. 12.1). Since better visible gravid females are at risk, a high predation mortality might have been the cause. Had DVM stopped too early? The 0^+ perch, decreased in numbers, had returned to the littoral zone and were feeding on different species. The decreased kairomone concentration did not enhance photobehaviour, thus DVM, anymore. In the mean time, also the high chlorophyll *a* concentration in the metalimnion had disappeared and starvation must have been the cause of the population collapse.

Let us turn our attention to the year 1991. The early summer was exceptional because temperature remained low for a long time (Chapter 9). The period of migration did not start as usual at the end of May, but between 26 June (177) and 3 July (184). See Fig. 12.6. The sequential changes in population parameters are more consistent than in 1989. Population size of *D. galeata x hyalina* increased and subsequently decreased during the pre-migration period (Fig. 12.1). A maximum of gravid females on day 143 was followed by a peak of juveniles 7 days later (day 150), while adults without eggs reached a maximum on day 163. All population parameters were high at the beginning but soon (day 160) decreased to very low values. Therefore, the retarded migration period started with a low birth rate (Fig. 12.5). However, a radical increase in all population parameters occurred during that period, notwithstanding the increased egg-development time. As in 1989, chlorophyll *a* concentration increased (Fig. 9.6 and Fig. 12.22) during the migration period and was certainly a stimulating factor of importance.

In conclusion, both years show that the arrival of large shoals of 0^+ perch predators does not necessarily lead to a collapse of the *Daphnia* population. A renewed algal growth was responsible for an increase in population size but DVM was needed to ascertain that daphnids were present in sufficient numbers to start population growth.

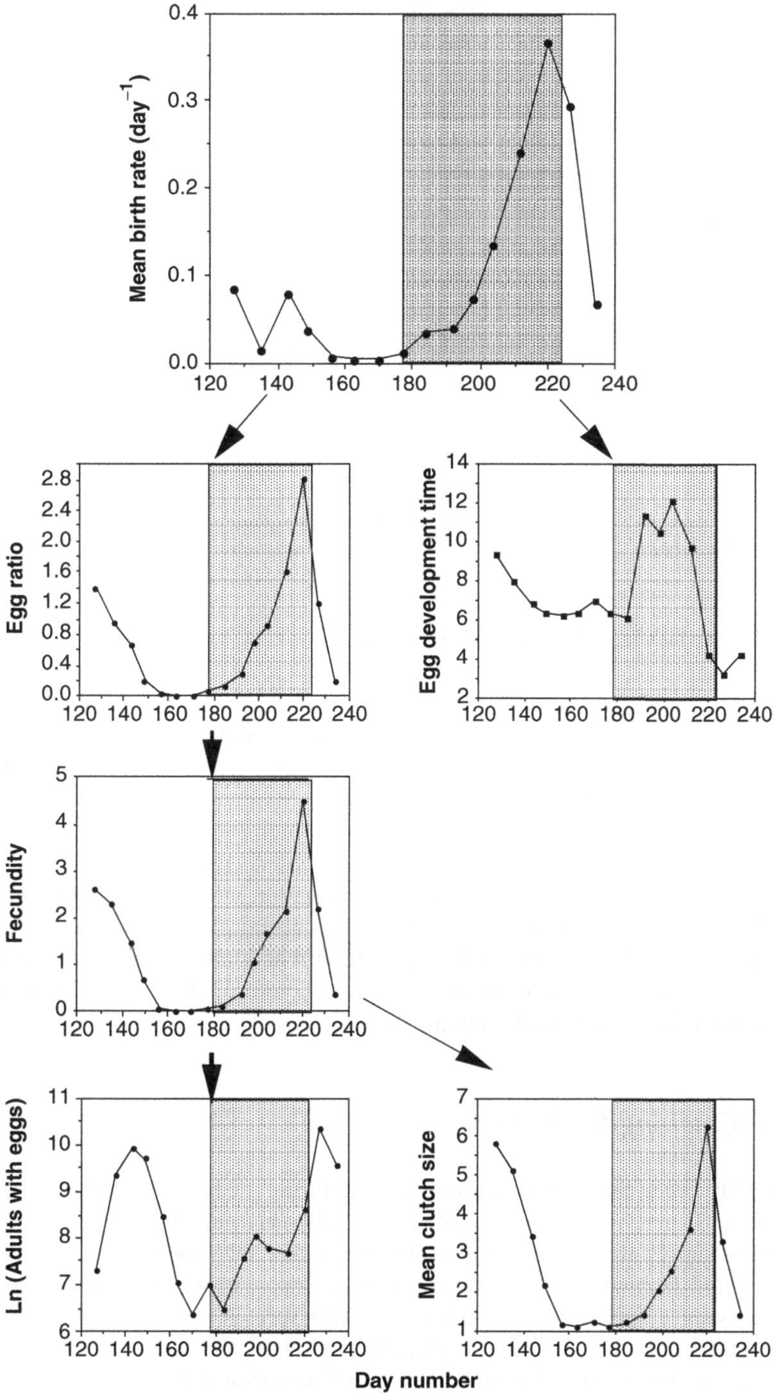

Fig. 12.5 As previous figure but now for the year 1991

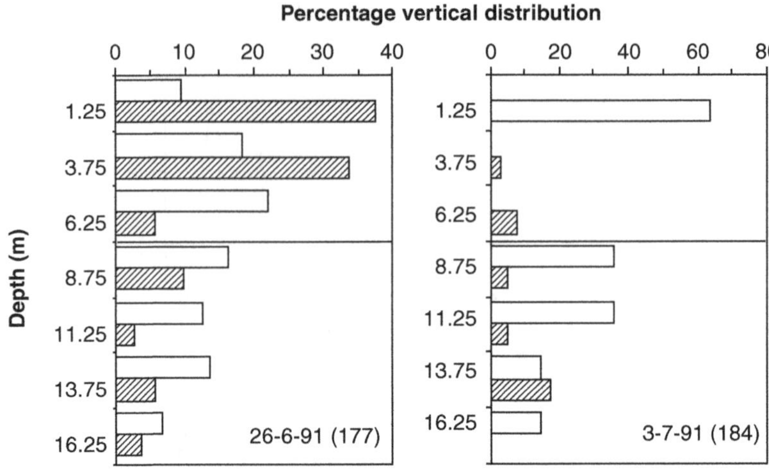

Fig. 12.6 The migration period in 1991 started within the week of 26 June to 3 July 1991. *Open bars* presents daytime distribution, *black bars* night-time distribution. The *horizontal line* indicates the depth of the thermocline

12.3 Quantifying 0^+ Fish Populations

During spring and early summer, large numbers of 0^+ perch often occur in lakes and they are a cause of mortality in zooplankton. These juvenile fish are considered the ultimate selecting factor through which DVM was evolved. Of course, selection still operates and maintains migration. These visually hunting fish are also a proximate factor and are conditional for the large amplitude migrations of copepods and cladocerans in a way as was discussed in previous chapters. Whether a proximate or an ultimate factor, of importance is quantity. If predation mortality is small compared to the birth rate, selection wanes. With a low concentration of fish kairomones in the water, the condition responsible for a strong reaction to changes in light intensity is small and zooplankton do not migrate. Therefore, the quantification of fish populations is important to understand DVM.

12.3.1 Catching 0^+ Perch

To catch 0^+ fish in lakes, fishing must be done during the hours of darkness because in daytime 0^+ perch aggregate in compact shoals which makes quantification difficult. After sunset, the majority leaves the littoral and sublittoral zones and disperse (Gliwicz and Jachner, 1992; Jachner, 1989; see also Fig. 12.7). Offshore–onshore migrations have been described for several fish species, for example, the golden shiner (Hall et al., 1979), bluegills (Werner and Hall, 1988), redbelly dace (Naud and Magnan, 1988) and yellow perch (Post and McQueen, 1988). According to Bohl (1980), this horizontal migration is correlated with the rate of relative decreases

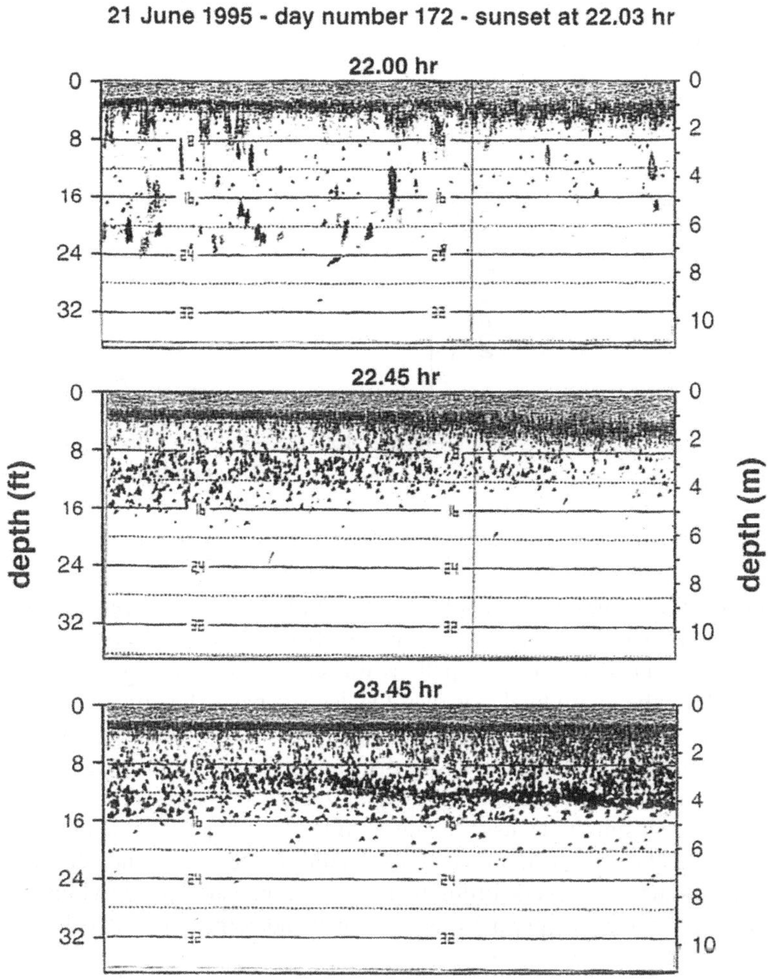

Fig. 12.7 Echograms of juvenile fish, mostly perch (*Perca fluviatilis*), taken around sunset in Lake Maarsseveen on 21 June 1995. Daytime shoals (2200 hours picture) are breaking up (2245 hours) and perch are leaving the littoral zone (2345 hours at left). From Flik et al. (1997)

in light intensity. Dispersion is not complete, however, and even at night a slight gradient in the 3.6–6.5 m depth range exists with numbers declining towards the open water (Duncan et al., 1998). Abundance is usually calculated from the number caught and the volume sampled (Jachner, 1989; Mehner et al., 1998a). This seems straightforward but catch efficiency of the net must be known. In the shallow Lake Tjeukemeer (the Netherlands), efficiency of the used trawl nets was calibrated by comparing the number of fish caught with that in a known area, encircled with a purse seine (Mooij et al., 1994). It was assumed that with that kind of netting all 0^+ fish were removed (Boersma et al., 1995). This method of calibration can only be used in shallow lakes but not in Lake Maarsseveen.

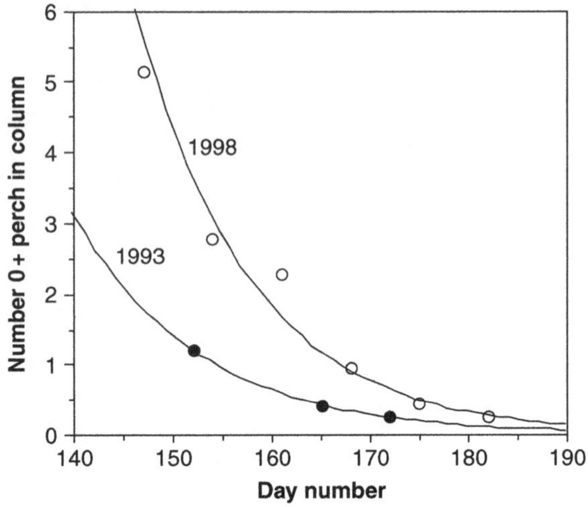

Fig. 12.8 The number of 0^+ perch in Lake Maarsseveen in 1993 (*black dots*) and 1998 (*open circles*). Lines are drawn according to the fitted functions $Y = 2.04 \times 10^5 \times$ e–0.079x ($R^2 = 0.999$) and $Y = 19.84 \times 10^5 \times$ e–0.087 × x for 1993 and 1998, respectively

Of the diverse fishing gears that were tested in Lake Maarsseveen, standard horizontal tows by two twin Bongo nets gave the best results. Tows between 3 and 5 m depth were made over the same path as was covered by the zooplankton torpedo nets, thus along the logitudinal axis over the deepest part. The efficiency of the nets decreased after the perch reached a total length of 30 mm (Aanen, 1993). The abundance of fish targets, with a similar size distribution as detected by dual-beam echo sounding in the lake (Duncan and Quiliiam, 1998), was comparable with the abundance of 0^+ perch simultaneously caught by Bongo nets up to that length. Numbers in the next size class of 30–42 mm had to be multiplied by 1.5 to meet the size spectrum of echo sounding targets. From the filtered volume and the individuals caught, the abundance of 0^+ perch per m^3 was calculated and compared with echo sounding data.

For 1998, the decrease in abundance of the 0^+ perch is depicted in Fig. 12.8. Numbers in the 0–6.5 m column under 1 m^2 decreased exponentially. The curves can be used for the period of about day 145 (25 May) until day 190 (9 July), which includes the period of migration. At day number 150 (30 May), t(0) was positioned, which is when DVM mostly starts in Lake Maarsseveen. The decrease function was calculated for two additional years giving an average of the exponent of 0.0812 with a standard error of 0.0034. I think a good approximation is obtained with an exponent of 0.08 day^{-1} and if N(0) is set at day 150 (or any other date with an empirically estimate of N), the equation becomes:

$$N(t) = N(t - 150)^* e^{-0.08*(t-150)}$$

in which t is day number and $N(t)$ is the number of 0^+ perch in a column under 1 m^2 and a depth of 0–6.5.

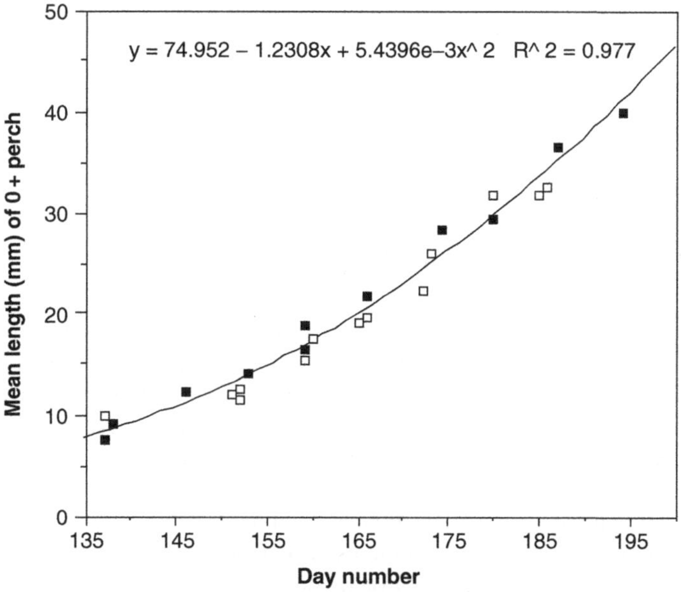

Fig. 12.9 Length increase of 0^+ perch (*P. fluviatilis*) in Lake Maarsseveen for the years 1994 (*white squares*) and 1995 (*black squares*). The quadratic function gives the best fit although it is obvious that the function does not hold outside the presented time interval

In 1994 and 1995, the development of total length of the 0^+ perch individuals is very similar (Fig. 12.9). This does not hold for all years, however. In 1993 length on day 150 was 20% and in 1998 even 33% higher.

To get an idea of the impact of 0^+ perch as predator of zooplankton the daily increase in biomass was estimated. Wet weight (Ww mg) as a function of total length

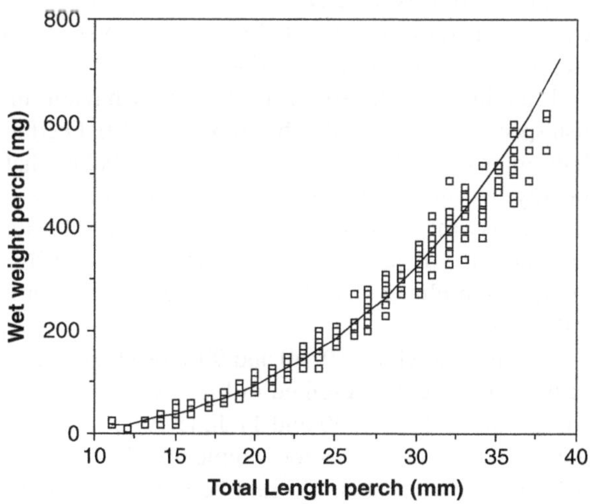

Fig. 12.10 Wet weight as a function of total length of 0^+ perch (*P. fluviatilis*) caught in Lake Maarsseveen in early summer of 1998. The function is $W = 0.0095 \times L^{3.0677}$ ($R^2 = 0.974$). (Van Gool and Ringelberg, 2002)

(Fig. 12.10) was determined by several authors and the decisive exponent is comparable in all cases, for example, $Ww = 0.02 \times L^{2.8}$ (Jachner, 1989), $Ww = 0.075 \times L^{3.064}$ (Mehner et al., 1995), $Ww = 0.006 \times L^{3.2}$ (Mooij et al., 1994; Ww in gram and length as fork length (cm), not total length (mm)) and the present determination produced $Ww = 0.01 \times L^{3.067}$.

12.3.2 Times of Feeding and Food Composition

As visually hunting predators, 0^+ fish need sufficient light to attack and catch zooplankton. The adaptive significance of diel vertical migration of zooplankton was based on this maxim. We must explore the range of light intensities between maximum and minimum feeding of 0^+ fish and describe this range in terms of time of day and depth. Combined with the vertical distribution of zooplankton, time and depth of feeding of 0^+ fish is then given. With a paper of Clark and Levy (1988) in mind, we might call that particular space–time a "feeding window".

In their model, the effect of light intensity on feeding of juvenile sockeye salmon (*Oncorhynchus nerka*) and on predation by piscivorous fish was combined to realise an "antipredation window" with a minimal value of the ratio between mortality risk and feeding rate.

With the "feeding window" emphasis is on feeding. Shortly after sunset the 0^+ perch have entered the open water from the littoral and are "waiting" for the migrating zooplankton to enter the epilimnion. Light intensity is still sufficiently high then for predation. Whether piscivorous predators are also involved and a reason for the daytime sojourn in the littoral zone is unknown for Lake Maarsseveen. For Lake Ros in Poland, Gliwicz and Jachner (1992) could neither find arguments for hiding of the 0^+ fish and called it "the ghost of predation past" that is a left-over behaviour from times visual predation on the juvenile perch had been high. Whether this historical argument holds is not easy to say: as zooplankton do, prey fish react to infochemicals mediated by predators (Chivers et al., 1996) and unfavourable behaviour might stop in the absence of relevant information.

Light intensity thresholds for feeding behaviour in several marine and freshwater fish species were compiled by Blaxter (1970). Feeding seems maximally possible at light intensities higher than 1–100 lux but below that intensity predation decreases rapidly.[3] A threshold, defined as the light intensity at which 10% of the maximum feeding is still possible, ranges from 0.1 to 0.01 lux. This agrees with the range presented for 0^+ perch from Lake Maarsseveen (Fig. 12.12). In the original figure, presented in Flik et al. (1997), 100% feeding was at 1 lux and 10% feeding at about 0.025 lux.

Isolumes at which 10, 50 and 90% of *D. galeata x hyalina* were eaten, as found in the experiments presented in Fig. 12.12, are used in a time-depth series around sunrise on 30 May 1990 and 11 June 1992 (Fig. 12.13). Obviously, a concept like the "feeding window" is too simple. On 30 May 1990, perch were able to predate on *Daphnia* from about 25 min before sunrise and continued to do so during the

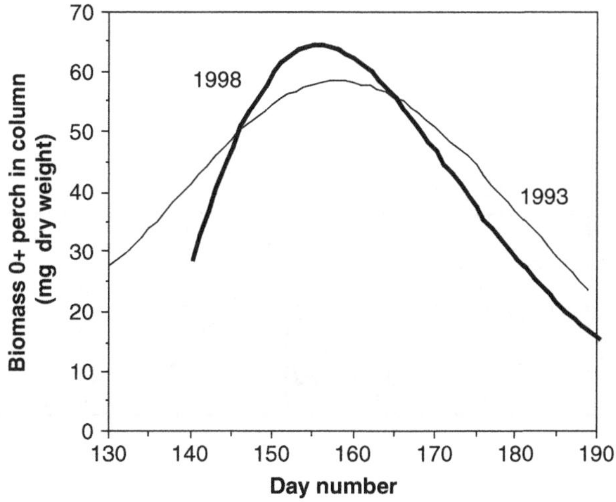

Fig. 12.11 Total biomass of 0^+ perch (*P. fluviatilis*) in the open water of Lake Maarsseveen during the early summers of 1993 and 1998. Values represent the calculated dry weight in a water column of 0–6.5 m under 1 m^2. For 1993, values are multiplied by 5 to fit both lines in one figure. Calculations are based on data presented in Fig. 12.8

Fig. 12.12 The relative predation rate of one 0^+ perch on 30 adult *D. galeata x hyalina* in 10 l filtered water from Lake Maarsseveen (average length 1.57 mm, s.d. = 0.197, $n = 10$). The concentration is low but often perceived in the lake. Observations were made in a dark room at a temperature of 18–20°C. After 15 min all remaining daphnids were removed and counted. The line is described by the equation $y = 1944.\ e^{1.7x}$

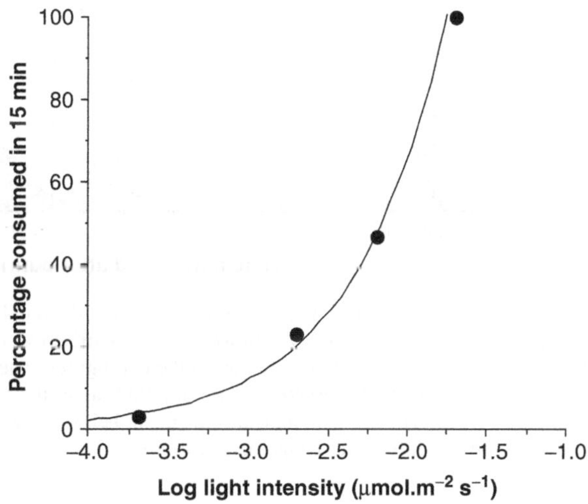

rest of the day. Unless, of course, they went into the littoral zone to spend the daylight hours, as they usually do. In 1992, *Daphnia* descended rapidly and at the time light intensity was sufficiently high for 0^+ perch to predate, *Daphnia* had left the epilimnion. The pattern of 1992 is similar to that of 1998 and, as suggested (see Fig. 12.11), the extensive DVM in both years could have been the result of the

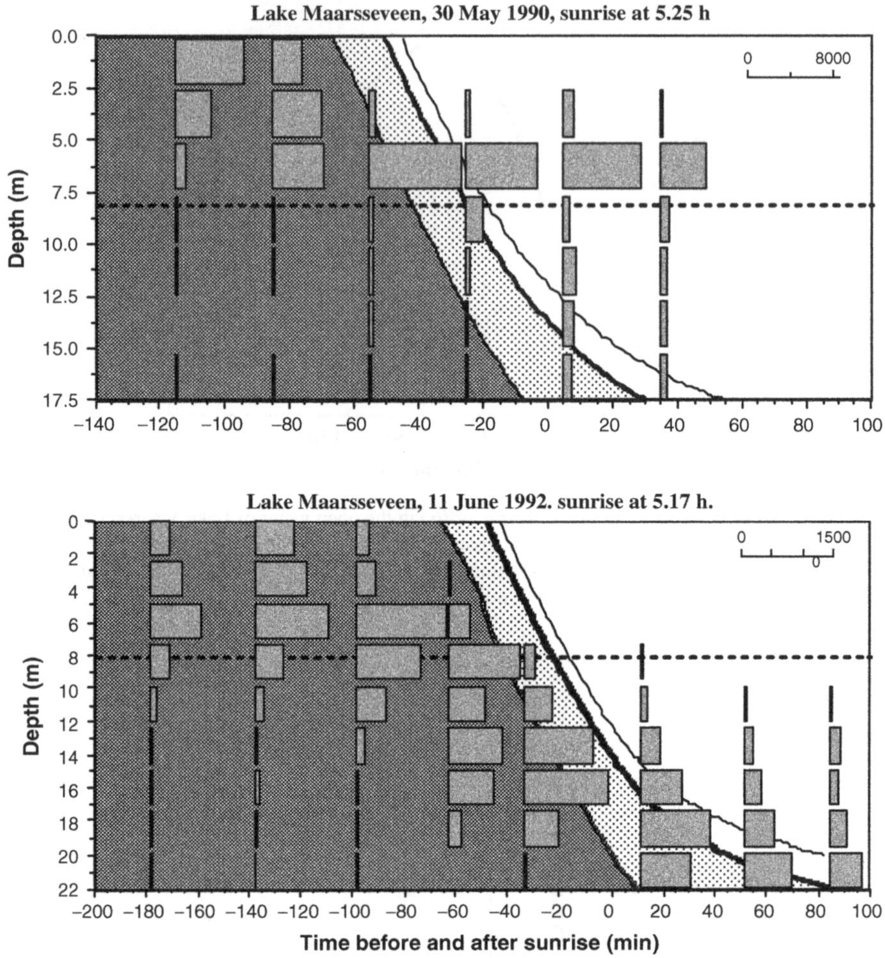

Fig. 12.13 Early morning downward migration of *Daphnia* in Lake Maarsseveen on 30 May 1990 (*upper figure*) and on 11 June 1992 (*lower figure*). Sunrise was on 0525 and 0517 hours local time, respectively. The depth scale is different in the two figures. The isolumes indicate light intensities at which 0^+ perch eat, 10, 50 (*thick line*) and 90% as in the experimental situation presented in Fig. 12.12. The *horizontal line* indicates the depth below which 0^+ perch were never caught in the lake (Slightly modified after Ringelberg, 1999b)

high abundance of perch. In most years in Lake Maarsseveen, migrating *Daphnia* and *Eudiaptomus gracilis* (see 21 June 1995 in Fig. 4 in Flik et al., 1997) descend slowly towards the metalimnion.

For many fish larvae only the smaller sized zooplankton, such as rotifers and nauplii, are available as food. In Lake Ros, the diet of perch <14 mm consisted almost exclusively of copepod nauplii. *Cyclops* and *Eudiaptomus* were dominant in the guts up to a perch length of about 20 mm and only 20% consisted of *Daphnia*.

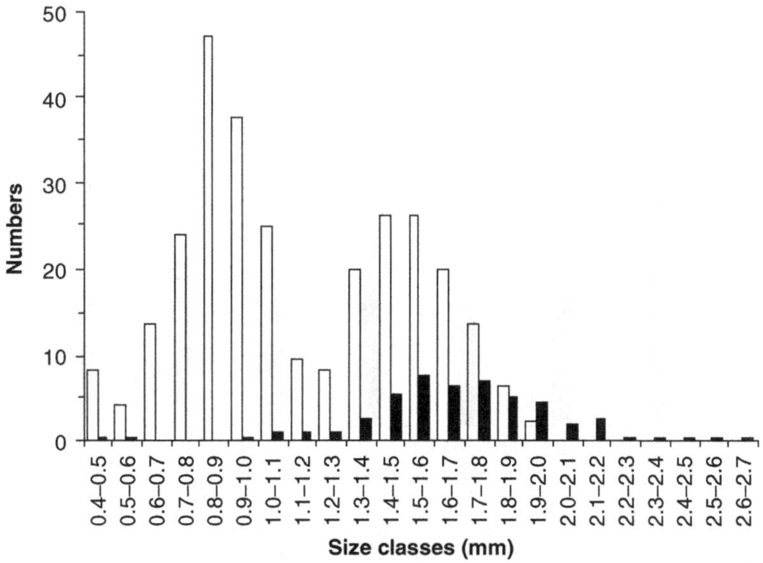

Fig. 12.14 Size distribution of *D. galeata x hyalina* in the open water of Lake Maarsseveen (*open bars*) and of daphnids in the gut of 0+ perch (*black bars*), both caught on 15 June 1993

With increasing length, *Daphnia* became dominant until a length of about 60 mm was reached (Jachner, 1989, 1991). The increase in gape width makes the change in catching and eating of larger prey species possible. As has been known for a long time (Brooks and Dodson, 1965), large sizes are preferred which is also nicely illustrated for *D. galeata x hyalina* in Lake Maarsseveen (Fig. 12.14).

A complicating factor is that the composition of the diet changes during evening, night and morning as illustrated for the night of 10–11 June 1992 (Fig. 12.15). At about sunset (2158 hours, local time), gut composition was still diverse but during the night *Daphnia* dominated nearly completely. In addition the number of adult daphnids carrying eggs increased relative to the number of females without eggs (Table 12.2). Against sunrise (0517 hours), *Cyclops* and *Eudiaptomus* were eaten again.

Table 12.2 The change in percentage of dominant species in the gut of 0+ perch during the night of 21/22 June 1995 in Lake Maarsseveen. Between brackets the percentage of *D. galeata x hyalina* with eggs as a percentage of total adult daphnids. Average total length of perch = 30.8 mm; standard error = 0.31; $n = 30$

Time (h)	*D. g x h* (ad.–) (%)	*D. g x h* (ad.+)	*Eudiaptomus* (%)	*Cyclops* (%)
23	31.3	1.8% (5.4%)	49.8	17.1
0	67.6	26.1% (27.8%)	2.3	4.0
01.30	60.4	38.2% (38.7%)	0.9	0.5

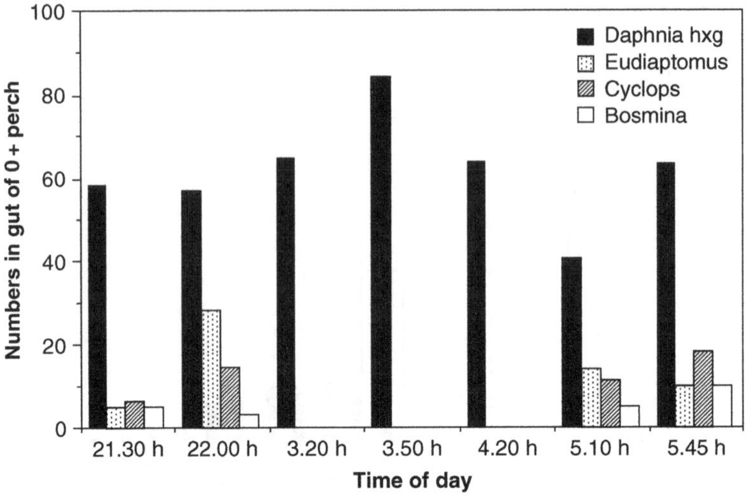

Fig. 12.15 Composition of gut content of 0^+ perch on different times of the night and early morning. Lake Maarsseveen, 10–11 June 1992; sunset 2158 hours, sunrise 0517 hours local time. Size of perch 25–30 mm total length

An interpretation of Fig. 12.15 is not easy to make. Perch left the littoral zone shortly before sunset, then individuals dispersed and predation started. Figure 12.15 suggests that at 2130 hours, eating had already started. Non-daphnid zooplankton in the gut increased (2200 hours) but had disappeared 5 h later. This suggests that evacuation time in the ambient situation was less than 5 h. On 21 June 1995 at 2300 hours, *Eudiaptomus* and *Cyclops* made up 75% of the gut contents but both species had disappeared 1 h later (Flik et al., 1997). At 0130 hours, *Daphnia* was the only prey species present, which narrows evacuation time to about 2 h or less. For yellow perch with a length of 25–30 mm length, Mills and Forney (1981) found in extensive experiments evacuation times of 1.5–2.0 h. Therefore, I think it is safe to conclude that the *D. galeata x hyalina* in the guts of 0^+ perch that were caught during the night were also digested during the nightly hours. Two hours after sunset the lower sensitivity limit of our photometer had been reached and light intensity had dropped to 5×10^{-3} lux (1×10^{-4} μmol. s^{-1} m^{-2}). This light intensity is a factor 2.5 below the light intensity at which perch were still able to catch a few *Daphnia* under laboratory conditions (Fig. 12.12). Nevertheless, I hesitate to conclude that 0^+ perch is not exclusively a visual predator because the angular light distribution in the experimental situation was certainly different from that in the lake. Snell's circle (Chapter 6) was not realised in the feeding experiments and the ability to fixate prey might have been different. An argument in favour for visually prey catching at night is that the percent of *Daphnia* with eggs increased between 2300 and 0130 hours at the expense of the less visible females without eggs. Reflexion of light from nearby city of Utrecht makes that the night sky over Lake Maarsseveen is not always really dark.

12.4 Comparing a Year with a "Weak" and a "Strong" Migration

In 1991 and 1992, *Daphnia* and *Eudiaptomus* went deeper than in previous years but in the following years amplitude was small again and the seasonal migration pattern often irregular. We had to wait until 1998 before migration depth suddenly increased again. I have chosen the 2 years 1993 and 1998 to compare what happened with *Daphnia* and 0^+ perch in Lake Maarsseveen during the early summer weeks. The pattern is irregular in 1993 and the difference between the mean population depth around noon and midnight does not exceed 2–3 m (Fig. 12.16). In 1998, DVM developed nicely and on our last day of observation, 4 July, the amplitude was on average 13 m. On that date, the migration period was clearly not over yet, although in most previous years it was by that time.

Perch abundance was different in the 2 years. On day 150 (30 May), when migration started as usual, the number of 0^+ perch in the column was 3.1 × larger in 1998 then in 1993 (Fig. 12.8 A). Length of the average individual was 4 mm larger in 1998 (Fig. 12.17). Spring temperature was about 3°C higher in 1998 and could have been responsible for the initial length difference. The difference in length remained the same, thus successive growth was comparable in both years. Both number and size were responsible for the difference in biomass (Fig. 12.11). Length increase as a function of the number of *Daphnia* consumed per daydid not differ much although the larger 0^+ perch of 1998 consumed more daphnids. The critical perch length of 15–20 mm was already reached on 20 May 1998 (day number 140), which was 8 days earlier than in 1993 but *D. galeata x hyalina* did not migrate. This does not conform to the earlier proposed hypothesis that *Daphnia* start migration when

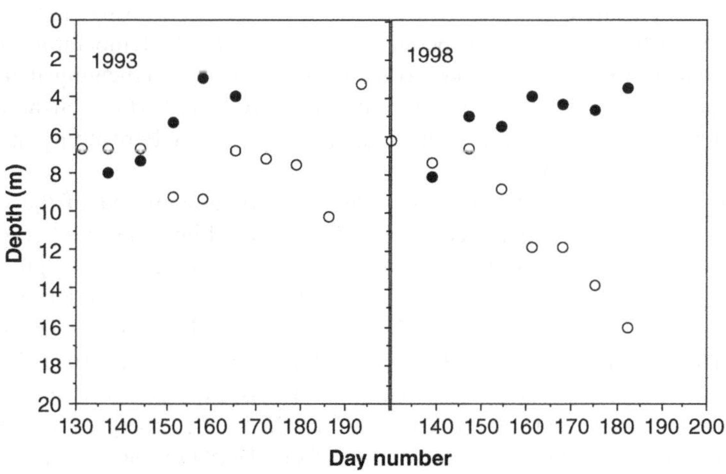

Fig. 12.16 Mean population depth of migrating *D. galeata x hyalina* in 1993 and 1998. *Black dots* represent depth at night and *open circles* around noon

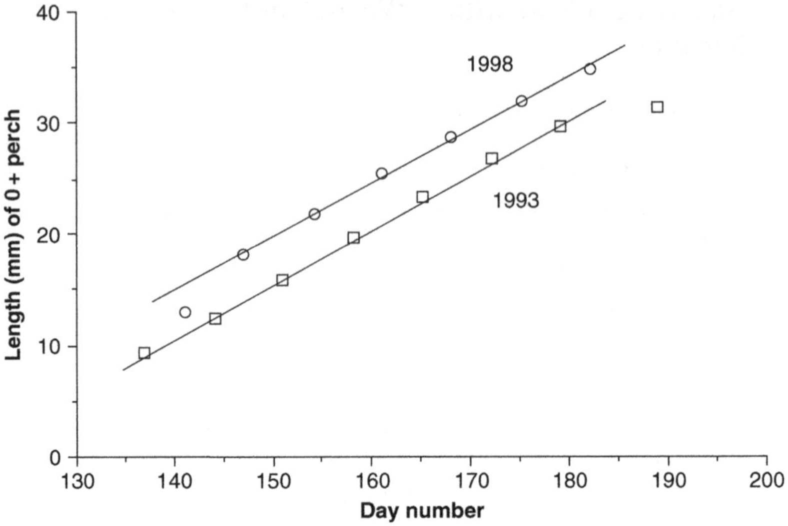

Fig. 12.17 Mean total length of 0^+ perch in Lake Maarsseveen in 1993 and 1998. Each point is the average of 50–300 measurements. Confidence limits were too small to insert in the picture. To compare the 5% confidence ranges on 7 June 1993 (day 158) and 9 June 1998 (day 160) are 19.23–21.64 and 22.49–26.29, respectively. The pseudo-linear increases in the growth curve are $y = -58.38 + 0.49 \times$ ($R^2 = 0.999$) for 1993 with the last point omitted, and $y = -51.56 + 0.48 \times$ ($R^2 = 0.998$) for 1998 with the first point omitted

the perch have reached a length of 15 mm or, rather, when the mouth gape is sufficiently large to eat daphnids. Of course, other factors are also of influence, for example, temperature. I suggest that on 20 May 1998 in the absence of a stable stratification, mixing of the upper water layers with a large part of the water column occurred, which diluted the fish kairomone to a concentration too low to induce DVM notwithstanding a large 0^+ perch biomass. On 10 June, temperature was constant to a depth of about 5 m, and lower than on 19 May, indicating that mixing had occurred earlier. Soon thereafter, a discontinuity layer had established confining mixing over a much smaller volume than in May, probably resulting in a higher kairomone concentration.

The change in 0^+ perch biomass is the result of growth and of the decrease in numbers due to mortality. Year class strength was higher in 1998 (Fig. 12.8). Biomass was expressed as dry weight and calculated from wet weight. (In the Appendix of this chapter, methods of calculation are presented.) The maximum in dry weight biomass, reached on 7 June 1993 (day 158), was a factor of 6.4 lower than the maximum in 1998, reached on 10 June (day 161). The food consumption that was needed to realise the biomass increase for a week was calculated assuming a food-growth efficiency of 25% (Vijverberg et al., 1990). A part of the amount consumed was derived from *Daphnia* and the rest predominantly from *E. gracilis.* The ratio *Daphnia*/*Eudiaptomus* depended on the size of the 0^+ perch but also on time of the day (Fig. 12.15). Total consumption and the part ascribed to *Daphnia* are presented in Fig. 12.18. In the course of

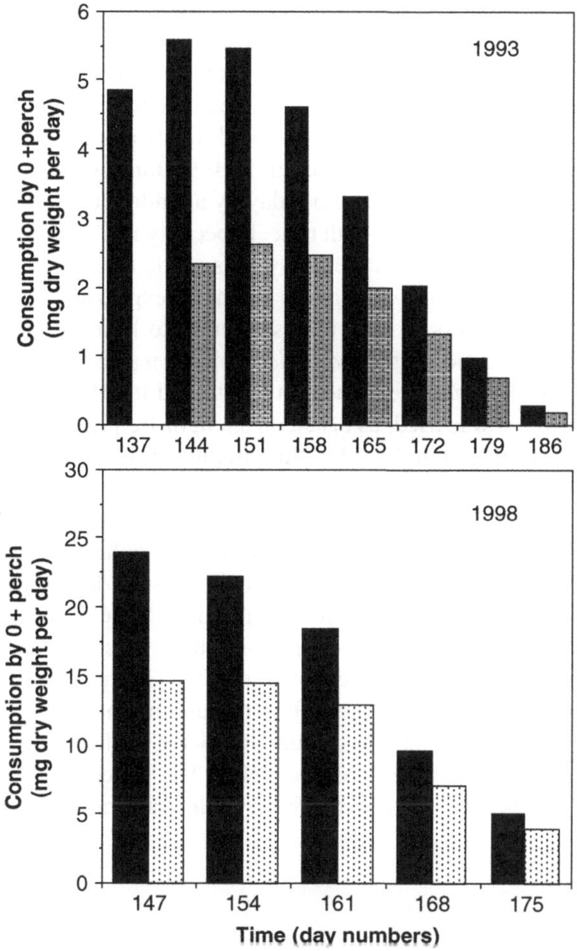

Fig. 12.18 Predation of 0^+ perch in Lake Maarsseveen in 1993. *Black bars*: total amount consumed in a water column of 0–7 m below 1 m^2; *grey bars:* the contribution of *Daphnia*. Calculations were made over 1-week intervals but consumption is expressed per day. See text for more extensive information about the calculations

the season the contribution of *Daphnia* in the diet gradually increased but the total consumption decreased rapidly as the number of 0^+ perch diminished. The daily food ration divided by body biomass decreased with time and increased with length. On day 137 (16 May 1993) it was 60%, in the first week of migration 31%, and it dropped to a mere 4% at the end of the migration period in the first week of July 1993. By that time perch had started to return to the littoral zone permanently and consumption in the open water became less important. A similar decrease in daily food ration with time was found by Mehner (2000) and Mills and Forney (1981). Daily ration per body mass (y) of 0^+ yellow perch (*P. flavescens*) as a function of length (l) is nearly the same for *P. fluviatilis* as found in the present study ($y = 186–115 \times \text{Log l}$, $R^2 = 0.726$ and $y = 165–104 \times \text{Log l}$, $R^2 = 0.986$, respectively).

In the first few weeks of the migration period, part of the daphnid population remained in the 0–6.5m depth layer even in daytime. These individuals did not migrate (see next paragraph) and were the first to be eaten. After 3 weeks, for instance, in 1998, no daphnids could be found anymore in the epilimnion during the day. In both years, the calculated number of consumed *Daphnia* decreased rapidly during these weeks although in 1993 more slowly and in smaller numbers (factor 30). The number eaten per day by an individual perch is comparable for both years, increasing steadily with time. Especially in 1998, the size of the 0^+ perch population rapidly declined in the first weeks of the migration period.

The consumed *Daphnia* biomass, expressed as dry weight, had to be changed into numbers of individuals in order to have a common currency for the study of the dynamics of the *Daphnia* population (see Appendix). The dry weight of an individual *Daphnia* depends on length and for 1993, an average length of *D. galeata x hyaline* was used, as based on a length-frequency distribution of daphnids in the gut of the 0^+ perch. In 1998, this simple approach was not possible because the average length of *D. galeata x hyalina* changed continuously during the observation period (Fig. 12.19). This can be attributed to the presence of fish kairomones as found by Machacek (1990, 1993) in life-history cultures and since then confirmed by several authors. A progressive smaller size at first reproduction resulting from shifts in clonal composition as described by Stibor and Lampert (2000) in the Schöhsee is another possible explanation, although size selective predation must have contributed to the change in length also. On the first two dates in Lake Maarsseveen, *Daphnia* was still present in the epilimnion and on day 154, these animals were already significantly smaller compared to those from deeper down (average length l = 1.60 and 1.82, respectively). The difference had disappeared 1 week later (l = 1.54 and 1.55, respectively). The kairomones-induced life-history effect might

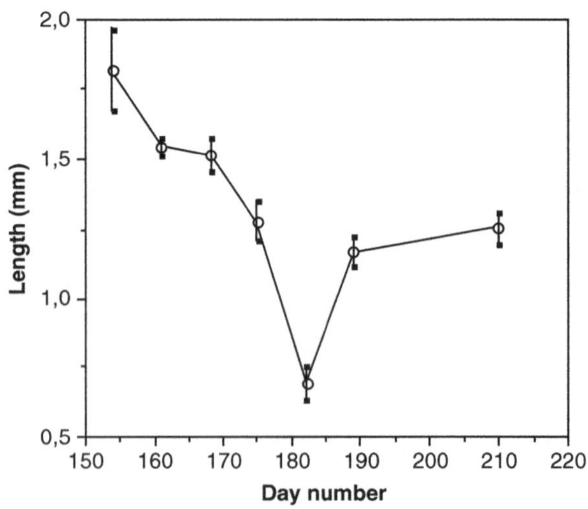

Fig. 12.19 Mean length of *D. galeata x hyalina* in the 12–20 m depth column, caught in daytime in Lake Maarsseveen in 1998. *Black dots* and *bars* indicate upper and lower 5% confidence limits of the mean. With the exception of day numbers 154 (3 June 1998) and 161 (10 June 1998), no *Daphnia* could be caught in the epilimnion. See text

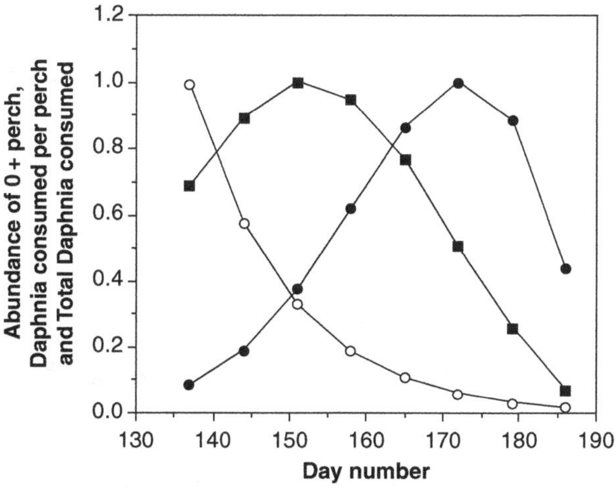

Fig. 12.20 Change in juvenile perch predation regime in Lake Maarsseveen in 1993. Number of 0^+ perch set to 1 (*open circles*) on day 137 (17 May) and the number of consumed *Daphhnia* per perch (*black dots*) is set to 1 on day 172 (21 June). Also the total number of consumed daphnids (*black squares*) is made relative to fit the figure

be the cause but, again, size selective predation cannot be excluded. After day 182 (1 July), the decrease in length of the migrating individuals stopped. By then, the number of 0^+ perch in the epilimnion had considerably decreased, so I assume that also the concentration of fish kairomones must have decreased.

The consequence of the changing length was that the calculation of the number of consumed *Daphnia* from biomass data had to be done for each date separately. In Fig. 12.20, the (relative) number of these consumed daphnids is compared with the (relative) number of 0^+ perch present in a water column of Lake Maarsseveen. The product of both or total number consumed is also inserted. The two basic lines cross at about day 151 (31 May) in 1993 and on 10 June (day 161) in 1998. Before these dates predation per individual perch was small and predation pressure was determined by perch numbers. Thereafter, the abundance of perch rapidly declined and total predation mortality of *Daphnia* decreased, although individual consumption increased. The date of crossing lines marks the maximum of predation and is a turning point in the relation between predator and prey.

The precise moment depends on the rate curve of perch mortality and on length growth of the fish and both are influenced by several factors as, for example, the initial size of the 0^+ perch population and the available zooplankton. The date of crossing lines coincides with the change in the dynamics of the *Daphnia* population as is visible, for example, in a suddenly increasing clutch size (Fig. 12.21) and a renewed growth of the population as was discussed in Section 12.2. I do not suggest a direct causal relation but a mark in an altered "set of necessary ecological conditions" of which the predation regime is a possible component.

In order to evaluate the role of 0^+ perch predation in the population dynamics of *D. galeata x hyalina*, mortality rate and birth rate were calculated.

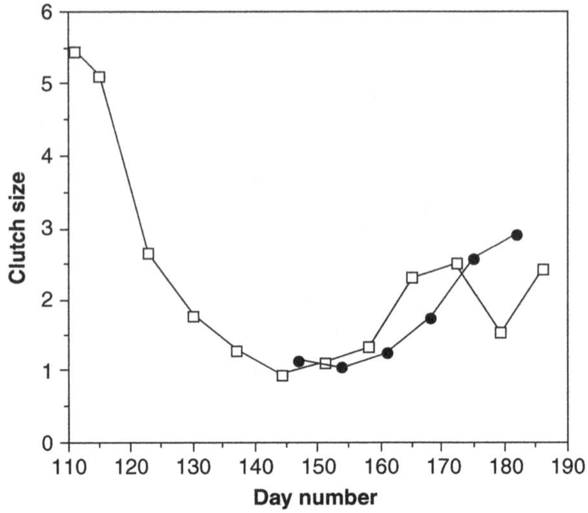

Fig. 12.21 Clutch sizes in 1993 (*open circles*) and in 1998 (*black dots*)

Calculation of mean birth rates (b) was according to the protocol described in Ringelberg et al. (1991b). Weighted egg ratios and weighted development times were used in the Paloheimo equation (Paloheimo, 1974) for number-depth distributions during the day and the night. Mortality rates (d) were estimated, as traditionally, from the intrinsic rates of population change (r) and birth rates (b) (see Fig. 12.3). However, total death rates do not equal predation rates (pr) , while these latter rates are of interest now.

Predation rates were calculated from the amount of food, necessary to realise the increase in 0^+ fish biomass, and the contribution of daphnids therein (Fig. 12.18). Total death rates and predation rates are thus independent estimates and the results are listed in Table 12.3. The calculation of dynamic parameters for a field population

Table 12.3 Parameters of the population of *D. galeata x hyalina* in Lake Maarsseveen

Day number	Birth rate	Predation rate	Pop. change	Death rate
1993				
137	0.0009	0.0008	–0.070	0.071
144	0.001	0.0028	–0.093	0.094
151	0.002	0.0077	–0.135	0.137
158	0.029	0.0270	–0.056	0.085
165	0.112	0.0280	–0.037	0.149
1998				
147	0.035	0.0026	–0.164	0.199

The predation rate is calculated from the part of the growth increase of the 0^+ perch contributed to consumed daphnids, while the death rate is calculated from the *Daphnia* birth rate and the intrinsic rate of population change.

is a precarious exercise, although birth rates and changes in population size are based on extensive sampling (see Ringelberg et al., 1991b). Nevertheless, I think the prudent conclusion is justified that predation mortality is a small part only of the total mortality in both years.

As a final conclusion of their review Mehner and Thiel (1999) wrote: "...the exclusive control of zooplankton dynamics by 0^+ juvenile fish predation in late spring or early summer (for example the induction of a midsummer decline of large cladocerans) could not beproven..." (p.178).

12.5 The Relevance of Life-History Experiments

As Fig. 12.19 revealed, *D. galeata x hyalina* do not only migrate but also change life-history traits under influence of fish kairomones. With both responses costs are involved and the simultaneous expression of different anti-predator traits could be non-adaptive in the opinion of De Meester et al. (1995) and De Meester and Pyanowska (1997). That might be justified, provided that each one alone suffices to reduce sufficiently predation mortality. If this is not so, selection will intervene. Anyhow, the example of Fig. 12.19 shows that DVM and a decrease in size do go together. Of course, it might be argued that size-selective predation was responsible for the decreasing average length in the population. Since the results of culture experiments (Reede and Ringelberg, 1995) showed that migrating and non-migrating clones from Lake Maarsseveen react to fish karomones with life-history adaptations (although to a different clone-specific extent), size-selective predation is not the only possible factor to explain the observed phenomenon in the field. The two anti-predation responses seem to be uncoupled because the sub-population that does not leave the epilimnion shows one but not the other. During the years of our study, non-migrating daphnids were always found in the epilimnion during the first weeks of the migration period.

Compared to behavioural changes, life-history responses need a long time to come to expression because a new generation has to appear before they might become effective. On the other hand, immediately after fish kairomones have been perceived individuals of *D. galeata x hyalina* start sinking. Triggered by the light intensity increases of the next dawn, Photobehaviour system 2 (PBS 2) is activated and morning descent starts. This short-time delay and the reversible character makes DVM superior as a strategy compared to life-history responses (Gabriel, 2006).

The juvenile daphnids, born from mothers that have been in water with fish kairomones, generally mature earlier, are smaller and the first clutch is larger although the eggs are smaller. The advantage of a large clutch size is obvious but the consequence of smaller juveniles might be a disadvantage. Smaller sized juveniles are less resistant to low food conditions than larger sized ones (Tessier and Consolatti, 1989; Reede, 1995). Generally, food conditions are worse in the hypolimnion compared to the epilimnion, thus a decrease in size and at the same time staying at a low food concentration for the better part of the day might be a bad strategy for survival (De Meester et al., 1995; Lass and Spaak, 2002). Nevertheless,

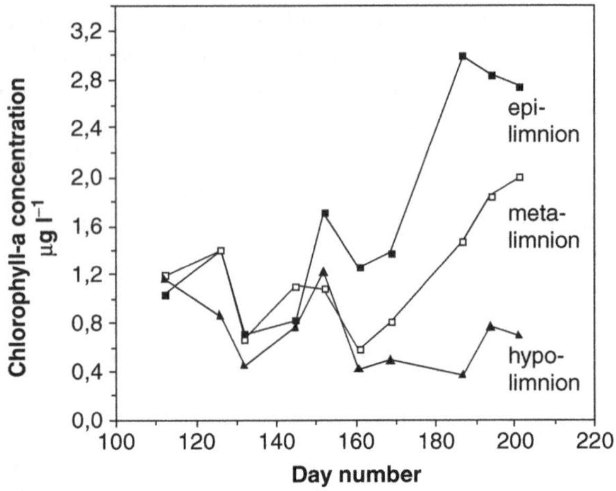

Fig. 12.22 Chlorophyll *a* concentration in Lake Maarsseveen in 1992

in Lake Maarsseveen both occurred, as is apparent from Fig. 12.19. Over about a month, the length of migrating adult daphnids continued to decrease. At the end of the migration period, when the abundance of 0^+ perch had decreased considerable, size of the daphnids did not decrease further. During the first week of the migration period, average length of non-migrating individuals of *D. galeata x hyalina* was significantly less than those caught in the hypolimnion (see previous paragraph). One week later, average length of the migrating daphnids had also decreased and the difference had disappeared. A life-history response alone was not enough to prevent predation sufficiently because after a few weeks, hardly any daphnids could be found in the epilimnion during the day. Culture experiments by Reede and Ringelberg (1995) showed that the difference in response could be ascribed to a higher sensitivity of the non-migrating animals for fish kairomones. Using a non-migrating clone, Reede (1995) determined age and size at maturity, the number of eggs in the first clutch and neonate size at six dilutions, indicated by the percentage of a "standardized fish water". The results are presented in Fig. 12.23. With increasing kairomone level, size at maturity decreased, which must have been due, at least in part, to the shorter juvenile growth period until maturity. The number of eggs increased rapidly with kairomone level and the size of the neonates decreased, although not proportionally.

Comparable experiments with a dilution range of "fish water", also using a clone from Lake Maarsseveen, were performed by Hülsmann et al., 2004. Their results differed from those of Reede (1995). For example, no significant effect of fish kairomone was found for size and age at maturity. On the other hand, both investigators described a significant size difference in the neonates. The unknown kairomone concentration makes the interpretation of different life-history experiments difficult. In addition, as the experimental results of both (and others, for example, Weber, 2001; Weetman and Atkinson, 2002) showed, the effect of kairomones depend on food concentration. Different combinations of environmental factors and also the

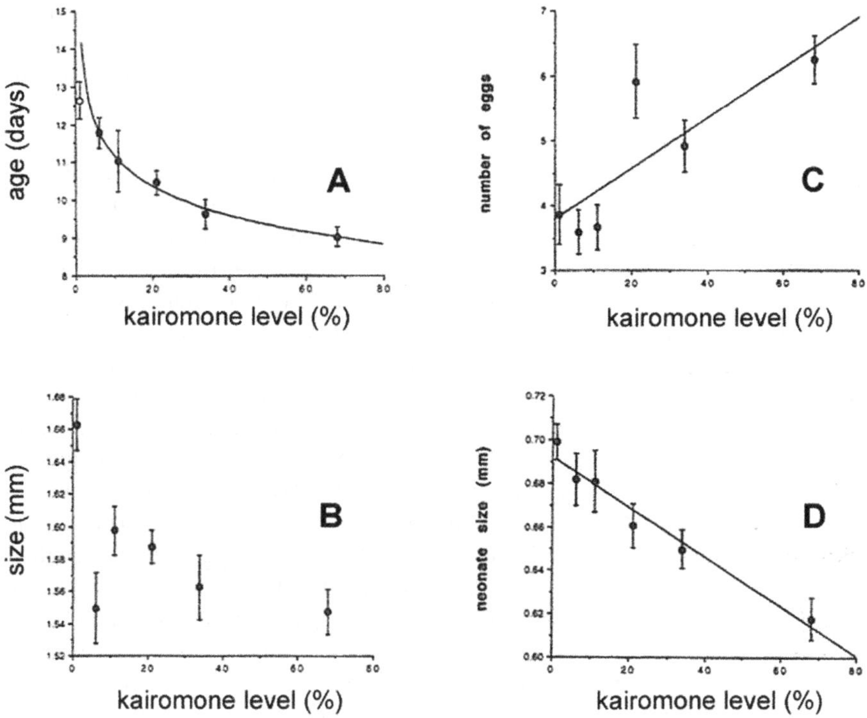

Fig. 12.23 Life-history experiments with a non-migrating clone of *D. galeata x hyalina* from Lake Maarsseveen in a series of different dilutions of "fish water" (kairomone level kl). (**a**) Age at maturity with the fitted line ($y = 13.4 - \log$ kl, $R^2 = 0.991$). (**b**) Size at maturity. (**c**) Size of the first clutch ($y = 3.76 + 3.93 \times 10^{-2}$ kl, $R^2 = 0.676$). (**d**) Neonate size ($y = 0.69 - 1.14 \times 10^{-3}$, $R^2 = 0.967$). Bars are 5% standard errors of the mean. (Modified from Reede, 1995)

culture history, for example, at what food concentrations mothers and grandmothers were grown, determine results (Reede, 1997). This makes the interpretation of life-history experiments to a specific field situation only possible in general terms even if animals from that situation are used. To make life of an ecologist even more complicated, the genetic composition of a *Daphnia* population in a lake is changing over the years and even seasonally (see next paragraph). Life-history experiments contribute to our understanding of the rich potential animals have to cope with environmental conditions but are a shot in the dark if we want to quantitatively understand what happens in a lake when fish predators appear.

The results of a multi-clonal experiment with *D. galeata x hyalina* and some clones of the parents by Reede and Ringelberg (1998) illustrate the complexity of the natural situation. The 15 clones studied in the paper were separated, a priori, into a migrating and a non-migrating group. This was based on the fact that the original females were caught around noon either in the epilimnion (0–7 m) or in the hypolimnion (15–25 m) during a period of seasonal migration in Lake Maarsseveen. Apart from the hybrid, two non-migrating *D. galeata* clones from shallow lakes in the Netherlands and one migrating *D. hyalina* clone (Stich and Lampert, 1984) from Lake Constance (Germany) were used. Characterisation of the clones was done with allozyme electrophoresis. It is unlikely that only two groups exist because not all animals migrate to the same depth (Ringelberg et al., 1991a), and this differential migration was correlated with different multilocus genotypes (Ringelberg et al., 2004). A considerable overlap in vertical distribution is present and this overlap is also present in the expression of life-history traits, for example, in the length at maturity of the used clones in the absence and the presence of fish kairomones. In the absence of fish kairomones, Reede and Ringelberg (1998) found that length at maturity did not differ for both groups but, if present, length of migrating clones was significantly larger ($P<0.001$). Life-history and behavioural responses blend into a continuous spectrum of combinations. This also holds for the *Daphnia* species from Lake Constance. The original separation of the non-migrating *D. galeata* and the migrating *D. hyalina* (Stich and Lampert, 1981) had to be reconsidered when it was found that in between these two extremes several hybrids occurred displaying an overlap in vertical distribution (Weider and Stich, 1992). Both predation prevention strategies are phenotypically plastic and are phenotypically induced. The seasonal recurrence of predation by 0^+ fish is not unpredictable but year class strength is, and thus the strength of predation pressure is unpredictable. Consider this spectrum of reaction norms of life history and of migration behaviour as "spreading of the risk" (Den Boer, 1968). This is not an evolutionary mechanism, but a phenomenological notion to appreciate the large diversification within the clonal community of *Daphnia* transcending the species concept. The composition of this clonal community is dynamic and changes yearly and within a season. I will discuss this in the next paragraph.

12.6 Complications Again: Genetic Differences

Over the last few decades, allozyme analysis made a better distinction possible between hybrids and parental species. Allozyme analysis also raised the veil of a high genetic diversity in lake *Daphnia* (Wolf and Mort, 1986; Wolf, 1987, 1988; Spaak and Hoekstra, 1993; Spaak, 1996; Spaak and Ringelberg, 1997; Boersma et al., 1998; Schwenk and Spaak, 1997). Although allozyme analysis has a poor resolution, it was possible to show a.o. in Lake Maarsseveen, that distinct genotypes have different migration patterns. It must be borne in mind that allozymes do not characterize these patterns. An allozyme type harbours several genotypes all having the particular allozyme in common. Some will migrate, some will not. Nevertheless, some insight in the complicated picture has emerged and will be discussed in short. For an extensive review on the genetic variation in *Daphnia*, see De Meester (1997).

In many lakes, hybrids have become dominant and even have replaced the parentals. Since the survey of Wolf and Mort (1986) in Lake Constance, the hybrid *D. galeata x hyalina* has increased in this lake according to Weider and Stich (1992). This hybrid has outnumbered the parental species also in Lake Maarsseveen (Spaak and Hoekstra, 1993). Although several genotypes of parents and hybrids coexist, a particular genotype is often dominant. The genotype composition of a *Daphnia* population complex changes with the seasons and over the years. This is possible because parthenogenetic species are capable of a fast reproduction, resulting in clonal lines.

Not all genotypes react with the same intensity to predator kairomones. With increasing knowledge, our appreciation of DVM becomes more complicated. This is illustrated by the famous description of the simple migration pattern of *D. hyalina* and a non-migrating *D galeata* in Lake Constance (Stich and Lampert, 1981). This was confounded since it became known that hybrids of the two species migrated to overlapping depths in between that of the two parental species (Weider and Stich, 1992).

Differences in migration behaviour of closely related clones may be small and mutual distances in distribution consequently not large. King and Miracle (1995) described for two clones of *D. longispina* in Spanish lake El Tobar the vertical differences in distribution. The relative frequencies over three depths, 0, 5 and 10 m, for two noon and two midnight samples are presented in Fig. 12.24. Although overlap was extensive, one clone was significantly higher in the water column than the other, in daytime as well as at night. Both clones are deeper in daytime than at night thus they migrate. However, do these two clones really differ in migration amplitude or do they start from different depths? A quick and dirty analysis of the day–night depth differences (2.73 and 1.80 m, respectively) suggests that amplitude is not significantly different (Mann–Whitney: P=0.059) and, therefore, photobehaviour is probably the same. The distinct difference in depth distribution of the two clones might be an adaptation to the typical environmental circumstances of lake El Tobar. A saline–freshwater interface separates the well-developed monimolimnion

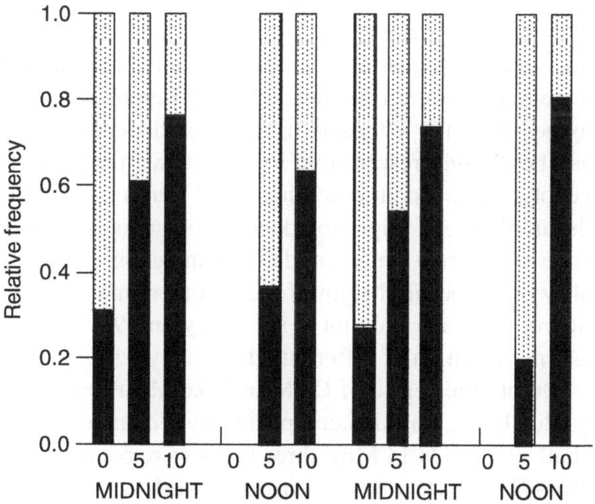

Fig. 12.24 Frequency of two clones of *D. longispina* in Lake El Tobar (Spain). For each time of day, these sample depths (m) are indicated below the *bars*. (Slightly modified after King and Miracle, 1995)

below 12 m from the freshwater mixolimnion. At the interface nutrients accumulate, chlorophyll *a* concentration peaks and light intensity rapidly declines (Vincente et al., 1993). In summer the two clones are present but the deep one disappeared during mixing in autumn (Miracle et al. 1993).

Also in Lake Maarsseveen hybrids of *D. galeata* and *D. hyalina* increased in dominance over the last decades. Before turning to the differential migration of the allozyme types, a short exposé is given to illustrate the dynamic character of the *D. galeata-hyalina* complex in the lake. *D. galeata* was abundantly present from June 1989 until December 1990 with a relative maximum of 56% in July 1990 (Spaak and Hoekstra, 1993). Thereafter, the species disappeared below our detection level in a few months and diverse hybrids took over. In 1991, this genetic variation in hybrids was largely lost and the entire population became almost fixed for one genotype. Next year on 25 May, before the period of migration started, two genotypes were dominant although other genotypes were still found. During the period of migration, the composition of the hybrid genotypes changed dramatically. One type, MFSF, became dominant for more than 80% while another one, MMMF, remained second best (Spaak and Ringelberg, 1997). When allozyme analysis was resumed in May 1997, the *Daphnia* population complex was dominated by the hybrids MMMF and MMFF. MFSF had been dominant in 1992 but was present now in very low percentages only. In the course of the summer, the hybrid MMMF that had left the field to MFSF 5 years earlier, now increased strongly and on 18 June reached a dominance of 80%. A same allozyme type might perform differently in another lake or at other times.

The environment changes for migrating animals. Food availability, temperature and the light climate differs. Consequently, the direction of selection becomes different. The new set of conditions may affect the competitive ability of the participating genotypes. Interclonal differences in life history of the hybrids were demonstrated (Reede and Ringelberg, 1998) and interclonal competition will contribute to a shift in the composition of the genotypes during a migration period. Although the presence of fish predators is the direct driving force of DVM, under this scenario predation itself is of minor importance. Spaak and Ringelberg (1997) remarked that the time needed for differential competition to become effective is longer than for a (passive) genotype-dependent predation. Whether competition can be thought responsible for a shift within 2 weeks is questionable but a direct, genotype-dependent predation by 0^+ perch is also difficult to imagine. However, if the amplitude of migration depends on genotype, and genotypes arrive in the danger zone at different times, predation can become genotype specific. Knowledge of the allozyme types in the gut of the predators in comparison with the relative composition in the water column is necessary. In 1997, we tried to study in more detail the ascent migration of different allozyme types.

Before the period of DVM in Lake Maarsseveen started in 1997, population size gradually decreased accompanied by a change in genetic composition (Ringelberg et al., 2004). On 14 May, three types were present in about equal numbers: the backcross *D. galeata x hyalina-hyalina*, MMMF and MMFF. On 4 June 1997, allozyme type MMFF migrated already, considering the mean depth of 11 m at noon and

6.5 m at night. The total number of adults had decreased by 27% notwithstanding an absolute increase of MMMF. On 11 June, the other allozyme types had followed because now 53% of the adults present in the epilimnion at night had descended to the metalimnion and the hypolimnion. The early start of MMFF might be an indication of a possible higher sensitivity for perch kairomones, for example. De Meester (1997) suggested that the onset of DVM in Lake Maarsseveen is a "combined effect of a phenotypically plastic response to the presence of fish kairomones and genotypic selection for clones showing the appropriate behavioural changes" (p. 81). The earlier start of MMFF compared to the other allozyme types seems to support this suggestion. Of the 47% of the population that did not migrate on 11 June nearly all were of the MMMF type. The descending daphnids consisted for 53% of MMMF, 27% of MMFF, 13% of MFSF and a diverse rest group. After 23 h on that day, the genetic diversity of the population in the epilimnion increased due to the ascent of a variety of genotypes from below. Within a week, the epilimnetic non-migrating subpopulation of MMMF was halved, paying the price for not migrating. In the guts of the 0^+ perch, MMMF dominated. Nevertheless, MMMF remained the dominant allozyme type in the water column. Most probably, MMMF consisted of multiple clone groups, comprising migrating and non-migrating genotypes. In 1992, MFSF replaced MMMF but that MMMF might have been different from the one becoming dominant in 1997. This example illustrates the inadequacy of allozyme analysis. In spite of uncertainties, analysis showed that the non-migrating part of the *Daphnia* population, observed for years in Lake Maarsseveen, is genetically different from the migrating part. A difference in migration pattern is illustrated in Fig. 12.25. The distribution is of 11 June and thus from before the large decrease of the epilimnetic subpopulation occurred. The short survey of the genetic diversity of the *Daphnia* clonal community in

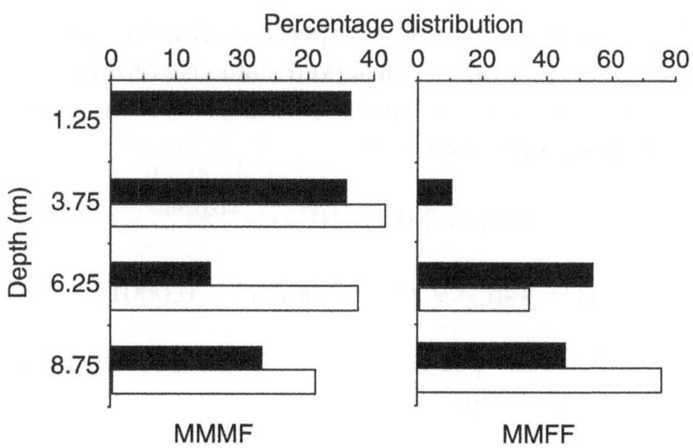

Fig. 12.25 The percentage vertical distribution of the allozyme types MMMF and MMFF at noon (*white bars*) and after 23 h (*black bars*) on 11 June 1997 in Lake Maarsseveen. At each time and depth, the allozyme types are represented by a percentage of the total allozyme community. The percentages are based on $n = 143$ (noon) and $n = 141$ (night) for MMMF, and $n = 64$ (noon) and $n = 31$ (night) for MMFF. (Modified after Ringelberg et al., 2004)

Lake Maarsseveen indicates that the erstwhile simple interpretation of migration patterns must be revised. This will be possible only if the variation in DVM is studied at a genetically more detailed level with the advanced techniques that are available now and, of course, in combination with the experimental analyses of photobehaviour.

Notes

1. Although predator-induced DVM is by far the most interesting, migrations with an amplitude too small to be of value for predation prevention also occur (Chapter 9). The migrating animals remain within the epilimnion and costs, if present at all, must be small. As adaptive significance, the prevention of damage by ultraviolet radiation was suggested. This kind of DVM seems to be a constitutive or a genetically fixed strategy. Photobehaviour system 1 was proposed to be responsible and this kind of phototaxis can be nearly always aroused, also in the absence of fish kairomones. It can be imagined that within a population the constitutive and the phenotypically strategy of migration is present in all individuals, although to a different extent. That would explain why in the beginning of the migration period in Lake Maarsseveen for many years it was observed that part of the population does not perform the extensive migration.
2. It might be thought difficult to determine precisely the start of migration. If diel vertical migration is absent, the ratio between mean day and night depth of a population category, for example, adults, must be one. For the pre-migration period in 1989 and 1993 this ratio was 1.056 for the adult part of the population up to day 144 (s.d. = 0.154, s.e. = 0.0265, $n = 34$). For the next sampling date, when migration was supposed to have started, the ratio was 1.810 (s.d. = 0.1621, s.e. 0.054, $n = 9$) and obviously different from the pre-migration period.
3. Blaxter (1970) presents his threshold light intensities in metre candles. These have changed in lux (ratio 1:1). The conversion of lux into mmol s^{-1} m^{-2} was at 1:0.02 (Instruction manual LI-COR (1982)).

Appendix

With this calculation of the predation by 0^+ perch on *Daphnia* in Lake Maarsseveen, I have kept as close as possible to observed data. Nevertheless, as always with this kind of exercise, assumptions had to be made. They are mentioned and can be valued. The following steps were made:

$$N(t) = 2.04 \times 10^5 \times e^{-0.079t} \tag{12.1}$$

$$TL(t) = 446 - 9.23t + 0.0622t^2 - 0.000132t^3 \tag{12.2}$$

The decrease in number of 0^+ perch in time (12.1) and the increase in total length can be found in the legend of Fig. 12.8.

$$W(t) = 0.0095 \times TL(t)^{3.0677} \tag{12.3}$$

$$\mathrm{Wdrywt}(t) = 2.0 \times 10^{-3} \times TL(t)^{3.0677} \tag{12.4}$$

The conversion of length into wet weight (12.3) is presented in Fig. 12.10. The conversion from wet weight to mg dry weight (12.4) was by a factor of 4.75. This value was based on a combination of 5.28 (original estimate), 4.17 Mehner {976}, 4.79 Vijverberg et al., (1990) and Mooy (1964):

$$\Delta W = W(t+7) - W(t) \tag{12.5}$$

Equation (12.5) calculates the difference in dry weight over 7 days which is the interval between samples of *Daphnia* made in the lake.

$$\Delta B = \frac{N(t+7)) \times \Delta W + 0.5((N(t) - N(t+7)) \times \Delta W}{(t+7) - t} \tag{12.6}$$

For the increase in biomass of the 0^+ perch population in the lake was assumed that half of the individuals had weight increases to the same amount but then died. The biomass increase is per day. To realise this increase in biomass a certain quantity of food had to be consumed. An efficiency of 25% as compiled from the literature, mentioned by Vijverberg et al., 1990, was used. Thus the total food consumed by the perch per day was 4ΔB. Of this, a certain proportion consists of *Daphnia,* depending on length of the perch and time of day. Based on gut analysis and Total Length, an estimate of the percentage daphnids was made:

$$\%\text{Daphnia} = -27.05 + 2.66TL \tag{12.7}$$

However, this estimate was made with perch caught in the lake between 2100 and 2230 hours and between 0500 and 0600 hours when other zooplankton species were also predated upon (Fig. 12.16). During the night, only daphnids were found in the gut. It was supposed that at 2 × 2.5 h a mixed diet was consumed and at 5 h only *Daphnia* were eaten. With an average length of the perch of 30 mm on day 165 in the centre of the period of calculation, according to (12.7), 53% daphnids were consumed during the evening and morning and 100% during the night or 76.5% over the whole period.

Length was calculated for the seasonal period from (12.2) and substituted in (12.7). The percentage *Daphnia* was increased with 76.5%. The dry weight consumed by the perch was calculated as

$$\Delta C = 4\Delta B \times \%\text{Daphnia} \tag{12.8}$$

Since ΔC is in mg dry weight per day while *Daphnia* population parameters are expressed in numbers, dry weight of an average length *Daphnia* was calculated from (12.9), compiled from Bottrell {61}

$$W(\text{mgdryweight}) = 4.7 \times 10^{-3} \times L^{2.70} \tag{12.9}$$

Average length of adult *D. galeata x hyalina* in Lake Maarsseveen was measured as $\bar{L}$= 1.514 mm, n = 160, s.e. = 0.0068 mm.

Chapter 13
Recapitulations and Considerations

Contents

13.1 The Importance of Studies of Behaviour

Our knowledge of behaviour at the base of migration is limited. Researchers of zooplankton are not often enthusiast students of behaviour. There are notable exceptions, of course, and we have encountered them in the previous chapters. In the "World of Plankton", a beautiful book by Sir Allister Hardy, the richness of the tiny organisms in the seas and oceans is described but the real world, as experienced by a copepod, a medusa or a chaetognath, is hardly touched upon. To touch upon that world, studies of behaviour are necessary.

Research never starts without preconceptions created by those before us. We are no tabula rasa. In the course of time some notions were largely forgotten and new ones became fashionable, as the present-day prejudices. They provide limited views on the real animal world, and I called them "windows" in Chapter 1. These windows are ways of thinking, frames of mind, not methods. Some have been moulded a long time ago and are antiquated because the art of observation is largely lost now the statistical processing of large data sets is supposed to lead to the truth. Consider the simple idea about a "Wirkwelt" and a "Merkwelt". The distinction draws attention to what the animal "feels" and what it perceives directly by means of its senses. It is a simple distinction but it led to the recognition of the help that animal functions have when vegetative needs cannot be solved directly by adequate behaviour. DVM

J. Ringelberg, *Diel Vertical Migration of Zooplankton in Lakes and Oceans*,
DOI 10.1007/978-90-481-3093-1_13,

is an example: without the perception of light changes and phototactic swimming, the avoidance of perceived predators would be impossible. I think some time must be devoted to the art of observation if we want to really understand the mechanisms leading to DVM.

The existence of adequate behaviour is often taken for granted in the migration literature. I mentioned "depth selecting behaviour", ascribed to *Daphnia,* but no author seems to be bothered by what factors and mechanisms selection of depth is realised, nor how that depth is maintained. Oscillating swimming is a possible solution for keeping to a certain depth.

Without knowledge of the mechanisms of individual behaviour, migration cannot be causally explained. Of course, not every ecologist has to analyse behaviour as detailed as is presented in Chapters 2, 3, 4 and 7. Moreover, the first principles of animal physiology are hardly taught anymore. The result is that mechanistic explanations have become scarce in the recent migration literature and hidden goal-directed explanations are offered instead. I will give an example. At the end of the clear water phase, food concentration was low in a reservoir. Young-of-the-year perch had already started to predate on *Daphnia* and Hulsmann and Wagner (2007) thought it strange that *Daphnia* did not migrate. A few weeks later when the algal concentration had increased, DVM started. The proposed explanation was that *Daphnia* could not risk the costs arising from an induced migration as long as food was scarce. This is not a causal explanation but "ultimate reasoning" . It might have an element of truth in it but, as far as we know, *Daphnia* is unable to assess "costs" and, what is necessary too, assess the "benefits" of a possible migration in order to compare both. This kind of explanation might prevent a continued analysis. Costs and benefits are important criteria for selection in the evolutionary history of the development of a physiological–behavioural mechanism. To substantiate the ultimate reasoning, costs and benefits have to be quantified and a change in ratio followed in time. In that way, insight in that other realm of biological explanation is obtained but it can never replace the causal explanation of behaviour. The "decision-making" mechanism as is presented in Chapter 4 is relevant in this case. In the previous chapters was stressed that in causal explanations of an ecological phenomenon nearly always more than one factor must be considered. A "set of necessary ecological conditions" renders the complexity. In the given example, predation was sufficient reason to look for an argument why migration did not occur. Therefore, also kairomone concentration must have been of influence but no attention was given to that factor. The concentration of food and kairomones might have been too low to enhance photobehaviour and facilitate migration. A comparable situation occurred in Lake Maarsseveen and was described in Chapter 9.

It seems useful now to present a comprehensive overview of the essential proximate aspects at the base of DVM. Figure 13.1 is a guide. Drifts are slow, interrupted displacements upwards or downwards. "Normal" swimming continues, but either the upward phase or the downward phase is a little lengthened. These drifts not only occur during prolonged sub-threshold changes in light intensity but also in the absence of these changes. In that case, it can be imagined that the internal oscillator is slightly out of phase (see Chapter 2) and perhaps under influence of an endogenous clock that is present in *Daphnia* too. For example, Harris (1963) found that in continuous light the position of *D. magna* in a tube changed in the course of the day. Ringelberg and Servaas (1971) determined the percentage reactions caused by relative decreases in light intensity over 24 h. The percentages were lower during the dark period but increased before light was normally switched on in the culture. Under continuous light, *D. magna* kept to this rhythm for at least 6 days, although the subjective night shifted about 3 h 50 min per day, an exceptional free running period but the same as the one found by Harris (1963). Ringelberg and Servaas (1971) assumed that an endogenous activity rhythm was responsible, activating the animals in time before dawn.

Supra-threshold relative increases or decreases in light intensity affect the internal oscillators by prolonging the downward or upward phase of "normal" swimming, respectively. This mechanism was dubbed photobehaviour system 1 (PBS 1) (Chapter 3). In the course of reactive swimming absolute intensity changes and thresholds must change. The change in relative intensity is an approximation of the real stimulus. It would be better to divide the change in absolute intensity ΔI not by I but by a function of I, thereby incorporating the dependence of the present stimulus on absolute light intensity.

I suggested that PBS 1 is genetically fixed in *Daphnia* and protects against damage of UVB radiation (Chapter 9). It can be imagined as first evolved and that PBS 2 evolved superimposed on PBS 1. Photobehaviour system 2 is essential to DVM. The rate of relative changes in light intensity accelerate and decelerate during dawn and dusk and these light effects lead to an enhancement of PBS 2 and was dubbed the phototactic coefficient (PC) in Chapter 4.

Of essential importance is that the concentration of fish kairomones and of algal food determine PC. In that way the "decision-making" mechanism leads to different migration amplitudes.

It must be realised that decision-making in *Daphnia* and other zooplankton species is not a conscious process. Although unknown, the central nervous system

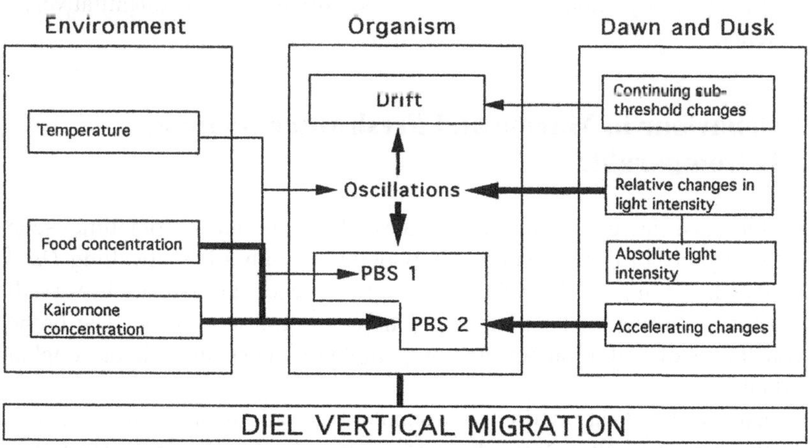

Fig. 13.1 A comprehensive scheme of the physiological–behavioural mechanisms at the base of DVM as developed for *Daphnia*

must have an important role in making a decision. The process must be considered an elaborate and complicated sensor-actor chain (Chapter 1). Since ultimate aspects of DVM have become hype, adapted and optimal behaviour is central in discussions of articles on DVM. However, I argued that one particular optimal behaviour does not exist and we can better leave the notion for what it is: not very realistic. If different genotypes "decide" to migrate with different amplitudes (Chapter 12), notwithstanding a similar environment for all of them, it is better to embrace the dynamic character of "decision-making" and refrain from "optimality".

In the presentation of photobehaviour and the light stimuli the effect of changes in wavelength was not discussed. Nevertheless, it is well-known that colour changes also cause phototactic swimming. A large, predominately older literature exists. Effects of colour changes in DVM were even proposed. For example, Smith and Baylor (1953) suggested that the red shift during sunset caused the ascent migration. It is true, that a spectral shift to longer wavelengths evokes a swimming of *Daphnia* in the direction of the light source. However, the red glow of the setting sun in the evening is invisible under water because the oblique incidence of the sun's rays are mostly reflected at the water surface and do not penetrate the water column. Below the water surface the spectral composition shifts to the blue instead. Interestingly, this latter effect was used by Frank and Case (1988) to suggest that shrimps used the blue shift in the evening to start the upward migration. Later, Frank and Widder (1996) withdrew the suggestion because they showed that the shift could not be seen sufficiently deep. It is improbable that diel shifts in wavelength play an important role in DVM.

The "decision-making" mechanism combines the proximate and the ultimate aspects of DVM. It evolved necessarily attuned to the need of adaptation within different concentrations of kairomones and food. I think that the discussions of ultimate aspects as it occurs in the literature at the present moment create no new aspects anymore. Experiments along the lines of decision-making promise a way out of the present cul-de-sac. The generated amplitudes can be used, for instance, in models of actual migrations, offering the possibility to estimate quantitatively costs and benefits.

13.2 Behaviour in Marine and Freshwater Biotopes Is Comparable

Marine biologists are used to the vast expanse of oceans and are sometimes sceptical about the validity of experimental results obtained with a lake-dwelling *Daphnia*. To us, the dimension of the pelagial biotope of lakes and oceans is very different, but that is not so for zooplankton. This is a question of scale. The properties and constraints of both biotopes are comparable on the small scale of zooplankton individuals.

In freshwater and the marine environment, migrations occur at the same time during the daily transition from darkness to light, and vice versa. These provide comparable stimuli all over the world. I think that photobehaviour, as described for *Daphnia,* is of universal validity in marine and freshwater zooplankton (see Chapter 11).

The vertical plane is where to look for in order to understand zooplankton life. In lakes and ocean, the water column is often structured by temperature and food concentration over distances that can be covered by swimming. The structure of vertical space differs over time and species-specific depth distributions are not stable.

Animals must orient in the vertical plane and thus far optical orientation has been the only described option. On first sight, optical orientation seems strange in an optically empty environment, but contrasts around the maximum angle of refraction at the water surface are universally present up to large depths. Contrast orientation was worked out with copepods and cladocerans from lakes but the tracking experiments with *Euphausia* can be interpreted in the same way as those with *Daphnia*. Optical body axis orientation is another common adaptation of marine and freshwater zooplankton (Chapter 7).

A critical aspect of uniformity and emptiness is the absence of hiding places as refuges for predators. Prey react to hydrodynamic disturbances made by swimming zooplankton predators. These tactile stimuli operate over small distances and on short notice. Jumps or immobility are effective escape behaviour. However, against predating fish they are less effective. These "escape" reactions are too weak to escape from fish with a superior swimming speed and visual acuity. A timely retreat is necessary. Diel vertical migration evolved not as *predator escape* behaviour but as a *predation prevention* strategy. DVM is a timely departure for greater depths, it is an "early warning" for the coming threat of predation.

If the proximate aspects of DVM are highly comparable in both biotopes, this does not hold for the adaptive significance of DVM. In lakes, protection against UVB radiation might play a role, but the prevention of predation by visually hunting fish holds a prominent place. Predation prevention has been described for estuaries, fjords and the like but to what extent it is an important reason for migration in oceans is not clear. The high complexity of the oceanic community makes it likely that competition also plays a role as was suggested for the species-rich assembly of euphausiids in the Atlantic Ocean near the Canary Islands (Chapter 10). Non-migrating species seemed to be vertically separated while in daytime the migrating ones were closer together below the depth of the non-migrating species but at night moved upwards, dispersing over the water column. Although difficult to study, this possible aspect of depth separation and migration is important for understanding the oceanic community.

13.3 Ontogenetic Developments

Ontogenetic developments of morphology and migration behaviour are fascinating. In Chapter 10, a few examples of differences in DVM by marine calanoid copepods were presented but ontogenetic differences are also present in freshwater cyclopoids and calanoids, for example, *Cyclops abyssorum* and *Eudiaptomus gracilis* (Einsle,

1987). In Chapter 11, it was described that only the final aquatic stage of phantom larvae (*Chaoborus*) performs DVM. Obviously morphology and DVM or better photobehaviour develops in concert and might be coupled. No systematic studies were made of the behaviour of nauplii, copepodites or adults of planktonic copepods, thus we must largely do with informed guesses and suggestions for future research.

I suppose that photobehaviour is basic to DVM also in the larval stages of copepods and that this behaviour, with the accompanying traits as perception of kairomones and food availability, evolved parallel with the changes in morphology. Because nauplii and the first copepodites of pelagic *Calanus* species do not migrate, photobehaviour has perhaps not developed yet. Nauplia of the non-pelagic brine shrimps (*Artemia*) react to changes in light intensity even with an enhanced response in the presence of fish kairomones (McKelvey and Forward, 1995; Forward and Hettler, 1992; Forward and Rittschof, 1999a) (Chapter 4) but the nauplius of this species might not be representative for those of pelagic copepods because they are larger and more vulnerable for predation. Of importance in the present context is the presence of photobehaviour as part of predator avoidance in these nauplia.

The third nauplius stage (N3) of *Calanus*, which is the first one feeding, moves upwards in the water column (Uye et al., 1990). This might be the light response indicated as "drift" in Fig. 13.1. Also N4–N6 and the non-migrating copepodite stages C1–C3 of *Calanus sinicus* and of *C. finmarchicus* (Dale and Kaartvedt, 2000) aggregate in the phytoplankton-rich euphotic zone and might show this drift behaviour. Starting with C4, real DVM occurred in *C. sinicus* but not in *C. finmarchicus* (Dale and Kaartvedt, 2000) and the amplitude increased in C5 to reach a maximum in adult females. Therefore, it is tentatively concluded that photobehaviour had developed. Huang et al. (1993) correlated the day depth of C1 through C5, adult females and males with the relative biomass of planktivorous fish but found a significant correlation for C5 and adults only. Therefore, it seems that the perception of fish smell (kairomones) is still absent in C4 but had developed in C5. A correlation between the chlorophyll *a* concentration and adult females was positive too but not with C5. These correlations suggest a stepwise development of the photobehaviour mechanisms that were described in Chapters 3 and 4. For example, the reaction to relative changes in light intensity (PBM 1) probably develops in C4, but the response to accelerations in light changes as dependent on food and fish kairomone concentration (PBM 2) had to wait until C5 and the adult stages had been reached. It must be stressed that these hypotheses about the development of mechanisms, basic to migration, are based on few correlations of DVM and environmental factors in the field. The development of photobehaviour can be tested experimentally, however. In photobehaviour experiments, attention must be given to a proper angular light distribution to which optical orientation and "normal" swimming is possible (Chapters 6 and 7). This will be more difficult to realise in experiments with copepods than with *Daphnia* whose orientation mechanism is more versatile. Examples of experimental apparatus can be found in the previously mentioned papers of Forward and co-workers. The set-up does not permit extensive vertical swimming but thresholds can be determined, which is the beginning of an exploration of the role light changes, food and kairomone concentrations play in the ontogenetic development of copepods.

In Chapter 10 was also noticed that the development of ontogenetic stages changed niches, population dynamics and food web structure. Survival was indicated as a chain of chances from egg to egg. Based on patterns of migrations, suggestions were made in the literature about these stage-specific adaptations but they are a far cry of the chances of survival. Of importance is the synchronisation of hatching and stage sequences with algal blooms. Changes in the environment, such as climatic temperature increases, might disturb synchronisations, thus population development. The adverse effects for food web relations go without saying because copepods in oceans are important links. We are at the beginning of the quantitative analysis of the functioning of food webs and have insufficient knowledge of the communication between participating taxa. Pelagic food webs are relatively easy to study quantitatively, especially those in freshwater and in semi-closed marine biotopes like Dabob Bay (see Chapter 10). A detailed study of ontogenetic development of behaviour and adaptive values of a freshwater species like *E. gracilis* is feasible if organised well. Interaction with *Daphnia* is an aspect of attention because it provides a model on which more difficult-to-access marine webs can be based.

13.4 Notes on the Chemical Nature of Infochemicals

If concentrations of infochemicals cannot be determined, their role cannot be studied further. Analysis of the chemical nature of the important fish kairomones did not lead to a consensus and even the precise origin remains controversial. Forward and co-workers (a.o. Forward and Rittschof, 1999; McKelvey and Forward, 1995; Cohen and Forward, 2003) found that mucus from the marine fish *Fundulus heteroclitus* and the non-visual predating ctenophore *Mnemiopsis leydyii,* enhanced photobehaviour in larvae of the crab *Rhithropanopeus harrisii* and the brine shrimp *Artemia franciscana* (see also Chapters 3 and 4). Analysis of the mucus and tests with commercially obtained mucopolysaccharides revealed that chondroitine A disaccharide and hyaluronic acid disaccharide, both common constituents of invertebrate as well as vertebrate mucus, had similar enhancing effects. Therefore, mucus-derived degradation products are likely candidates for kairomones in marine systems.

The source of kairomones in freshwater is less clear. Von Elert and Pohnert (2000) studied photoresponses of *D. magna* using kairomone water from pikes (*Esox lucius*) and sticklebacks (*Gasterosteus aculeatus*). The extent of the reaction depended on the extent of dilution as had been shown for kairomones from cyprinid fishes. Chemical analyses showed similarity of active substances over taxonomic different freshwater fishes but the essential factor was not found. Mucus did not seem to contain that factor, although Ringelberg and Van Gool (1998) had found that mucus from juvenile perch, treated with antibiotics, enhanced photoresponses significantly less than mucus from untreated perch. They suggested that bacterial breakdown products were probably responsible.

Von Elert and Stibor (2006) made a subsequent attempt at characterisation, using Size at First Reproduction as criterion. Interestingly, HPLC revealed that active fractions were different from those that activate photoresponses. However, the exact structure of fish kairomone remained elusive.

It would be surprising if kairomones mediated by different freshwater fish would differ much from kairomones produced by marine fish. All have a mucus envelop with a comparable composition and the mucus of the marine fish harbours the responsible kairomones, as did the mucus of a taxonomic totally different coelenterate. There is one observation indicating that also in freshwater fish mucus might be a source. Therefore, it would be interesting to test mucus from freshwater fish again using photobehaviour as test criterion because that reaction to relative changes in light intensity is at the base of DVM where kairomones play a crucial role.

13.5 Information Exchanges

Kairomones from different species might be present simultaneously in the water column. Non-specific infochemicals elicit non-specific prey defences, as different species of freshwater fish do. An increase in population size of one predator must lead to a higher kairomone concentration and possibly to an increased prey response, for instance, a larger amplitude of migration. The resulting decrease in prey availability might drive another predator to a critically low density. There is no knowledge about the reality of these subtle predator–prey interactions but the consequences for population dynamics and community composition can be imagined.

Predators mediating different kairomones, eliciting predator-specific defences with conflicting strategies, may result in risk-enhancement, for example, 0^+ perch and *Chaoborus*. Normal DVM evoked by the fish is in conflict with the reverse migration induced by the phantom larvae. The behavioural effect has not been studied as far as I know but life-history effects in *D. galeata* proved to be in conflict (Weber and Declerck, 1997) as can be expected because fish prefer large-sized and *Chaoborus* larvae small-sized prey. If it is supposed that these kairomones have linear, additive effects with equal effective strengths, life-history adaptations would be nullified leaving the prey unprotected. In that case, multiple predator assemblages would be risk enhancing. This did not occur in the culture experiments by Weber and Declerck (1997) because the *Chaoborus* effect seemed to be neutralised or overruled by the perch kairomone. Since the *D. galeata* originated from a lake with heavy fish predation, the authors suggested that selection for genotypes with a higher sensitivity for fish kairomones could have been responsible. The assumption cannot be rejected, of course, but nothing can be said about conflicting effects of species-specific kairomones. More in line with the original hypothesis of departure is that a higher (effective) concentration of the fish kairomone had been responsible. The experiment must be repeated with the determination of separate effects of the (regrettably unknown) kairomone concentration before combined effects are studied.

Is the simultaneous occurrence of multiple predators in lakes a real problem? Most invertebrate predators are preyed upon by fish and reach a high abundance after the 0^{+} fish have left the open water zone. The higher diversity in the marine environment may present a different picture. Osgood and Frost (1994b) described that copepodites C3 and C4 of *Metridia pacifica* performed a reverse migration. The authors suggested that the adaptive significance was to escape from the invertebrate predators *Euchaeta elongata* and *Sagitta elegans* migrating normally to evade visually predating fish. This observation and story calls for an experimental investigation into the development of photobehaviour. The role of simultaneously present predator-specific kairomones and behaviour is of high physiological and ecological interest.

Infochemicals also play a role between herbivores and their algal food. Hessen and Van Donk (1993) and Lampert et al. (1994a) found that *D. magna* released an infochemical that enlarged the number of cells in colonies (coenobia) of *Scenedesmus subspicatus* and *S. acutus*, respectively. Subsequently, several species of *Scenedesmus* were extensively tested by Lürling (1999) using water that had been incubated with different *Daphnia* species. Enlarged coenobia were not always induced, not even in the two species originally studied.

The *Daphnia* factor does not only induce colony formation. In *Staurastrum spec*, Wiltshire et al. (2003) observed that mucus excretions made individual cells clump together into large aggregates. Planktonic desmids often have mucilaginous envelopes (Coesel, 1994, 1997). Diatoms, like *Asterionella formosa* and *Synedra tenuis,* enlarge the number of cells in aggregates when the *Daphnia* factor is present (Lürling, 1999) but also in these taxa the effects were variable. Obviously, the phenotypic induction of colonies and other effects is widespread under algae but not universally present.

Grazing is necessary to make "*Daphnia* water" active (Lampert et al., 1994). Van Donk et al. (1997) wrote that "the inducing factor probably originates from the daphnids digestive system" but this was rebutted by Von Elert and Franck (1999) who thought that the chemical cue originated from non-digestive metabolism rather than from the digestion of alga. They even suggested that bacteria in the gut were involved, since antibiotics added to the incubation water prevented that coenobia were formed. History seems to be repeated. Also fish kairomones were thought for a long time to be the product of digestion of *Daphnia* in the guts of 0^{+} fish, until Ringelberg and Van Gool (1998) found that phototactic swimming was significantly less enhanced after 0^{+} perch were treated with an antibiotic.

The *Daphnia* factor elicits a conditional anti-predation response in algae and, therefore, these infochemicals can be called kairomones. Whether the morphological change is effective depends on the size of algae in relation to the size of the grazer. Hessen and Van Donk (1993) observed that four- and eight-cell coenobia of *S. subspicatus* reduced even grazing in the large *D. magna*. In contrast, Lampert et al. (1994) mentioned that grazing on coenobial *S. acutus* was only reduced for the small *D. cucullata*. Coesel (1994) observed that removal of the mucus envelop in *Cosmarium abreviatum* facilitated ingestion in *D. galeata x hyalina*. These few examples show without doubt that algal colonies reduce grazing but Wiltshire et al.

(2003) doubted whether the ecological relevance of colony induction is really high because *Scenedesmus* is not a common species in lake phytoplankton and not the preferred food for pelagic *Daphnia*. The same holds for Desmidiaceae and Coesel (1994) doubted that mucus sheaths are evolved for the reduction of grazing.

On the other hand, the large group of flagellates, Cryptomonads and Chrysophytes, are common and abundantly present in the pelagic area of lakes and also favoured as food by zooplankton. The single cells form no colonies, nor do they change cell walls in the presence of *Daphnia* but these motile cells perform reverse diel vertical migrations swimming upwards very early in the morning and moving downwards in late afternoon or early evening (Jones, 1988; Soeder, 1967; Sommer, 1985; Salonen and Rosenberg, 2000). Amplitude is often restricted to a few metres but the day–night difference of the mean population depth in *Cryptomonas marsonii* might be 8 m. (Soeder, 1967), or 5–7.5 m in *C. ovata* (Tilzer, 1973). Active swimming is involved because in most cases the rate of going down is faster than can be accounted for by passive sinking. Jones (1993) even suggested that (relative) changes in light intensity are possible stimuli triggering the migrations. Multiple advantages characterise these reverse migration of flagellates. As autotrophes, they have to be in sunlit surface layers. But why go down during the night? In some cases it is evident that the uptake of nutrients is an ultimate reason for migrating downwards: *Gonyostemum semen* depleted soluble reactive phosphorus (SRP) near an anoxic bottom in a short time according to Salonen and Rosenberg (2000). As a third adaptive goal was mentioned the avoidance of predation by herbivores that perform a normal DVM. Hansson (1996) on finding that *G. semen* disappeared to the bottom in the presence of *Daphnia* suggested that this was an active avoidance reaction to grazers. Also Smolander and Arvola (1988) supported the grazing avoidance hypothesis for *C. marssonii*. They found that the largest amplitude occurred in July when the population density of zooplankton was highest and an SRP chemocline was even crossed. Jones (1991, 1993) likewise thought that an anoxic hypolimnion offers an excellent refuge against grazing.

The analogy with normal migrations of zooplankton is obvious. The nightly descent of phytoplankton is comparable to the daytime descent in zooplankton, being a strategy to prevent mortality by predators. In both cases the indispensable habitat is left: the availability of light for photosynthesis in algae is as essential as algal food is for zooplankton. The reverse migrations seem to be phenotypically induced and phototaxis might be the basic behavioural mechanism as is casually mentioned by Sommer (1985) and emphasised by Jones (1993).

Of importance is that *Daphnia* infochemicals induce anti-grazing responses in algal species too.

13.6 Infochemicals and the Pelagic Food Web

The given examples of chemical information transfer in plankton makes one suspect that infochemicals of all kinds play an important role in the communication highway of the pelagic area of lakes. Visual or sound stimuli do not work and tactile stimuli, although important, have a short effective range and no early warning function. Many simultaneously present chemical substances with a signal function cannot be a problem at the physiological level because the chemical senses and the central nervous system must have evolved in a way that only relevant infochemicals are reacted to.

Infochemicals are important for population development and food webs. A traditional food web is a construct of predator–prey links representing the pathways of material flow in the community. The quantity of these flows depends on the abundance of food and activity of consumer and these activities are to a large extent determined by information. Therefore, a network of information can be thought superimposed on the material food web. As far as I know the idea of an information network was first proposed by Patten and Odum (1981) but not developed and never picked up. Simple illustrations were given by Ringelberg (1987a,1997). In Fig. 13.2, an example of a tritrophic linear food web (black arrows) with the governing infochemical network in red is presented. For the time being, examples have to be simple. Increasing (+) and decreasing (–) infochemical effects are indicated. Van Gool and Ringelberg (1996) demonstrated that the number of *D. galeata x hyalina* swimming into the leg of a Y-tube with filtered water from a culture of *S. acuminatus* or *Oscillatoria limnetica* was significantly larger than the number in the control leg with clean water. They concluded that *Daphnia* could perceive the odour of algae. The experiments were repeated by Roozen and Lurling (2001) but the results could not be confirmed. Probably, this was due to the experimental technique used by these authors.[1] In Fig. 13.2, the allomone (terminology according to Dicke and Sabelis (1988)) representing enhanced feeding is inserted.

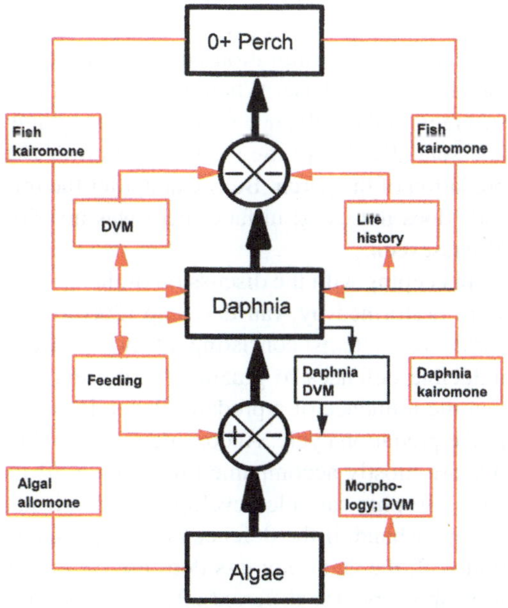

Fig. 13.2 Representation of a traditional tritrophic food chain (*black*) between algae, *Daphnia* and 0^{+} perch. The + and – signs in the cross-divided circles indicate the increasing or decreasing effects on the transfer of matter. Superimposed is the information network, consisting of infochemicals (*red lines* and *arrows*) and the mechanisms or phenomena that are triggered by the infochemicals in *squares*

The effect of the information flows is determined by the concentration of the information carrying chemical substances and this is to a large extent determined by the biomass of the populations. The material-information network looks like a cybernetic system (Chapter 1) with (in this case) three negative feedbacks and one positive feedback. For example, an increase in 0^+ perch biomass (Chapter 12) leads to stronger migration responses of the *Daphnia* population and a decrease in material flow (or more simply said, less food) for the perch. As the predator biomass declines, less fish kairomone will be produced and the response of the *Daphnia* will decrease too. Negative feedback loops stabilise the system but the response times have to be short. Response times of life-history changes are long in comparison with photobehaviour leading to DVM and therefore are less effective as was already concluded in Chapter 12. However, the Perch-*Daphnia* subsystem will not reach equilibrium on the short time scale of the migration period because the development times of the participating biological units are too different. Especially the growth of individual perch is independent of the subsystem. However, on a time scale of years DVM might have a stabilising effect because in the absence of migration the daphnid population would be reduced to a very low level and followed by a mass starvation of the 0^+ perch, thus recruitment would be very poor.

In the *Daphnia*-Algae subsystem a positive feedback is included as realised by the allomone that amplifies feeding of *Daphnia*. In general, positive feedbacks are inherently destabilising. Also in the present case this can be easily imagined. Better algal food detection will lead to higher grazing and large fluctuations in the populations of both algae and daphnids with the eventual disappearance of the daphnids. A counteracting and stabilising effect is expected from the change in morphology and algal migration. In addition, a negative effect on the transfer of matter towards *Daphnia* and thus a strong positive effect on the algal population is realised by DVM of the herbivore. In the previous chapter the recovery of the *Daphnia* population in the course of the migration period was ascribed to a renewed algal growth ending the clear water phase. When the migration period starts, the top down cascade gets strong with the influence of the perch reaching down to the algal level although this is essentially due to the strong reaction of *Daphnia* and not because of the heavy predation of the perch. Before and after the migration period a strong tritrophic cascade does not exist; in fact, there was no tritrophic system of importance in Lake Maarsseveen.

In keeping with the discussion of the trophic web are interesting experiments that were performed by Van der Stap (2007) with tri- and bi-trophic algal-herbivore-carnivore systems consisting of two rotifer species and algae with or without inducible defences to grazing. Of present interest is the conclusion of the author that the influence of a predator depends on the time after the prey defence appears in the predation cycle. According to her, induced defences becoming rapidly effective are mostly accompanied by a trade-off in algal food uptake preventing that a strong trophic cascade develops and this leads to an increase in algal biomass. On the other hand, if the defence becomes effective late in the predation cycle with no trade-off, the algal biomass does not change. The terminology differs from the one used above but the different effect of short and long response times in the feedback loop can be recognised. Van der Stap (2007) seems a bit surprised by this "somewhat counterintuitive situation that herbivore defenses may either prevent or facilitate a trophic cascade" (p. 82). It all depends on the presence of a trade-off with regard to food uptake. Translating to the Lake Maarsseveen situation, the rapidly developing predator-induced defence DVM has an evident trade-off with the algal food uptake and the algal population changes, whether due to the indirect effect of the introduced fish or not. The change in life history also presented in Fig. 13.2 develops slowly and does not affect food uptake. Now the trophic cascade is not blocked and the algal population should not be affected.

Obviously, we deal with two different approaches or two different windows (Chapter 1) of looking at food web relations. A natural situation is, of course, more complicated than an experimental one. The merit of experiments is in simplification because it makes understanding easier. The web of Fig. 13.2 is, however, also a simplification. Not only the algae but also *Daphnia* (which, in previous chapters, was called a clone collective consisting of different genotypes with different migration patterns and, therefore, different positions in the food web) are unspecified. In the large Plöner Plankton Towers, Spaak and Boersma (2006) studied the influence of planktivorous fish and their kairomones on assemblages of 20 clones of the *D. galeata-hyalina-cucullata* complex as characterised by allozyme analysis. One important result was that the instantaneous rates of clonal increase depended on the intensity of predation. Continuously changing predation pressure facilitated the co-occurrence of taxa in the *Daphnia* complex. In Lake Maarsseveen, various clones migrated to a different day depth (Fig. 12.25) leading to a time-sequential appearance in the danger zone of predation in the evening. We deal with clone-dependent predation contributing to clonal diversity during the period of predation as Spaak and Boersma (2006) observed in the Plankton Towers.

13.7 Finally

How to continue research? Not in the same way. Ecological investigations easily become anecdotal and then do not comfortably contribute to a consistent body of knowledge. Migration research is an example. There are many species and many biotopes with different properties, even in the relatively simple aquatic environment. For some time, we have thought that the species interacted in a way that the community or the ecosystem was more than a mere collection of selfish items. But the holistic approach did not provide sufficient answers to believe that the right road was taken. Research continued to drift away in different directions, topics and species. We need a specific subject around which different aspects of interest come together. Diel vertical migration can have such pivotal function around which physiological, behavioural and ecological problems can be formulated. I tried to illustrate this in the book.

New roads of research are announced. Information transfer directs population development and the study of species-specific relations that rule the community must become more sophisticated. Genetic analysis will become of more importance in DVM research. The application of the powerful techniques must serve to provide insight into the central topic and not drift away.

Descriptions in general terms of vertical distributions and migrations in the field are not necessary anymore and must be replaced by sampling programmes designed to solve special problems.

However, above all, experiments must be done. Techniques must be developed in order to better study behaviour not only with *Daphnia* but especially with marine calanoids and euphausiids.

Note

1. Roozen and Lurling (2001) succeeded in making two laminar flows of test water from the two arms into the common leg. Van Gool and Ringelberg (1996) were unable to realise that and at the junction of the two arms turbulent mixing could not be prevented. Coloured water from one arm mixed into clouds with clear water from the other arm. Roozen and Lurling (2001) placed test animals at the far end of the common leg from where they swam upstream towards the junction of the Y-tube. Chance probably determined whether swimming started in the left or the right laminar stream and on reaching the junction reorientation was impossible. Van Gool and Ringelberg (1996) on the other hand trapped the daphnids for some time phototactically within the turbulent mixing zone at a small distance from the junction where odour gradients must have been present. Upon liberation, taking away the optical trap, most daphnids swam into the arm with the algal odour.

References

Aanen, D.K. 1993. Population dynamics of *Daphnia galeata x hyalina* and the presence of juvenile perch (*Perca fluviatilis*) in Lake Maarsseveen. Masters Thesis, Univ. of Amsterdam. (In Dutch).

Aarseth, K.A. and Schram, T.A. 1999. Wavelength-specific behaviour in *Lepeotheirus salmonis* and to ultraviolet and visible light in laboratory experiments (Crustacea: Copepoda). Mar. Ecol. Prog. Ser. 186:211–217.

Aarseth, K.A. and Schram, T.A. 2002. Suspectibility to ultraviolet radiation in *Calanus finmarchicus* and Lepeophtheirus salmonis and the adaptive value of external filtering (Crustacea: Copepoda). J. Plankton Res. 24:661–679.

Alonso, C., Rocco, V., Barriga, J.P., Battini, M.A. and Zagarese, H. 2004. Surface avoidance by freshwater zooplankton: Field evidence on the role of ultraviolet radiation. Limnol. Oceanogr. 49:225–232.

Alverdes, F. 1924. Das Verhalten positiv phototaktischer Daphnien gegen eine doppelte Lichtquelle. Z. Wissenschaftliche Zoologie 123:359–371.

Andersen, V. and Sardou, J. 1992. The diel migrations and vertical distributions of zooplankton and micronekton in the Northwestern Mediterranean Sea. 1. Euphausiids, mysids, decapods and fishes. J. Plankton Res. 14:1129–1154.

Angel, M.V. 1968. The thermocline as an ecological boundary. Sarsia 34:229–312.

Angel, M.V. 1977. Studies on Atlantic Halocyprid Ostracods: Vertical distribution of the species in the top 1000 m in the vicinity of 44o N, 13o W. J. Mar. Biol. Assoc. U.K. 57:239–252.

Angel, M.V. 1993. Biodiversity of the pelagic ocean. Conserv. Biol. 7:760–772.

Angel, M.V. 2000. Pelagic biodiversity. In: Ormond, R.F.G., Gage, J.D., and Angel, M.V. (Eds.) Marine biodiversity: Patterns and processes, pp. 35–68. Cambridge Univ. Press, Cambridge.

Angel, M.V. and Fasham, M.J.R. 1973. Sond cruise, 1965: Factor and cluster analyses of the plankton results, a general summary. J. Mar. Biol. Ass. U.K. 53:185–231.

Angel, M.V. and Fasham, M.J.R. 1974. Sond cruise 1965: Further factor analyses of the plankton data. J. Mar. Biol. Ass. U.K. 54:879–894.

Angeli, N., Pinel-Alloul, B., Balvay, G. and Menard, I. 1995. Diel pattern of feeding and vertical migration in daphnids and diaptomids during the clear water phase in Lake Geneva (France). *Hydrobiologia* 300/301:163–184.

Anonymous 1995. The Wordsworth dictionary of biology. Wordsworth Editions Ltd, Ware, England. pp. 1–324.

Arkett, S.A. 1985. The shadow response of a hydromedusan (*Polyorchis penicillatus*): Behavioral mechanism controlling diel and ontogenetic vertical migration. Biol. Bull. 169:297–312.

Atkins, W.R.G. and Poole, H.H. 1933. The photo-electric measurement of penetration of light of various wavelengths into the sea and the physiological bearing of the results. Phil. Trans. Roy. Soc. Lond. B 222:129.

Backus, R.H., Clarke, R.C. and Wing, A.S. 1965. Behaviour of certain marine organisms during the solar eclipse of July 20, 1963. Nature 205:989–991.

DOI 10.1007/978-90-481-3093-1_BM2,

Baerends, G.P. 1957. The ethological analysis of fish behaviour. In: Brown, M.E. (Ed.) The physiology of fishes. Vol. II, pp. 229–269. Academic Press, New York.

Bainbridge, R. 1952. Underwater observations on the swimming of marine zooplankton. J. Mar. Biol. Assoc. U.K. 31:107–112.

Bainbridge, R. 1961. Migrations. In: Waterman, T.H. (Ed.) The physiology of crustacea, pp. 431–463. Academic Press, New York.

Bainbridge, R. and Waterman, T.H. 1958. Turbidity and the polarized light orientation of the Crustacean *Mysidium*. J. Exp. Biol. 35:487–493.

Baker, A.de C. 1970. The vertical distribution of euphausiids near Fuerteventura, Canary Islands (Discovery Sond Cruise, 1965). J. Mar. Biol. Ass. U.K. 50:301–342.

Baker, K.S. and Smith, R.C. 1982. Bio-optical classification and model of natural waters. Limnol. Oceanogr. 27:500–509.

Baldwin, R.W. 1993. Invertebrate drift in a subarctic Alaskan Stream in relation to photoperiod, light intensity and rate of relative light change. Master Thesis, University of Wyoming, Laramie, Wyoming. pp. 1–123.

Balino, B.M. and Aksnes, D.L. 1993. Winter distribution of the sound scattering layers, zooplankton and mircronekton in Masfjorden, western Norway. Mar. Ecol. Progr. Ser. 102:35–50.

Banse, K. 1964. On the vertical distribution of zooplankton in the sea. Prog. Oceanogr. 2:53–125.

Barange, M. 1990. Vertical migration and habitat partitioning of six euphausiid species in the northern Benguela upwelling system. J. Plankton Res. 12:1223–1237.

Barlow, Jr., R.B., Powers, M.K. and Kass, L. 1988. Vision and mating in *Limulus*. In: Atema, J., Fay, R.R., Popper, A.N. and Tavolga, W.N. (Eds.) Sensory biology of aquatic animals, pp. 419–434. Springer-Verlag, New York.

Bary, B.McK. 1967. Diel vertical migrations of underwater scattering, mostly in Saanich Inlet, British Columbia. Deep-Sea Res. 14:35–50.

Bauer, V. 1908. Uber die reflectorische Regulierung von Schwimmbewegungen bei Mysiden. Z. Allgem. Physiol. 2:20–27.

Baumbach, P. 1922. Studien über das Zooplancton des Chiemsees. Arch. Naturgesch. Berlin 88A:177–203.

Bayley, I.A.E. 1986. Aspects of diel vertical migration in zooplankton, and its enigma variations. In: DeDecker, P. and Williams, W.D. (Eds.) Limnology in Australia, Monogr. Biol. Vol. 61, pp. 349–368. Dr. W. Junk Publishers, Dordrecht.

Baylor, E.R. and Hazen, W.E. 1962. The analysis of polarized light in the eye of *Daphnia*. Biol. Bull. 123:233–242.

Baylor, E.R. and Smith, F.E. 1957. Diurnal migration of plankton crustaceans. In: Scheer, B. (Ed.) Recent advances in invertebrate physiology, pp. 21–35. University of Oregon Publications, Eugene, Oregon.

Bayly, I.A.E. 1963. Reversed diurnal vertical migration of planktonic Crustacea in inland waters of low hydrogen ion concentration. Nature 200:704.

Blachowiak-Samolyk, K., Kwasniewski, S., Richardson, K., et al. 2006. Arctic zooplankton do not perform diel vertical migration (DVM) during periods of midnight sun. Mar. Ecol. Prog. Ser. 308:101–116.

Black, A.W. 1993. Predation-induced phenotypic plasticity in *Daphnia pulex*: Life history and morphological responses to *Notonecta* and *Chaoborus*. Limnol. Oceanogr. 38:986–996.

Blackwell, R. 1971. Luminance difference thresholds. In: Autrum, H., Jung, R., Loewenstein, W.R., et al. (Eds.) Handbook of sensory physiology;VolVII/4 visual psychophysics, pp. 78–100. Springer-Verlag, Berlin.

Blaxter, J.H.S. 1970. Light, animals, fishes. In: Kinne, O. (Ed.) Marine biology, pp. 213–285. John Wiley and Sons, London.

Blumstein, D.T. and Bouskila, A. 1996. Assessment and decision making in animals: A mechanistic model underlying behavioral flexibility can prevent ambiguity. Oikos 77:569–576.

Blumthaler, M., Ambach, W., et al. 1997. Increase in solar UV radiation with altitude. J. Photochem. Photobiol. 39:130–134.

Boden, B.P. and Kampa, E.M. 1958. Lumiere, bioluminescence et migrations de la couche diffusante profonde en Mediterrane occidentale. Vie et Milieu 9:1–10.

Boersma, M., van Tongeren, O.F.R. and Mooij, W.M. 1995. Seasonal pattern in the mortality of *Daphnia* species in a shallow lake. Can. J. Fish. Aquat. Sci. 53:18–28.

Boersma, M., Spaak, P. and de Meester, L. 1998. Predator-mediated plasticity in morphology, life history and behavior in *Daphnia*: The uncoupling of responses. Am. Nat. 152:237–248.

Bogorov, B.G. 1946. Peculiarities of diurnal vertical migrations of zooplankton in polar seas. Mar. Res. 6:25–32.

Bohl, E. 1980. Diel pattern of pelagic distribution and feeding in planktivorous fish. Oecologia (Berlin) 44:368–375.

Bohn, G. 1906. Les tropismes, les réflexes et l'intelligence. Année Psychol. 12:137–156.

Bolhuis, J.J. 2005. Function and mechanism in neuroecology: Looking for clues. Anim. Biol. 55:457–490.

Bollens, S.M. and Frost, B.W. 1989a. Predator induced vertical migration in a planktonic copepod. J. Plankton Res. 11:1047–1065.

Bollens, S.M. and Frost, B.W. 1989b. Zooplanktivorous fish and variable diel vertical migration in the marine planktonic copepod *Calanus pacificus*. Limnol. Oceanogr. 34:1072–1083.

Bollens, S.M. and Frost, B.W. 1991. Diel vertical migration in zooplankton: Rapid individual response to predators. J. Plankton Res. 13:1359–1365.

Bollens, S.M., Frost, B.W, Thorenson, D.S. and Watts, J.S. 1992. Diel vertical migration in zooplankton: Field evidence in support of the predator avoidance hypothesis. Hydrobiologia 234:33–39.

Bollens, S.M., Osgood, K.E., Frost, B.W. and Watts, S.D. 1993. Vertical distributions and susceptibilities to vertebrate predation of marine copepods *Metridia lucens* and *Calanus pacificus*. Limnol. Oceanogr. 38:1827–1837.

Bollens, S.M., Frost, B.W. and Cordell, J.R. 1994. Chemical, mechanical and visual cues in the vertical migration behavior of the marine planktonic copepod *Acartica hudsonica*. J. Plankton Res. 16:555–564.

Borradaile, L.A., Eastham, L.E.S., Pots, F.A. and Saunders, J.T. 1955. The invertebrata. Cambridge University Press, Cambridge. Ed. 2. pp. 1–725.

Bourne, G.H. 1960. Structure and function of muscle. Vol. II. Academic Press, New York.

Bove, G.E. and Stewart, K.M. 2002. Zooplankton migration under ice. Verh. Internat. Verein. Limnol. 28:1134–1139.

Boxshall, G.A. 1998. Mating biology of copepod crustaceans. Phil. Trans. Roy. Soc. Lond. 353: 1–815.

Brancelj, A. and Blejec, A. 1994. Diurnal vertical migration of *Daphnia hyalina* Leydig,1860 (Crustacea: Cladocera) in Lake Bled (Slovenia) in relation to temperature and predation. Hydrobiologia 284:125–136.

Brehm, V. 1938. Die Rotfarbung von Hochgebirgsseeorganismen. Biol. Rev. 13:307–318.

Bright, T., Ferrari, F., Martin, D. and Franceschini, G.A. 1972. Effects of a total solar eclipse on the vertical distribution of certain oceanic zooplankters. Limnol. Oceanogr. 17:296–302.

Brinton, E. 1967. Vertical migration and avoidance ability of euphausiids in the California Current. Limnol. Oceanogr. 12:451–483.

Brooks, J.L. and Dodson, S.I. 1965. Predation, body size, and composition of plankton. Science 150:28–35.

Buchanan, C. and Haney, J.F. 1980. Vertical migration in zooplankton in the Arctic: A test of environmental controls. In: Kerfoot, W.C. (Ed.) Evolution and ecology of zooplankton communities. Chapter 7:69–79. Univ. Press of New England, Hanover, New Hampshire.

Bullock, T.H. 1988. The comparative neurology of expectation: Stimulus acquisition and neurobiology of anticipated and unanticipated input. In: Atema, J., Fay, R.R., Popper, A.N. and Tavolga, W.N. (Eds.) Sensory biology of aquatic animals, pp. 269–284. Springer-Verlag, New York.

Bullock, T.H. 1996. Neuroethology of zooplankton. In: Lenz, P.H., Hartline, D.K., Purcell, J.E. and Macmillan, D.L. (Eds.) Zooplankton: Sensory ecology and physiology, pp. 1–16. Overseas Publishers Association, Amsterdam.
Buskey, E.J. 1984. Swimming pattern as an indicator of the roles of copepod sensory systems in the recognition of food. Mar. Biol. 79:165–175.
Buskey, E.J.A., Smith, R.C. and Swift, E. 1989. Photosensitivity of the oceanic copepods *Pleuromamma xiphias* and its relationship to light penetration and daytime depth distribution. Mar. Ecol. Prog. Ser. 55:207–216.
Buskey, E.J., Coulter, C. and Strom, S. 1993. Locomotory patterns of microzooplankton: Potential effects on food selectivity of larval fish. Bull. Mar. Sci. 53:29–43.
Calaban, M.J. and Makarewicz, J.C. 1982. The effect of temperature and density on the amplitude of vertical migration of *Daphnia magna*. Limnol. Oceanogr. 27:262–271.
Caruthers, J.W., Thompson, R.C. and Novarini, J.C. 1972. The response of deep scattering layers in the Gulf of Mexico to a total solar eclipse. Deep-Sea Res. 37:337–338.
Cellier, S., Rehailia, M., Berthon, J-L. and Buisson, B. 1998. Le role de l'oeil dans les rythmes migratoires de *Daphnia magna* et *Daphnia longispina* (Cladocères). Ann. Limnol. 34:159–164.
Chaston, I. 1969. The light threshold controlling the vertical migration of *Chaoborus punctipennis* in a Georgia impoundment. Ecology 50:916–920.
Chivers, D.P., Brown, G.E. and Smith, R.J.F 1996. The evolution of chemical alarm signals: Attracting predators benefits alarm signal senders. Am. Nat. 148:649–659.
Clark, C.W. and Levy, D.A. 1988. Diel vertical migration by juvenile sockeye salmon and the antipredation window. Am. Nat. 131:271–290.
Clarke, G.L. 1930. Change of phototropic and geotropic signs in *Daphnia* induced by changes of light intensity. J. Exp. Biol. 7:109–131.
Clarke, G.L. 1932. Quantitative aspects of the change in phototropic signs in *Daphnia*. J. Exp. Biol. 9:180–211.
Clarke, G.L. 1933. Diurnal migration of plankton in the Gulf of Maine and its correlation with changes in submarine irradiation. Biol. Bull. 65:402–435.
Clarke, G.L. 1934a. Further observations on the diurnal migration of copepods in the Gulf of Maine. Biol. Bull. (Woods Hole) 67:432–455.
Clarke, G.L. 1934b. The diurnal migration of Copepods in St. Georges Harbor, Bermuda. Biol. Bull. (Woods Hole) 67:456–459.
Clarke, G.L. and Backus, R.H. 1956. Measurements of light penetration in relation to vertical migration and records of luminescence of deep-sea animals. Deep-Sea Res. 4:1–14.
Clarke, G.L. and Backus, R.H. 1964. Interrelations between the vertical migration of deep scattering layers, bioluminescence and changes in daylight in the sea. Bull. Inst. Oceanogr. Monaco. 64:1–36.
Clarke, G.L. and Wertheim, G.K. 1956. Measurement of illumination at great depths and at night in the Atlantic Ocean by means of a new bathyphotometer. Deep-Sea Res. 3:189–205.
Coesel, P.F.M. 1994. The ecological significance of a cellular mucilaginous envelope in planktic desmids. Algol. Stud. 73:65–74.
Coesel, P.F.M. 1997. The edibility of *Staurastrum cheatoceras* and *Cosmarium abbreviatum* (desmidiaceae) for *Daphnia galeata/hyalina* and the role of desmids in the aquatic food web. Aquatic Ecol. 31:73–78.
Cohen, J.J. and Forward, R.B. Jr. 2002. Spectral sensitivity of vertically migrating marine copepods. Biol. Bull. (Woods Hole) 203:307–314.
Cohen, J.H. and Forward, Jr., R.B. 2003. Ctenophore kairomones and modified aminosugar disaccharides alter the shadow response in a larval crab. J. Plankton Res. 25:203–213.
Cohen, J.J. and Forward, R.B. Jr. 2005. Diel vertical migration of the marine copepod *Calanopia americana*. 1. Twilight DVM and its relation to the diel light cycle. Mar. Biol. 147:387–398.
Conover, R.J. and Huntley, M. 1991. Copepods in ice-covered seas – Distribution, adaptations to seasonally limited food, metabolism, growth patterns and life cycle strategies in polar seas. J. Mar. Systems 2:1–41.

Conover, R.J., Bedo, A.W., Herman, A.W., Head, E.J.H., Harris, L.R. and Horne, E.P.W. 1988. Never trust a copepod – some observations on their behavior in the Canadian arctic. Bull. Mar. Sci. 43:650–662.
Consi, T.R., Macagno, E.R. and Necles, N. 1987. The oculomotor system of *Daphnia magna*. The eye muscles and their motor neurons. Cell Tissue Res. 247:515–523.
Consi, T.R., Passani, M.B. and Macagno, E.R. 1990. Eye movements in *Daphnia magna*: Regions of the eye are specialised for different behaviors. J. Comp. Physiol. 166:411–420.
Cowgill, U.M., Williams, D.M. and Esquivel, J.B. 1984. Effects of maternal nutrition on fat content and longevity of neonates of *Daphnia magna*. J. Crust. Biol. 4:173–190.
Cronin, T.W. 1982. Estuarine retention of larrae of the crab *Rhithropanopeus harrisii*. Estuarine Coastal Shelf Sci. 15:201–220.
Cronin, T.W. 1988. Vision in marine invertebrates. In: Atema, J., Fay, R.R., Popper, A.N. and Tavolga, W.N. (Eds.) Sensory biology of aquatic animals, pp. 403–418. Springer-Verlag, New York.
Cronin, T.W. and Forward, Jr., R.B. 1980. The effect of starvation on phototaxis and swimming of larvae of the crab *Rhithropanopeus harrisii*. Biol. Bull. 158:283–294.
Cunningham, L. 1972. Vertical migrations of *Daphnia* and copepods under the ice. Limnol. Oceanogr. 17:301–304.
Cushing, D.H. 1951. The vertical migration of planktonic Crustacea. Biol. Rev. 26:158–192.
Cuthill, I.C. 2005. The study of function in behavioural ecology. Anim. Biol. 55:399–417.
Daan, N. and Ringelberg, J. 1969. Further studies on the positive and negative phototactic reaction of *Daphnia magna* Straus. Neth. J. Zool. 19:525–540.
Dagg, M.J., Frost, B.W. and Walser, J. 1989. Copepod diel migration, feeding, and the vertical flux of pheopigments. Limnol. Oceanogr. 34:1062–1071.
Dagg, M.J., Frost, B.W. and Newton, J.A. 1997. Vertical migration and feeding behavior of *Calanus pacificus* females during a phytoplankton bloom in Dabob Bay, U.S. Limnol. Oceanogr. 42:974–980.
Dale, T. and Kaartvedt, S. 2000. Diel pattern in stage-specific vertical migration of *Calanus finmarchicus* in habitats with midnight sun. ICES J. Mar. Sci. 57:1800–1818.
Damkaer, D.M., Dey, D.B. and Heron, G.A. 1981. Dose/dose-rate responses of shrimp larvae to UV-B radiation. Oecologia 48:178–182.
Davis, A.D., Weatherby, T.M., Hartline, D.K. and Lenz, P.H. 1999. Myelin-like sheaths in copepod axons. Nature 398:571.
Dawidowicz, P. and Loose, C.J. 1992. Metabolic costs during predator-induced diel vertical migration of *Daphnia*. Limnol. Oceanogr. 37:1589–1595.
Dawidowicz, P., Pyanowska, J. and Ciechomski, K. 1990. Vertical migration of *Chaoborus* larvae is induced by the presence of fish. Limnol. Oceanogr. 35:1631–1637.
Delcomyn, F. 1980. Neural basis of rhythmic behavior in animals. Science 210:492–498.
De Lange, H.J. 1999. Effects of Ultraviolet-B radiation on phytoplankton – zooplankton interactions. Thesis Wageningen Agricultural University, Wageningen pp. 1–187.
De Meester, L. 1989. An estimation of the heritability of phototaxis in *Daphnia magna* Straus. Oecologia 78:142–144.
De Meester, L. 1991. Analysis of the phototactic behaviour of *Daphnia magna* clones and their sexual descendants. Hydrobiologia 225:217–227.
De Meester, L. 1992. Influence of fish-mediated chemicals on phototactic behaviour of different *Daphnia magna* clones. (Abstract).
De Meester, L. 1993a. On the genetical ecology of phototactic behaviour in *Daphnia magna* (Crustacea: Cladocera). Academia Analecta 55:145–188.
De Meester, L. 1993b. The vertical distribution of *Daphnia magna* genotypes selected for different phototactic behaviour: Outdoor experiments. Arch. Hydrobiol. Beih. Ergebn. Limnol. 39: 137–156.
De Meester, L. 1993c. Genotype, fish-mediated chemicals, and phototactic behavior in *Daphnia magna*. Ecology 74:1467–1474.

De Meester, L. 1995. Life history characteristics of *Daphnia magna* clones differing in phototactic behaviour. Hydrobiologia 307:167–175.

De Meester, L. 1997. Neutral markers, ecologically relevant traits, and the structure of genetic variation in *Daphnia*. Aquatic Ecol. 31:79–87.

De Meester, L. and Pyanowska, J. 1997. On the trait-specificity of the response of *Daphnia* genotypes to the chemical presence of a predator. In: Lenz, P.H., Hartline, D.K., Purcell, J.K. and Macmillan, D.L. (Eds.) Zooplankton: Sensory Ecology and Physiology, pp. 407–417. Gordon and Breach, Amsterdam.

De Meester, L. and Weider, L.J. 1999. Depth selection behavior, fish kairomones, and the life histories of *Daphnia hyalina x galeata* hybrid clones. Limnol. Oceanogr. 44:1248–1258.

De Meester, L., Weider, L.J. and Tollrian, R. 1995. Alternative antipredator defences and genetic polymorphism in a pelagic system. Nature 378:483–485.

De Meester, L., Dawidowicz, P., van Gool, E. and Loose, C.J. 1998. Ecology and evolution of predator-induced behavior of zooplankton: Depth selection and diel vertical migration in *Daphnia*. In: Harvell, C.D. and Tollrian, R. (Eds.) Consequences of inducible defences for population biology, Princeton University Press, New York.

De Robertis, A., Schell, C. and Jaffe, J.S. 2003. Acoustic observations of the swimming behavior of the euphausiid *Euphausia pacifica* Hansen. ICES J. Mar. Sci. 60:885–898.

De Wit, Y. 1990. De invloed van verschillende combinaties van temperatuur, voedselconcentratie en predatiedruk op het verticale migratiegedrag van *Daphnia hyalina* Leydig. Masters Thesis, Univ. of Amsterdam.

Den Boer, P.J. 1968. Spreading of risk and stabilisation of animal numbers. Acta Biotheoretica 18:165–194.

Dennett, D.C. 1995. Darwin's dangerous idea. Allen Lane The Penguin Press, London. pp. 1–587.

Dicke, M. and Sabelis, M.W. 1988. Infochemical terminology: Based on cost-benefit analysis rather than origin of compounds? Funct. Ecol. 2:131–139.

Diffey, B.L. 1991. Solar ultraviolet radiation effects on biological systems. Phys. Med. Biol. 36:299–328.

Digby, P.S.B. 1961. The vertical distribution and movements of marine plankton under midnight sun conditions in Spitsbergen. J. Anim. Ecol. 30:9–25.

Dill, L.M. 1987. Animal decision making and its ecological consequences: The future of aquatic ecology and behaviour. Can. J. Zool. 65:803–811.

Dini, M.L. and Carpenter, S.R. 1988. Variability in *Daphnia* behavior following fish community manipulations. J. Plankton Res. 10:621–635.

Dini, M.L., Soranno, P.A., Scheuerell, M. and Carpenter, S.R. 1993. Effects of predators and food supply on diel vertical migration of *Daphnia*. In: Carpenter, S.R. and Kitchell, J.F. (Eds.) The trophic cascade in lakes, pp. 153–172. Cambridge University Press, Cambridge.

Doall, M.H., Colin, S.P., Strickler, J.R. and Yen, J. 1998. Locating a mate in 3D: The case of *Temora longicornis*. Phil. Trans. Roy. Soc. Lond 353:681–689.

Dodson, S.I. 1988. The ecological role of chemical stimuli for the zooplankton: Predator avoidance behavior in *Daphnia*. Limnol. Oceanogr. 33:1431–1439.

Dodson, S. 1990. Predicting diel vertical migration of zooplankton. Limnol. Oceanogr. 35: 1195–1201.

Dodson, S.I. 1996. Optimal swimming behavior of zooplankton. In: Lenz, P.H., Hartline, D.K., Purcell, J.E. and Macmillan, D.L. (Eds.) Zooplankton: Sensory ecology and physiology, pp. 365–375. Overseas Publishers Association, Amsterdam.

Dodson, S. and Ramcharan, C. 1991. Size-specific swimming behavior of *Daphnia pulex*. J. Plankton Res. 13:1367–1379.

Dodson, S.I., Hanazato, T. and Gorski, P.R. 1995. Behavioral responses of *Daphnia pulex* exposed to carbaryl and *Chaoborus* kairomone. Environ. Toxicol. Chem. 14:43–50.

Dodwell, P.C. (ed.) 1971. Perceptual processing. Appleton-Century-Crofts, New York. pp. 41–71.

Downing, A.C. 1974. The hydraulic suspension of the *Daphnia* eye – a new kind of universal joint? Vision Res. 14:647–652.

Dumont, H.J., Guisez, Y, Carels, I. and Verheye, H.M. 1985. Experimental isolation of positively and negatively phototactic phenotypes from a natural population of *Daphnia magna* Straus.: A contribution to the genetics of vertical migration. Hydrobiologia 126:121–127.

Duncan, A. Kubecka, J. and Quiliiam, B. 1998. An acoustic study of fish in Lake Maarsseveen June 1998. Internal Report to Project SNIFFS. Centre for Limnology, Nieuwersluis, The Netherlands.

Dusenbery, D.B. 1992. Sensory ecology. W.H. Freeman and Company, New York. pp. 1–549.

Duval, W. and Geen, G.H. 1975. Diel rhythms in the feeding and respiration of zooplankton. Verh. Internat. Verein. Limnol. 19:518–523.

Eckert, F. 1935. Die positiv phototaktische Einstellreaktion des Komplexauges von *Daphnia pulex* im Zweilichterversuch. Lotos (Prag) 83:40–69.

Einsle, U. 1969. Untersuchungen zur Vertikalwanderung planktischer Crusataceen im Bodensee-Obersee. Geschichte des Bodensees und seiner Umgebung 87:177–187.

Einsle, U. 1987. Zur Vertikalwanderung planktischer Copepoden im Bodensee-Obersee. Schweiz. Z. Hydrobiol. 49:303–315.

Elster, U. 1969. Untersuchungen zur Vertikalwanderung planktischer Crusataceen im Bodensee-Obersee. Schr. Ver. Gesch. Bodensee 87:177–187.

Esterly, C.E. 1911. Diurnal migrations of *Calanus finmarchicus* in the San Diego region during 1909. Int. Rev. Hydrobiol. 4:140.

Esterly, C.O. 1917. The occurrence of a rhythm in the geotropism of two species of plankton Copepods when certain recurring external conditions are absent. University of California Publications of Zoology 16:393.

Esterly, C.O. 1919. Reactions of various plankton animals with reference to their diurnal vertical migrations. Univ. Czlif. Publ. Zool. 19:1–83.

Ewald, W.F. 1910a. Ueber Orientierung, Lokomotion und Lichtreaktionen einiger Cladoceren und deren Bedeutung für die Theorie der Tropismen. Biol. Centralblatt 33:49–63.

Ewald, W.F. 1910b. Ueber Orientierung, Lokomotion und Lichtreaktionen einiger Cladoceren und deren Bedeutung für die Theorie der Tropismen. Biol. Centralblatt 33:379–399.

Ewald, W.F. 1910c. Ueber Orientierung, Lokomotion und Lichtreaktionen einiger Cladoceren und deren Bedeutung für die Theorie der Tropismen. Biol. Centralblatt 30:1–16.

Fahrenbach, W.H. 1964. The fine structure of a nauplius eye. Z. für Zellforschung 62:182–197.

Farran, G.P. 1947. Vertical distribution of plankton (Sagitta, Calanus and Metridia) off the south coast of Ireland. Proc. Roy. Irish Acad. 51:121.

Fields, D.M. and Yen, J. 1996. The escape behavior of *Pleuromamma xiphias* in response to a quantifiable fluid mechanical disturbance. In: Lenz, P.H., Hartline, D.K., Purcell, J.E. and Macmillan, D.L. (Eds.) Zooplankton: Sensory ecology and physiology, pp. 323–339. Overseas publishers Association, Amsterdam.

Fischer, J. and Visbeck, M. 1993. Seasonal variation of the daily zooplankton migration in the Greenland Sea. Deep-Sea Res. 40:1547–1557.

Fleminger, A. 1967. Distributional atlas of calanoid copepods in the California current region part 2. Calif. Mar. Res. Comm. CalCofi Atlas 7, 213 p.

Flik, B.J.G. and Ringelberg, J. 1993. Influence of food availability on the initiation of diel vertical migration (DVM) in lake Maarsseveen. Arch. Hydrobiol. Beih. Ergebn. Limnol. 39:57–65.

Flik, B.J.G. and Vijverberg, J. 2003. Contrasting migration behaviour of *Daphnia pulicaria* and *D. galeata x hyalina*, in avoidance of predation of 0+perch (*Perca fluviatilis*). Hydrobiologia 491:289–299.

Flik, B.J.G., Hallegraeff, G.M. and Lingeman, R. 1973. Limnological notes on Lac Pavin (Auvergne, France). Annales de la station Biologique de Bess-en-Chandesse 7:119–146.

Flik, B.J.G., Aanen, D.K. and Ringelberg, J. 1997. The extent of predation by juvenile perch during diel vertical migration of *Daphnia*. Arch. Hydrobiol. Beih. Ergebn. Limnol. 49:51–58.

Fortier, M., Fortier, L., Hattori, H., Saito, H. and Legendre, L. 2001. Visual predators and the diel vertical migration of copepods under Arctic sea ice during the midnight sun. J. Plankton Res. 23:1263–1278.

Forward, Jr., R.B. 1974. Negative phototaxis in crustacean larvae: Possible functional significance. J. Exp. Mar. Biol. Ecol. 16:11–17.

Forward, R.B., Jr. 1985. Behavioral responses of larvae of the crab *Rhithropanopeus harrisii* (Brachyura: Xanthidae) during diel vertical migration. Mar. Biol. 90:9–18.

Forward, Jr., R.B. 1986. A reconsideration of the shadow response of a larval crustacean. Mar. Behav. Physiol. 12:99–113.

Forward, Jr., R.B. 1988. Diel vertical migration: Zooplankton photobiology and behavior. Oceanogr. Mar. Biol. Annu. Rev. 26:361–393.

Forward, Jr., R.B. 1993. Photoresponses during diel vertical migration of brine shrimp larvae: Effect of predator exposure. Arch. Hydrobiol. Beih. Ergebn. Limnol. 39:37–44.

Forward, Jr., R.B. and Cronin, T.W. 1980. Tidal rhythms of activity and phototaxis of an estuarine crab larva. Biol. Bull. 158:295–303.

Forward, Jr., R.B. and Hettler, W.F.J. 1992. Effects of feeding and predator exposure on photoresponses involved in diel vertical migration of brine shrimp larvae. Limnol. Oceanogr. 37:1261–1270.

Forward, Jr., R.B. and Rittschof, D. 1999a. Brine shrimp larval photoresponses involved in diel vertical migration: Activation by fish mucus and modified amino sugars. Limnol. Oceanogr. 44:1904–1916.

Forward, Jr., R.B. and Rittschof, D. 1999b. Alteration of photoresponses involved in diel vertical migration of a crab larva by fish mucus and degradation products of mucopolysaccharides. Limnol. Oceanogr. (submitted).

Forward, Jr., R.B., Cronin, T.W. and Stearns, D.E. 1984. Control of diel vertical migration: Photoresponses of a larval crustacean. Limnol. Oceanogr. 29:146–154.

Forward, Jr., R.B., Tankersley, R.A., De Vries, M.C. and Rittschof, D. 1996. Sensory physiology and behavior of Blue Crab (*Callinectes sapidus*) postlarvae during horizontal transport. In: Lenz, P.H., Hartline, D.K., Purcell, J.E. and Macmillan, D.L. (Eds.) Zooplankton: Sensory ecology and physiology, pp. 225–240. Overseas Publishers Association, Amsterdam.

Fraenkel, G.S. and Gunn, D.L. 1961. The orientation of animals. Dover publications, New York. pp. 1–376.

Franceschini, G.A., Bright, T.J., Caruthers, J.W., El-sayed, S.Z. and Vastano, A.C. 1970. Solar eclipse of March 7, 1970: Effects on migration of marine organisms in the Gulf of Mexico. Nature 226:1155–1156.

Frank, T.M. and Case, J.F. 1988. Visual spectral sensitivities of bioluminescent deep-sea crustaceans. Biol. Bull. 175:261–273.

Frank, T.M. and Widder, E.A. 1996. UV light in the deep-sea: In situ measurements of downwelling irradiance in relation to the visual threshold sensitivity of UV-sensitive crustaceans. In: Lenz, P.H., Hartline, D.K., Purcell, J.E. and Macmillan, D.L. (Eds.) Zooplankton: Sensory ecology and physiology, pp. 185–193. Overseas Publishers Association, Amsterdam.

Frank, T.M. and Widder, E.A. 1999. Comparative study of the spectral sensitivities of mesopelagic crustaceans. J. Comp. Physiol. 185:225–265.

Franz, V. 1912. Zur Frage der vertikalen Wanderungen der Plantontiere. Arch. Hydrobiol. 7: 493–499.

Friedman, M.M. and Strickler, J.R. 1975. Chemoreceptors and feeding in calanoid copepods (Arthropoda: Crustacea). Proc. Nat Acad. Sci. USA 72:4185–4188.

Friesen, W.O. 1989. Neural control of leach swimming movements. In: Jacklet, J.W. (Ed.) Cellular and neuronal oscillators, pp. 269–316. Marcel Dekker, New York.

Friesen, W.O. and Block, G.D. 1984. What is a biological oscillator? Am. J. Physiol. 246: R847–R851.

Frisch, K. von and Kupelwieser, H. 1913. Ueber den Einfluss der Lichtfarbe auf die phototaktischen Reaktionen niederer Krebse. Biol. Zentrallblat 33:517–552.

Frost, B.J. 1975. Eye movements in *Daphnia pulex* (De Geer). J. Exp. Biol. 62:175–187.

Frost, B.W. 1988. Variability and possible adaptive significance of diel vertical migration in *Calanus pacificus*, a planktonic marine copepod. Bull. Mar. Sci. 43:675–694.

Frost, B.W. 2005. Diatom blooms and copepod responses: Paradigm or paradox. Prog. Oceanogr. 67:283–285.

Fulmer, J.H. and Bollens, S.M. 2005. Responses of the chaetognath, *Sagitta elegans*, and larval Pacific hake, *Merluccius productus*, the spring diatom and copepod bloom in a temperate fjord (Dabob bay, Washington). Prog. Oceanogr. 67:442–461.

Gabriel, W. 1993. Models in diel vertical migration. Arch. Hydrobiol. Beih. Ergebn. Limniol. 39:123–136.

Gabriel, W. 2006. Selective advantage of irreversible and reversible phenotypic plasticicty. Arch. Hydrobiol. 167:1–20.

Gabriel, W. and Thomas, B. 1988. Vertical migration of zooplankton as an evolutionarily stable strategy. Am. Nat. 132:199–216.

Gaedke, U. and Schweitzer, A. 1993. The first decade of oligotrophication in Lake Constance. I. The response of phytoplankton biomass and cell size. Oecologia 93:268–275.

Gal, G., Loew, E.R., Rudstam, L.G. and Mohammadian, A.M. 1999. Light and diel vertical migration: Spectral sensitivity and light avoidance by Mysis relicta. Can. J. Fish. Aquat. Sci. 56:311–322.

Geller, W. 1986. Diurnal vertical migration of zooplankton in a temperate great lake (L. Constance): A starvation avoidance mechanism. Arch. Hydrobiol. 74(Suppl.):1–60.

George, D.G. 1983. Interrelations between the vertical distribution of *Daphnia* and chlorophyll a in two large limnetic enclosures. J. Plankton Res. 5:457–475.

Gerritsen, J. 1978. Instar-specific swimming patterns and predation of planktonic copepods. Verh. Internat. Verein. Limnol. 20:2531–2536.

Gerritsen, J. 1980. Adaptive responses for encounter problems. In: Kerfoot, W.C. (Ed.) Evolution and ecology of zooplankton communities, pp. 52–62. University Press of New England, Hanover.

Gerritsen, J. 1982. Behavioral response of *Daphnia* to rate of temperature change: Possible enhancement of vertical migration. Limnol. Oceanogr. 27:254–261.

Gerritsen, J. and Strickler, J.R. 1977. Encounter probabilities and community structure in zooplankton: A mathematical model. J. Fish. Res. Bd. Can. 34:73–82.

Ghan, D., Hyatt, K.D. and McPhail, J.D. 1998. Benefits and costs of vertical migration by the freshwater copepod *Skistodiaptomus oregonensis*: Testing hypotheses through population comparison. Can. J. Fish. Aquat. Sci. 55:1338–1349.

Gibbons, M.J., Barange, M. and Pillar, S.C. 1991. Vertical migration and feeding of *Euphausia lucens* (Euphausiacea) in the Southern Benguela. J. Plankton Res. 13:473–486.

Gicklhorn, J. 1931. Beobachtungen an den lateralen frontalorganen von *Daphnia magna* nach elektiver vitalfarbung. Protoplasma 12:725–739.

Giguere, L.A., Delage, A., Dill, L.M. and Gerritsen, J. 1982. Predicting encounter rates for zooplankton: A model assuming a cylindrical encounter field. J. Fish. Res. Bd. Can. 39: 237–242.

Giroud, C. and Balvay, G. 1999. L'eclipse solaire du 11 Aout 1999 et la migration de quelque crustaces planctonique dans le Lac Leman. Archs Sci. Geneve 52:199–208.

Gliwicz, Z.M. and Jachner, A. 1992. Diel migration of juvenile fish: A ghost of predation past or present? Arch. Hydrobiol. 124:385–411.

Gliwicz, Z.M. and Pijanowska, J. 1988. Effect of predation and resource depth distribution on vertical migration of zooplankton. Bull. Mar. Sci. 43:695–709.

Golubitsky, M., Stewart, I., Buono, P-L. and Collins, J.J. 1999. Symmetry in locomotor central pattern generators and animal gaits. Nature 401:693–695.

Gorski, P.R. and Dodson, S.I. 1996. Free-swimming *Daphnia* pulex can avoid following Stoke's law. Limnol. Oceanogr. 41:1815–1821.

Grad, G. and Williamson, C.E. 2001. Zooplankton survival and reproduction responses to damaging UV radiation: A test of reciprocity and photoenzymatic repair. Limnol. Oceanogr. 46:584–591.

Greene, C.H., Widder, E.A., Tamse, A. and Johnson, G.E. 1992. The migration behavior, fine structure, and bioluminescent activity of krill sound-scattering layers. Limnol. Oceanogr. 37:650–658.

Greve 1995. Lecture at the symposium "Zooplankton: Sensory ecology and physiology", January, 8–12, Honolulu.

Grigorova, V.K. and Kornilova, L. 1996. Microgravity effect on the vestibulo-ocular reflex is dependent on otolith and vision contributions. Aviat. Space Environ. Med. 67:947–954.

Grillner, S. 1975. Locomotion in vertebrates: Central mechanism and reflex interaction. Physiol. Rev. 55:247–304.

Gulati, R.D. and DeMott, W.R. 1997. The role of food quality for zooplankton: Remarks on the state-of-the-art, perspectives and priorities. Freshw. Biol. 38:753–768.

Guldner, F-H. and Wolff, J.R. 1970. Ueber die Ultrastruktur de Komplexauges von *Daphnia pulex.* Z. für Zellforschung 104:259–274.

Hailman, J.P. and Jaeger, R.G. 1996. A model of phototaxis and its evaluation with anuran amphibians. Behaviour 56:215–249.

Hairston, N.G. 1979. The adaptive significance of color polymorphism in two species of *Diaptomus* (Copepoda). Limnol. Oceanogr. 24:15–37.

Hall, D.J., Werner, E.E., Giliam, J.F., Mittelbach, G.G., Howard, D. and Doner, C.G. 1979. Diel foraging behavior and prey selection in the Golden Shiner (*Notemigonus crysoleucas*). J. Fish. Res. Bd. Can. 36:1029–1039.

Halsband-Lenk, C., Pierson, J.J. and Leising, A.W. 2005. Reproduction of *Pseudocalanus newmani* (Copepoda: Calanoida) is deleteriously affected by diatom blooms – A field study. Prog. Oceanogr. 67:332–348.

Hamner, W.M. 1996. Predation, cover, and convergent evolution in epipelagic oceans. In: Lenz, P.H., Hartline, D.K., Purcell, J.E. and Macmillan, D.L. (Eds.) Zooplankton: Sensory ecology and physiology, pp. 17–35. Overseas Publishers Association, Amsterdam.

Han, B-P. and Straskraba, M. 1998. Modeling patterns of zooplankton diel vertical migration. J. Plankton Res. 20:1463–1487.

Haney, J.F. 1993. Environmental control of diel vertical migration behaviour. Arch. Hydrobiol. Beih. Ergebn. Limnol. 39:1–17.

Haney, J.F. and Hall, D.J. 1975. Diel vertical migration and filter-feeding activities of *Daphnia*. Arch. Hydrobiol. 75:413–441.

Haney, J.F., Beaulieu, T.R., Berry, R.P., et al. 1983. Light intensity and relative change as factors regulating stream drift. Arch. Hydrobiol. 97:73–88.

Haney, J.F., Craggy, A., Kimball, K. and Weeks, F. 1990. Light control of evening vertical migrations by *Chaoborus punctipennis* larvae. Limnol. Oceanogr. 35:1068–1078.

Hansson, L-A. 1996. Behavioural response in plants: Adjustment in algal recruitment by herbivores. Proc. Roy. Soc. Lond B 263:1241–1244.

Hansson, L-A. 2004. Plasticity in pigmentation induced by conflicting threats from predation and UV radiation. Ecology 85:1005–1016.

Hardy, A.C. and Bainbridge, R. 1954. Experimental observations on the vertical migrations of plankton animals. J. Mar. Biol. Assoc. U.K. 33:409–448.

Harris, J.E. 1953. Physical factors involved in the vertical migration of plankton. Quart. J. Microscop. Sci. 94:537–550.

Harris, J.E. 1963. The role of endogenous rhythms in vertical migration. J. Mar. Biol. Assoc. U.K. 43:153–166.

Harris, J.E. and Mason, P. 1956a. Vertical migration in eyeless *Daphnia*. Proc. Roy. Soc. B145:280–290.

Harris, J.E. and Wolfe, U.K. 1956b. A laboratory study of vertical migration. Proc. Roy. Soc. B144:329–354.

Hart, R.C. 1977. Feeding rhythmicity in a migratory copepod (*Pseudodiaptomus hessei* (Mrzak)). Freshw. Biol. 7:1–8.

Hart, R.C. and Allanson, B.R. 1976. The distribution and diel vertical migration of *Pseudocalanus hessei* (Mrzak) (Calanoida: Copepoda) in subtropical lake in southern Africa. Freshw. Biol. 6:183–198.

Hartline, H.K. and Ratliff, F. 1972. Inhibitory interaction in the retina of *Limulus*. In: Fuortes, M.G.F. (Ed.) Handbook of sense physiology VII/2; physiology of photoreceptor organs, pp. 381–444. Springer-Verlag, Berlin.

Hartline, D.K., Lenz, P.H. and Herren, C.M. 1996. Physiological and behavioral studies of escape responses in calanoid copepods. In: Lenz, P.H., Hartline, D.K., Purcell, J.E. and Macmillan, D.L. (Eds.) Zooplankton: Sensory ecology and physiology, pp. 341–354. Overseas Publishers Asociation, Amsterdam.

Hartmann, J. and Löffler, H. 1989. Tiefenverteilung der Fische des Bodensees. Osterreichs Fischerei 42:236–240.

Hattori, H. and Saito, H. 1997. Diel changes in vertical distribution and feeding activity of copepods in ice-covered Resolute Passage, Canadian Arctic, in spring 1992. J. Mar. Systems 11:205–219.

Haury, L.R. and Weihs, D. 1976. Energetically efficient swimming behavior of negatively buoyant zooplankton. Limnol. Oceanogr. 21:797–803.

Hays, G.C., Warner, A.J. and Tranter, P. 1997. Why do the two most abundant copepods in the North Atlantic differ so markedly in their diel vertical migration behaviour? J. Sea Res. 38:85–92.

Hays, G.C., Kennedy, H. and Frost, B.W. 2001. Individual variability in diel vertical migration of a marine copepod: Why some individuals remain at depth while others migrate. Limnol. Oceanogr. 46:2050–2054.

Heberdey, R.F. and Kupka, E. 1941. Das Helligkeitsunterscheidungsvermögen von *Daphnia pulex*. Z. Vergl. Physiol. 29:541–582.

Hebert, P.D.N. and McWalter, D.B. 1983. Cuticular pigmentation in Arctic *Daphnia*: Adaptive diversification of asexual lineages? Am. Nat. 122:286–291.

Heckey, R.E. and Fee, E.J. 1981. Primary production and rates of algal growth in Lake Tanganyika. Limnol. Oceanogr. 26:532–547.

Heerebout, G.R. 1970. A classification system for isolated brackish inland waters, based on median chlorinity and chlorinity fluctuation. Neth. J. Sea Res. 4:494–503.

Hess, C. von 1910. Neue Untersuchungen über den Lichtsinn bei wirbellosen Tieren. Pflüg. Arch. Ges. Physiol. 136:282–367.

Hessen, D.O. 1993. DNA-damage and pigmentation in alpine and arctic zooplankton as bioindicators of UV-radiation. Verh. Internat. Verein. Limnol. 25:482–486.

Hessen, D.O. 1994. *Daphnia* responses to UV-light. Arch. Hydrobiol. Beih. Ergebn. Limnol. 43:185–195.

Hessen, D.O. and Van Donk, E. 1993. Morphological changes in *Scenedesdus* induced by substances released from *Daphnia*. Arch. Hydrobiol. 127:129–140.

Heywood, K.J. 1996. Diel vertical migration of zooplankton in the Northeast Atlantic. J. Plankton Res. 18:163–184.

Hobaek, A. and Wolf, H.G. 1991. Ecological genetics of Norwegian *Daphnia*. II. Distribution of *Daphnia longispina* in relation to short-wave radiation and water color. Hydrobiologia 225:228–243.

Hogan, J.A. 2005. Causation: The study of behavioural mechanisms. Anim. Biol. 55:323–341.

Holmes, R.W. 1970. The Secchi disk in turbid water. Limnol. Oceanogr. 15:688–694.

Horner, R.A., Postel, J.R., Halsband-Lenk, C, Pearson, J.J., Pohnert, G. and Wichard, T. 2005. Winter-spring phytoplankton blooms in Dabob Bay, Washington. Progr. Oceanogr. 67:286–313.

Hu, V.J.H. 1978. Relationships between vertical migration and diet in four species of euphausiids. Limnol. Oceanogr. 23:296–306.

Huang, C., Uye, S. and Onbe, T. 1992. Ontogenetic diel vertical migration of the planktonic copepod *Calanus sinicus* in the Inland Sea of Japan. II. Late fall and early spring. Mar. Biol. 113:391–400.

Huang, C., Uye, S. and Onbe, T. 1993. Ontogenetic vertical migration of the planktonic copepod *Calanus sinicus* in the Inland Sea of Japan III. Early summer and overall seasonal pattern. Mar. Biol. 117:289–299.

Hülsmann, S. and Wagner, A. 2007. Multiple defense strategies of *Daphnia galeata* against predation in a weakly stratified reservoir. Hydrobiologia 594:187–199.

Hülsmann, S., Vijverberg, J., Boersma, M. and Mooij, W.M. 2004. Effects of infochemicals released by gape-limited fish on life history traits of *Daphnia*: A maladaptive response? J. Plankton Res. 26:535–543.

Hutchinson, G.E. 1957. A treatise on limnology. Vol. 1. John Wiley & Sons, London. 1015p.

Hutchinson, G.E. 1967. A treatise on limnology. Vol. II New York: John Wiley & Sons. 1115p.

Hutchinson, G.E. 1978. An introduction to population ecology. Yale University Press, New Haven. pp. 1–260.

Ianora, A., Miralto, A., Poulet, S.A., et al. 2004. Aldehyde suppression of copepod recruitment in blooms of a ubiquitous plankton diatom. Nature 429:403–407.

Isaacs, J.D. 1977. The life of the open ocean. Nature 267:778–780.

Jachner, A. 1989. Growth of fry of three fish species from pelagial of a mesotrophic lake. Pol. Arch. Hydrobiol. 36:359–371.

Jachner, A. 1991. Food and habitat partitioning among juveniles of three fish species in the pelagic of a mesotrophic lake. Hydrobiologia 226:81–89.

Jander, R. and Waterman, T.H. 1960. Sensory discrimination between polarized light and light intensity patterns by arthropods. J. Cellular Comp. Physiol. 56:137.

Janssen, J. 1981. Searching for zooplankton just outside Snell's window. Limnol. Oceanogr. 26:1168–1171.

Jerlov, N.G. and Nygard, K. 1969. A quanta and energy meter for photosynthetic studies. Univ. Copenhagen Inst. Phys. Oceanogr. Rep. 10.

Johnson, W.H. 1938. The effect of light on the vertical movements of *Acartia clausi* (Giesbrecht). Biol. Bull. (Woods Hole) 75:106–118.

Johnson, G.H. and Jakobsen, P.J. 1987. The effect of food limitation on vertical migration in *Daphnia longispina*. Limnol. Oceanogr. 32:873–880.

Johnson, C.R., O'Brien, W.J., MacIntyre, S. 2007. Vertical and temporal distribution of two copepods species, *Cyclops scutifer* and *Diaptomus probilofensis*, in 24 arctic light. J. Plankton Res. 29:275–289.

Jones, R.I. 1988. Vertical distribution and diel migration of flagellated phytoplankton in a small humid lake. Hydrobiologia 161:75–87.

Jones, R.I. 1991. Advantages of diurnal vertical migrations to phytoplankton in sharply stratified, humic forest lakes. Arch. Hydrobiol. 120:257–266.

Jones, R.I. 1993. Phytoplankton migrations: Patterns, processes and profits. Arch. Hydrobiol. Beih. Ergebn. Limnol. 39:67–77.

Jones, R.H. and Flynn, K.J. 2005. Nutritional status and diet composition affect the value of diatoms as copepod prey. Science 307:1457–1459.

Kaartvedt, S., Webjorn, M., Knutsen, T. and Skjoldal, H.R. 1996. Vertical distribution of fish and krill beneath water of varying optical properties. Mar. Ecol. Prog. Ser. 136:51–58.

Kampa, E.M. and Boden, B.P. 1954. Submarine illumination and the twilight movement of a sonic scattering layer. Nature 174:869–872.

Kats, L.B. and Dill, L.M. 1998. The scent of death: Chemosensory assessment of predation risk by prey animals. Ecoscience 5:361–394.

Kerfoot, W.C., Kellogg, D.L. and Strickler, J.R. 1980. Visual observations of life zooplankters: Evasion, escape and chemical defenses. In: Kerfoot, W.C. (Ed.) Evolution and ecology of zooplankton communities, pp. 10–27. University Press of New England, Hanover, New Hampshire.

Kersting, K. 1981. Annual and diel oxygen and temperature regime of the Lakes Maarsseveen. Hydrobiol. Bull. 15:10–28.

King, C.E. and Miracle, M.R. 1995. Diel vertical migration by *Daphnia longispina* in a Spanish lake: Genetic sources of distributional variation. Limnol. Oceanogr. 40:226–231.

Kipling, R. 1902. Just so stories. Macmillan, London.

Kirk, J.T.O. 1983. Light and photosynthesis in aquatic ecosystems. Cambridge University Press, Cambridge. pp. 401.

Kleiven, O.T., Larsson, P. and Hobaek, A. 1996. Direct distributional response in *Daphnia pulex* to a predator kairomone. J. Plankton Res. 18:1341–1348.

Klotzsche, K. 1913. Beitrage zur Kenntnis des feineren Baues der Cladoceren (*Daphnia magna*). Jenaische Z. für Naturwissenschaften 50:601–647.

Knight-Jones, E.W. and Morgan, E. 1966. Responses of marine animals to changes in hydrostatic pressure. Oceanogr. Mar. Biol. Annu. Rev. 4:267–299.

Koestler, A. 1964. The sleepwalkers. Hutchinson & Co., London. pp. 1–624.

Kosobokova, K.N. 1978. Diurnal vertical distribution of *Calanus hyperboreus* Kroeyer and *Calanus glacialis* Jaschnov in the central Polar basin. Okeanologiya 18:722–728.

Kouwenberg, J.H.M., Browman, H.I., Cullen, J.J., Davis, R.F., St-Pierre, J.-F. and Runge, J.A. 1999a. Biological weighting of ultraviolet (280–400 nm) induced mortality in marine zooplankton and fish. I. Atlantic cod (*Gadus morhua*) eggs. Mar. Biol. 134:269–284.

Krebs, J.R. and Davies, N.B. 1993. An introduction to behavioural ecology. Blackwell, Science, Oxford. Ed. 3. pp. 1–420.

Kubecka, J. 1991. A case of an extremely low share of the 0-age group on the total biomass, production and ration of the perch (*Perca fluviatilis*) population. Acta Soc. Zool. Bohemoslov. 55:114–119.

Kubecka, J. and Bohm, M. 1991. The fish fauna of the Jordan reservoir, one of the oldest man-made lakes in central Europe. J. Fish. Biol. 38:935–950.

Kuhn, P., Browman, H., McArthur, B. and St-Pierre, J-F. 1999. Penetration of ultraviolet radiation in the waters of the estuary and Gulf of St. Lawrence. Limnol. Oceanogr. 44:710–716.

Lagergren, R. and Stenson, J.A.E. 2000. Chemical cues from the invertebrate predator *Leptodora kindtii* affect the development of cyclomorphic traits in *Eubosmina coregoni gibbera*. J. Plankton Res. 22:1213–1219.

Lampert, W. 1989. The adaptive significance of diel vertical migration of zooplankton. Funct. Ecol. 3:21–27.

Lampert, W. 1993. Ultimate causes of diel vertical migration of zooplankton: New evidence for the predator-avoidance hypothesis. Arch. Hydrobiol. Beih. Ergebn. Limnol. 39:79–88.

Lampert, W., Fleckner, W., Rai, H. and Taylor, B.E. 1986. Phytoplankton control by grazing zooplankton: A study on the spring clear-water phase. Limnol. Oceanogr. 31:478–490.

Lampert, W., Rothhaupt, K.O. and von Elert, E. 1994a. Chemical induction of colony formation in a green alga (*Scenedesmus acutus*) by grazers (*Daphnia*). Limnol. Oceanogr. 39:1543–1550.

Lampert, W., Tollrian, R. and Stibor, H. 1994b. Chemische Induktion von Verteidigungs-mechanismen bei Süsswassertieren. Natur & Wissenschaften 81:375–382.

Land, M.F. 1980. Eye movements and the mechanism of vertical steering in euphausiid crustacea. J. Comp. Physiol. 137:255–265.

Langford, R.R. 1938. Diurnal and seasonal changes in the distribution of limnetic Crustacea of Lake Nipissing. Univ. Toronto Stud. Biol. 45:1.

LaRow, E.J. 1968. A persistent diurnal rhythm in *Chaoborus*. I. The nature of the rhythmicity. Limnol. Oceanogr. 13:250–256.

Larsson, P. and Dodson, S. 1993. Chemical communication in planktonic animals. Archiv für Hydrobiologie 129:129–155.

Lass, S. and Spaak, P. 2002. Chemically induced anti-predator defences in plankton. Hydrobiologia.

Laurion, I., Vincent, W.F. and Lean, D.R.S. 1997. Underwater ultraviolet radiation: Development of spectral models for northern high latitude lakes. Photochem. Photobiol. Sci. 65: 107–114.

Laurion, I., Ventura, M., Catalan, J., Psanner, R. and Sommaruga, R. 2000. Attenuation of ultraviolet radiation in mountain lakes: Factors controlling the among- and within lake variability. Limnol. Oceanogr. 45:1274–1288.

Leder, H. 1915. Untersuchungen über den feineren Bau des Nervensystems der Cladoceren. Arbeiten der zoologische Institut Wien 20:1–96.

Leech, D.M. and Williamson, C.E. 2000. Is tolerance to UV radiation in zooplankton related to body size, taxon, or lake transparency? Ecol. Appl. 10:1530–1540.

Leech, D.M. and Williamson, C.E. 2001. In situ exposure to ultraviolet radiation alters the depth distribution of *Daphnia*. Limnol. Oceanogr. 46:416–420.

Leech, D.M., Williamson, C.E., Moeller, R.E. and Hargreaves, B.R. 2005. Effects of ultraviolet radiation on the seasonal vertical distribution of zooplankton: A database analysis. Arch. Hydrobiol. 162:445–464.

Legand, M., Bourret, P., Fourmanoir, P., Grandperrin, R., Gueredrat, J.A., Michel, A. and Raneurel, P. 1972. Rélations trophiques et distributions verticales en milieu pélagique dans l'Ocean Pacifique intertropical. Cahiers Office de la Recherche Scientifique et Technique Outre-Mer 10:303–393.

Leibold, M.A., Tessier, A.J. and West, C.T. 1994. Genetic, acclimatization, and ontogenetic effects on habitat selection behavior in *Daphnia pulicaria*. Evolution 48:1324–1332.

Leising, A.W., Pierson, J.J., Halsband-Lenk, C., Horner, R. and Postel, J. 2005. Copepod grazing during spring blooms: Does *Calanus pacificus* avoid harmful diatoms? Prog. Oceanogr. 67:384–405.

Lenz, P.H. 1980. Ecology of an alkali adapted variety of *Artemia* from Mono Lake, California, USA. In: Persoone, G., Sorgeloos, P., Roels, O. and Jaspers, E. (Eds.) The brine shrimp artemia, pp. 79–96. Universa Press, Belgium.

Lenz, P.H., Hartline, D.K. and Davis, A.D. 2000. The need for speed. I. Fast reactions and myelinated axons in copepods. J. Comp. Physiol. 186:337–345.

Levine, D.M. and Sulkin, S.D. 1984. Nutritional singnificance of long-chained polyunsaturated fatty acids of the zoeal development of the brachyuran crab *Eurpanopeus depressus* (Smith). J. Exp. Mar. Biol. Ecol. 81:211–223.

Levins, R. 1968. Evolution in changing environments. Princeton University Press, Princeton, New Yersey. pp. 1–120.

Lima, S.L. 1998. Stress and decision making under the risk of predation: Recent developments from behavioral, reproductive, and ecological perspectives. Adv. Study Beh. 27:215–290.

Lima, S.L. and Bednekoff, P.A. 1999. Temporal variation in danger drives antipredator behavior: The predation risk allocation hypothesis. Am. Nat. 153:649–659.

Lima, S.L. and Dill, L.M. 1990. Behavioral decisions made under the risk of predation: A review and prospectus. Can. J. Zool. 68:619–640.

Lincoln, R.J. 1971. Observations of the effects of changes in hydrostatic pressure and illumination on the behaviour of some planktonic crustaceans. J. Exp. Biol. 54:677–688.

Loeb, J. 1893. Ueber künstliche Umwandlung positiv heliotropischer Thiere in negativ Heliotropische und umgekehrt. Arch. Gesamte Physiol. Mensch Tiere 54:81–107.

Loeb, J. 1904. The control of heliotropic reactions in fresh water crustaceans by chemicals, especially CO2. Univ. Calif. Publ. Physiol. 2:1–3.

Longhurst, A.R. 1976. Vertical migration. In: Cushing, D.H. and Walsh, J.J. (Eds.) The ecology of the seas, pp. 116–137. Blackwell Scientific Publications, Oxford.

Longhurst, A.R., Sameoto, D. and Herman, A. 1984. Vertical distribution of Arctic zooplankton in summer eastern Canadian Archipelago. J. Plankton Res. 6:137–168.

Loose, C.J. 1993a. *Daphnia* diel vertical migration behavior: Response to vertebrate predator abundance. Arch. Hydrobiol. Beih. Ergebn. Limnol. 39:29–36.

Loose, C.J. 1993b. Lack of endogenous rhythmicity in *Daphnia* diel vertical migration. Limnol. Oceanogr. 38:1837–1841.

Loose, C.J. and Dawidowicz, P. 1994. Trade-offs in diel vertical migration by zooplankton: The costs of predator avoidance. Ecology 75:2255–2263.

Loose, C.J., von Elert, E. and Dawidowicz, P. 1993. Chemical-induced diel vertical migration in *Daphnia*: A new bioassay for kairomones exuded by fish. Arch. Hydrobiol. 126:329–337.

Lorke, A., McGinnis, D.F., Spaak, P. and Wuest, A. 2004. Acoustic observations of zooplankton in lakes using Doppler current profiler. Freshw. Biol. 49:1280–1292.

Lüning, J. 1992. Phenotypic plasticity of *Daphnia pulex* in the presence of invertebrate predators: Morphological and life history responses. Oecologia 92:383–390.

Lurling, M. 1999. The smell of water. Wageningen Agricultural University, Wageningen.

Macagno, E.R. 1977. Abnormal synaptic connectivity following UV-induced cell death during *Daphnia* development. In: Lash, J.W. and Burger, M.M. (Eds.) Cell and tissue interactions. Raven Press, New York.

Macagno, E.R. 1984. Formation of ordered connections in the visual system of *Daphnia magna*. BioScience 34:308–312.

Macagno, E.R., Lopresti, V. and Levinthal, C. 1973. Structure and development of neural connections in isogenic organisms: Variations and similarities in the optic system of *Daphnia magna*. Proc. Nat Acad. Sci. USA 70:57–61.

Machacek, J. 1990. The response of *Daphnia* to the fish-produced chemical stimuli. Annual Report 1990. Hydrobiol. Inst. Czechoslovak Acad. Sci. 31:45–49.

Machacek, J. 1993. Comparison of the response of *Daphnia galeata* and *Daphnia obtusa* to fish-produced chemical substance. Limnol. Oceanogr. 38:1544–1550.

Manca, M., De Bernardi, R. and Savia, A. 1986. Effects of fluctuating temperature and light conditions on the population dynamics and life strategies of migrating and non-migrating *Daphnia* species. Mem. Ist. Ital. Idrobiol. Pallanza 44:177–202.

Mann, C.G. and Swift, E. 1988. Swimming behavior of the ctenophore *Mnemiopsis leidyi*. Bull. Mar. Sci. 43:848.

Markager, S. and Vincent, W.F. 2000. Spectral light attenuation and the absorption of UV and blue light in natural waters. Limnol. Oceanogr. 45:642–650.

Mauchline, J. 1980. The biology of mysids and euphausiids. Advances in marine biology. Vol. 18. Academic Press, London. 681 pp.

McAllister, C.D. 1971. Some aspects of nocturnal and continuous grazing by planktonic herbivores in relation to production studies. J. Fish. Res. Bd. Can. 248:1–281.

McKelvey, L.M. and Forward, Jr., R.B. 1995. Activation of brine shrimp nauplii photoresponses involved in diel vertical migration by chemical cues from visual and non-visual planktivores. J. Plankton Res. 17:2191–2206.

McLaren, I.A. 1963. Effects of temperature on growth of zooplankton, and the adaptive value of vertical migration. J. Fish. Res. Bd. Can. 20:685–727.

McLaren, I.A. 1974. Demographic strategy of vertical migration by a marine copepod. Am. Nat. 108:91–102.

Mehner, T. 2000. Influence of spring warming on the predation rate of underyearling fish on *Daphnia* – a deterministic simulation approach. Freshw. Biol. 45:253–263.

Mehner, T. and Thiel, R. 1999. A review of predation impact by 0+ fish on zooplankton in fresh and brackish waters of the temperate northern hemisphere. Env. Biol. Fish. 56:169–181.

Mehner, T., Schultz, H. and Herbst, R. 1995. Interaction of zooplankton dynamics and diet of 0+ perch (*Perca fluviatilis* L.) in the top-down manipulated Bautzen Reservoir (Saxony, Germany) during summer. Limnologica 25:1–9.

Mehner, T., Hülsmann, S., Worischka, S., Plewa, M. and Benndorf, J. 1998a. Is the midsummer decline of *Daphnia* really induced by age-0 fish predation? Comparison of fish consumption and *Daphnia* mortality and life history parameters in a biomanipulated reservoir. J. Plankton Res. 20:1797–1811.

Mehner, T., Plewa, M., Hülsmann, S. and Worischka, S. 1998b. Gape-size dependent feeding of age-0 perch (*Perca fluviatilis*) and age-0 zander (*Shizostedium lucioperca*) on *Daphnia galeata*. Arch. Hydrobiol. 142:191–207.

Mikkelsen, P.M. 1981. Studies on euphausiacean crustaceans from the Indian River region in Florida. 1. *Stylocheiron longicorne* species-group, with emphasis on reproductive morphology. Proc. Biol. Soc. Wash. 94:1174–1204.

Miller, C.B. 1970. Some environmental consequences of vertical migration in marine zooplankton. Limnol. Oceanogr. 15:727–741.

Miller, W.H., Ratliff, F. and Hartline, H.K. 1961. How cells receive stimuli. Sci. Am. 205: 222–238.

Mills, C.E. 1981. Diversity of swimming behaviors in hydromedusae as related to feeding and utilization of space. Mar. Biol. 64:185–189.

Mills, E.L. and Forney, J.L. 1981. Energetics, food consumption, and growth of young yellow perch in Oneida Lake, New York. Trans. Am. Fish. Soc. 110:479–488.

Miracle, M.R., Armengol-Diaz, J. and Dasi, M.J. 1993. Extreme meromixis determines strong differential planktonic vertical distributions. Verh. Internat. Verein. Limnol. 25:705–710.

Moloney, C.L. and Gibbons, M.J. 1996. Sampling and analysis of gut contents in relation to environmental variability and diel vertical migration by herbivorous zooplankton. J. Plankton Res. 18:1535–1556.

Mooij, W.M., Lammens, E.H.R.R. and Van Densen, W.L.T. 1994. Growth rate of 0+ fish in relation to temperature, body size, and food in shallow eutrophic Lake Tjeukemeer. Can. J. Fish. Aquat. Sci. 51:516–526.

Moore, H.B. 1955. Variations in temperature and light response within a plankton population. Biol. Bull. 108:175–181.

Morris, D.P., Zagarese, H., Williamson, C.E. 1995. The attenuation of solar UV radiation in lakes and the role of dissolved organic carbon. Limnol. Oceanogr. 40:1381–1391.

Mück, P. and Lampert, W. 1984. An experimental study of the importance of food conditions for the relative abundance of calanoids copepods and cladocerans. Arch. Hydrobiol. 66(Suppl.): 157–179.

Muller, J. and Seitz, A. 1993. Habitat partitioning and differential vertical migration of some *Daphnia* genotypes in a lake. Arch. Hydrobiol. Beih. Ergebn. Limnol. 39:167–174.

Naud, M. and Magnan, P. 1988. Diel onshore – offshore migrations in northern redbelly dace, *Phoxinus eos* (Cope), in relation to prey distribution in a small oligotrophic lake. Can. J. Zool. 66:1249–1253.

Nauman, E. 1921. Untersuchungen ueber das Verteilungsproblem des limnische Biosestons. 1. Die allgemeinen reizphysiologischen Verteilungsbedingungen des heliophilen biosestons. Kungl. Svenska Vetenskapsakademiens Handlingar 61:1–28.

Nauwerck, A. 1963. Die Beziehungen zwischen Zooplankton und Phytoplankton im See Erken. Symbolae Botanicae Uppsalienses 18:1–163.

Nauwerck, A. 1993. Migration strategies of crustacean zooplankton in lake Mondsee. Arch. Hydrobiol. Beih. Ergebn. Limnol. 39:109–122.

Nilsson, D.E. and Odselius, R. 1983. Regionally different optical systems in the compound eye of the water-flea *Polyphemus* (Cladocera,Crustacea). Proc. Roy. Soc. Lond B 217:163–175.

Odselius, R. and Nilsson, D.E. 1983. Regionally different ommatidial structures in the compound eye of the water-flea *Polyphemus* (Cladocera,Crustacea). Proc. Roy. Soc. Lond B 217:177–189.

Ohman, M.D. and Wood, S.N. 1996. Mortality estimation for planktonic copepods: *Pseudocalanus newmani* in a temporate fjord. Limnol. Oceanogr. 41:126–135.

Ohman, M.D., Frost, B.W. and Cohen, E.H. 1983. Reverse diel vertical migration – an escape from invertebrate predators. Science 220:1404–1407.

Onsrud, M.S.R. and Kaartvedt, S. 1998. Diel vertical migration of the krill *Meganyctiphanes norvegica* in relation to physical environment, food and predators. Mar. Ecol. Prog. Ser. 171:209–219.

Onsrud, M.S.R., Kaartvedt, S., Rostad, A. and Klevjer, T.A. 2004. Vertical distribution and feeding patterns in fish foraging on the krill *Meganyctiphanes norvegica*. ICES J. Mar. Sci. 61: 1278–1290.

Orcutt, J.D. and Porter, K.G. 1983. Diel vertical migration by zooplankton: Constant and fluctuating temperature effects on life history parameters of *Daphnia*. Limnol. Oceanogr. 28:720–730.

Osgood, K.E. and Frost, B.W. 1994a. Comparative life histories of three calanoid copepods in Dabob Bay, Washington. Mar. Biol. 118:627–636.

Osgood, K.E. and Frost, B.W. 1994b. Ontogenetic diel vertical migration behaviors of the marine planktonic copepods *Calanus pacificus* and *Metridia lucens*. Mar. Ecol. Prog. Ser. 104:13–25.

Paanakker, J.E. and Hallegraeff, G.M. 1978. A comparative study on the carotenoid pigmentation of the zooplankton of Lake Maarsseveen (Netherlands) and of Lac Pavin (France). I. Chromatographic characterization of carotenoid pigments. Comp. Biochem. Physiol. 60B: 51–58.

Paloheimo, J.E. 1974. Calculation of instantaneous birth rate. Limnol. Oceanogr. 19:692–694.

Park, S., Chandra, S., Muller-Navarra, D. and Goldman, C.R. 2004. Diel and vertical variability of seston food quality and quantity in a small subalpine oligomesotrophic lake. J. Plankton Res. 26:1489–1498.

Parks, T.S. 1968. Calanoid copepods from the central North Pacific Ocean. Fish. Bull. 66:527–572.

Parma, S. 1971. *Chaoborus flavicans* (Meigen) (Diptera, Chaoboridae): An autecological study. Ph D. Thesis, University of Groningen, pp. 1–128.

Patten, B.C. and Odum, E.P. 1981. The cybernetic nature of ecosystems. Am. Nat. 118:885–219.

Pearre, Jr., S. 1973. Vertical migration and feeding in *Sagitta elegans* Verrill. Ecology 54:300–314.

Pearre, S., Jr. 1979. Problems of detection and interpretation of vertical migration. J. Plankton Res. 1:29–44.

Pearre, S. Jr. 2003. Eat and run? The hunger/satiation hypothesis in vertical migration: History, evidence and consequences. Biol. Rev. 78:1–79.

Pennisi, E. 2000. In nature, animals that stop and start win the race. Science 288:83–85.

Peters, R.H. 1987. Metabolism in *Daphnia*. In: Peters, R.H. and De Bernardi, R. (Eds.) Daphnia, pp. 193–243. Memorie dell' Istituto Italiano di Idrobiologia, Verbania, Pallanza.

Pianka, E.R. 1974. Niche overlap and diffuse competition. Proc. Natl. Acad. Sci. USA 71: 2141–2145.

Pierson, J.J., Halsband-Lenk, C. and Leising, A.W. 2005a. Reproductive success of *Calanus pacificus* during diatom blooms in Dabob Bay, Washington. Prog. Oceanogr. 67:314–331.

Pierson, J.J., Leising, A.W., Halsband-Lenk, C. and Ferm, N. 2005b. Vertical distribution and abundance of *Calanus pacificus* and i in relation to chlorophyll a concentrations in Dabob Bay, Washington. Prog. Oceanogr. 67:349–365.

Popper, K.R. 1960. The poverty of historicism. Routledge & Kegan Paul, London. Ed.2. pp. 1–166.

Post, J.R. and McQueen, D.J. 1988. Ontogenetic changes in the distribution of larval and juvenile yellow perch (*Perca flavescens*): A response to prey or predators? Can. J. Fish. Aquat. Sci. 45:1820–1826.

Price, H.J. 1989. Swimming behavior of krill in response to algal patches: A mesocosm study. Limnol. Oceanogr. 34:649–659.

Pyanowska, J. 1993. Diel vertical migration in zooplankton: Fixed or inducible behaviour? Arch. Hydrobiol. Beih. Ergebn. Limnol. 39:89–97.

Radl, E. 1901. Ueber den Phototropism einiger Arthropoden. Biol. Zentralblatt Leipzig 21:75–86.

Radl, E. 1903. Untersuchungen über den Phototropismus der Tiere. Wilhelm Engelmann, Leipzig. pp. 1–188.

Ramcharan, C.W. and Sprules, W.G. 1989. Preliminary results from an inexpensive motion analyzer for free-swimming zooplankton. Limnol. Oceanogr. 34:457–462.

Ratliff, F., Hartline, H.K. and Lange, D. 1966. The dynamics of lateral inhibition in the compound eye of *Limulus* In: Bernhard, C.G. (Ed.) The functional organisation of the compound eye, pp. 399–425. Pergamon Press, New York.

Reede, T. 1995. Life history shifts in response to different levels of fish kairomones in *Daphnia*. J. Plankton Res. 17:1661–1667.

Reede, T. 1997. Effects of neonate size and food concentration on the life history responses of a clone of the hybrid *Daphnia hyalina x galeata* to fish kairomones. Freshw. Biol. 37: 389–396.

Reede, T. and Ringelberg, J. 1995. The influence of a fish exudate on two clones of the hybrid *Daphnia galeata x hyalina* . Hydrobiologia 37:207–212.

Reede, T. and Ringelberg, J. 1998. Differential life history responses of several pelagic *Daphnia* clones differing in migratory behaviour. Aquatic Ecol. 32:245–253.

Rhode, S.C., Pawlowski, M. and Tollrian, R. 2001. The impact of ultraviolet radiation on the vertical distribution of zooplankton. Nature 412:69–71.

Rice, A.L. 1964. Observations on the effects of changes of hydrostatic pressure on the behaviour of some marine animals. J. Mar. Biol. Assoc. U.K. 44:163–175.

Rice, A.L. 1965. The orientation of pressure responses of some marine crustacea. Proceedings of Symposium on Crustacea 3:1124–1131.

Richards, S.A, Possingham, H.P. and Noye, J. 1996. Diel vertical migration: Modelling light-mediated mechanisms. J. Plankton Res. 18:2199–2222.

Rimet, M. 1960a. La phototaxie polyphasique de *Daphnia pulex* de Geer. Journal Physiologie 52:769–781.

Rimet, M. 1960b. Persistances rythmiques dans le phototropisme de *Daphnia pulex* de Geer. Annales de biologique (Paris) 64:189–198.

Ringelberg, J. 1964. The positively phototactic reaction of *Daphnia magna* Straus: A contribution to the understanding of diurnal vertical migration. Neth. J. Sea Res. 2:319–406.

Ringelberg, J. 1969. Spatial orientation of planktonic crustaceans. 2. The swimming behaviour in a vertical plane. Verh. Internat. Verein Limnol. 17:814–847.

Ringelberg, J. 1976. Aquatische Oecologie. Bohn, Scheltema & Holkema, Utrecht. pp 1–240 (in Dutch).

Ringelberg, J. 1987a. Ecosystemen: Concreet, abstract of fictie? Vakblad voor Biologen 67:45–49.

Ringelberg, J. 1987b. Light induced behaviour in *Daphnia*. In: Peters, R.H. and De Bernardi, R.,(Eds.) Daphnia, pp. 285–323. Memorie dell' Istituto Italiano di Idrobiologia, Verbania, Pallanza.

Ringelberg, J. 1991. Enhancement of the phototactic reaction in *Daphnia hyalina* by a chemical mediated by juvenile perch (*Perca fluviatilis*). J. Plankton Res. 13:17–25.

Ringelberg, J. 1993. Phototaxis as a behavioural component of diel vertical migration in a pelagic *Daphnia*. Arch. Hydrobiol. Beih. Ergebn. Limnol. 39:45–55.

Ringelberg, J. 1995a. An account of a preliminary mechanistic model of swimming behaviour in *Daphnia*: Its use in understanding diel vertical migration. Hydrobiologia 307:161–165.

Ringelberg, J. 1995b. Changes in light intensity and diel vertical migration: A comparison of marine and freshwater environments. J. Mar. Biol. Assoc. U.K. 75:15–25.

Ringelberg, J. 1995c. Is diel vertical migration possible without a rhythmic signal? Comments on a paper by Bollens et al. (1994). J. Plankton Res. 17:653–655.

Ringelberg, J. 1997. Some suggestions for future cladoceran research. Hydrobiologia 360: 291–294.

Ringelberg, J. 1999a. The photobehaviour of *Daphnia* ssp. as a model to explain diel vertical migration in zooplankton. Biol. Rev. 74:397–423.

Ringelberg, J. 1999b. The role of infochemicals in lake plankton crustacean behavior and predator-prey relations. In: Von Vaupel Klein, J.C. and Schram, F.R. (Eds.) The biodiversity crisis and crustacea: Proc. 4th crustacean congres, Amsterdam 1998, pp. 161–175. A.A. Balkema, Rotterdam.

Ringelberg, J. and Flik, B.J.G. 1984. The mortality effect of ultraviolet radiation in a translucent and in a red morph of *Acanthodiaptomus denticornis* (Crustacea, Copepoda) and its possible ecological relevance. Hydrobiologia 112:217–222.

Ringelberg, J. and Flik, B.J.G. 1994. Increased phototaxis in the field leads to enhanced diel vertical migration. Limnol. Oceanogr. 39:1855–1864.

Ringelberg, J. and Servaas, H. 1971. A circadian rhythm in *Daphnia magna*. Oecologia (Berlin) 6:289–292.

Ringelberg, J. and Van Gool, E. 1995. Migrating *Daphnia* have a memory for fish kairomones. Mar. Freshw. Behav. Physiol. 26:249–257.

Ringelberg, J. and Van Gool, E. 1998. Do bacteria, not fish, produce 'fish kairomone'? J. Plankton Res. 20:1847–1852.

Ringelberg, J. and van Gool, E. 2003. On the combined analysis of proximate and ultimate aspects in diel vertical migration (DVM) research. Hydrobiologia 491:85–90.

Ringelberg, J., Kasteel, J.V. and Servaas, H. 1967. The sensitivity of *Daphnia magna* Straus to changes in light intensity at various adaptation levels and its implication in diurnal vertical migration. Z. Vergl. Physiol. 56:397–407.

Ringelberg, J., Flik, B.J.G. and Buis, R.C. 1975. Contrast orientation in *Daphnia magna* and its significance for vertical plane orientation in the pelagic biotope in general. Neth. J. Zool. 25:454–475.

Ringelberg, J., Keyser, A.L. and Flik, B.J.G. 1984. The mortality effect of ultraviolet radiation in a translucent and in a red morph of *Acanthodiaptomus denticornis* (Crustacea, Copepoda) and its possible ecological relevance. Hydrobiologia 112:217–222.

Ringelberg, J., Flik, B.J.G., Lindenaar, D. and Royackers, K. 1991a. Diel vertical migration of *Daphnia hyalina* (sensu latiori) in Lake Maarsseveen: Part 1. Aspects of seasonal and daily timing. Arch. Hydrobiol. 121:129–145.

Ringelberg, J., Flik, B.J.G., Lindenaar, D. and Royackers, K. 1991b. Diel vertical migration of *Daphnia hyalina* (sensu latiori) in Lake Maarsseveen: Part 2. Aspects of population dynamics. Arch. Hydrobiol. 122:385–401.

Ringelberg, J., Flik, B.J.G., Lindenaar, D. and Royackers, K. 1991c. Diel vertical migration of *Eudiaptomus gracilis* during a short summer period. Hydrobiol. Bull. 25:77–84.

Ringelberg, J., Flik, B.J.G., Aanen, D.K. and van Gool, E. 1997. Amplitude of vertical migration (DVM) is a function of fish biomass, a hypothesis. Arch. Hydrobiol. Beih. Ergebn. Limnol. 49:71–78.

Ringelberg, J. and Lingeman, R. 2003. A coupled oscillator model describes normal and strange zooplankton swimming behaviour. Neth. J. Zool. 52:225–241.

Ringelberg, J., van Gool, E. and Brehm, M. 2004. Density and depth variations of *Daphnia* multilocus genotypes during a summer period in Lake Maarsseveen. Aquatic Ecol. 38:525–535.

Roger, C. 1971. Distribution verticals des euphausias (Crustacean) dans les courants quatoriaux de l'Océan Pacifique. Mar. Biol. 10:134–144.

Roger, C. 1975. Rythmes nutritionels et organisation trophique d'une population de crustaces pelagiques (Euphausiacea). Mar. Biol. 32:365–378.

Roozen, F. and Lurling, M. 2001. BehaviouraL response of *Daphnia* to olfactory cues from food, competitors and predators. J. Plankton Res. 23:797–808.

Rose, M. 1925. Contribution a l'étude de la biologie du plankton; le problème des migrations verticales journalières. Arch. Zool. expérimentale et génerale 64:387–542.

Russell, F.S. 1934. The vertical distribution of marine macroplankton. XII Some observations on the vertical distribution of *Calanus finmarchicus* in relation to light intensity. J. Mar. Biol. Ass. U.K. 19:569–585.

Salonen, K. and Rosenberg, M. 2000. Advantages from diel vertical migration can explain the dominance of *Gonyostomum semen* (Raphidophyceae) in a small, deeply-stratified humid lake. J. Plankton Res. 22:1841–1853.

Sameoto, D.D. 1976. Distribution of sound scattering layers caused by euphausiids and their relationship to chlorophyll a concentrations in the Gulf of St. Lawrence estuary. J. Fish. Res. Bd. Can. 33:681–687.

Sameoto, D.D. 1980. Relationships between stomach contents and vertical migration in *Meganyctiphanes norvegica, Thysanoessa raschil* and *T. inermis* (Crustacea Euphausiacea). J. Plankton Res. 2:129–143.

Sameoto, D.D. 1984. Vertical distribution of zooplankton biomass and species in Northeastern Baffin Bay related to temperature and salinity. Polar Biol. 2:213–224.

Sasaki, T., Okami, N., Watanabe, S. and Oshiba, N. 1958. Angular distribution of submarine daylight intensity in horizontal plane. J. Scientific Res. Inst. Japan 52:71–77.

Sasaki, T., Watanabe, S., Oshiba, G. and Okami, N. 1960. Measurements of perpendicular and horizontal angular distributions of submarine daylight by means of a new remote control instrument. Records Oceangr. Works Japan 4 (special number):197–205.

Sasaki, T., Watanabe, S., Oshiba, G., Okami, N. and Kajihara, M. 1962. On the instrument for measuring angular distribution of underwater radiance. Bull. Jap. Soc. Sci Fisheries 28: 489–496.

Sauberer, F. 1962. Empfehlungen für die Durchführung von Strahlungsmessungen an und in Gewässern. Mitt. Int. Ver. Limnol. 11:1–77.

Savage, R.E. 1926. The plankton of a herring ground. Fish. Invest. Lond. 9. Ser. 2, 9:1–35.

Schallek, W. 1942. The vertical migration of the copepod *Acartia tonsa* under controlled illumination. Biol. Bull. 82:112–126.

Scheffer, D., Robert, P. and Medioni, J. 1958. Réactions oculo-motrices de la Daphnie (*Daphnia pulex* De Geer) en response à des lumières monochromatiques d'égale energie. Sensibilité visuelle et sensibilité dermatoptique. Comtes rendue de Sociéte Biologique, Strasbourg 152:1000–1004.

Schloss, A. and Haney, J.F. 2002. Direct observations of the activity responses of mayfly nymphs to relative light change and light intensity. Arch. Hydrobiol. 154:353–374.

Schloss, A. and Haney, J.F. 2006. Clouds, shadows or twilight? Mayfly nymphs recognise the difference. Freshw. Biol. 51:1079–1089.

Schröder, R. 1959. Die Vertikalwanderungen des Crustaceen Planktons der Seen des südlichen Schwarzwaldes. Arch. Hydrobiol. 25(Suppl.):1–43.

Schröder, R. 1962. Vertikalverteilung des Zooplanktons und Thermokline. Arch. Hydrobiol. 25(suppl.):401–410.

Schulz, H. 1928. Ueber die Bedeutung des Lichtes im Leben niederer Krebse (nach Versuchen an Daphniden). Z. Vergl. Physiol. 7:488–551.

Schwenk, K. and Spaak, P. 1997. Ecology and genetics of interspecific hybridization in *Daphnia*. In: Streit, B., Stadler, T. and Lively, C.M. (Eds.) Evolutionary ecology of freshwater animals, pp. 199–229. Birkhauser Verlag, Basel.

Segal, E. 1970. Light. In: Kinne, O. (Ed.) Marine ecology, pp. 159–211. Wiley-Interscience, London.

Siebeck, O. 1960. Untersuchungen über die Vertikalwanderungen planktischer Crustacean unter Berücksichtigung der Strahlungsverhältnisse. Int. Rev. Ges. Hydrobiol. 45:381–454.

Siebeck, O. 1965. Orientierungsphysiologische Untersuchungen mit Planktoncrustaceen unter Berücksichtigung raumlicher Strahlungsverteilungen. Verh. Dtsch. Zool. Ges. 28:319–328.

Siebeck, O. 1968. Uferflucht und optische Orientierung pelagischer Crustaceen. Arch. Hydrobiol. 35(suppl.):1–118.

Siebeck, O. 1969. Spatial orientation of planktonic crustaceans. 1. The swimming behaviour in a horizontal plane. Verh. Internat. Verein. Limnol. 17:831–847.

Siebeck, O. 1978. Ultraviolet tolerance of planktonic crustaceans. Verh. Internat. Verein. Limnol. 20:2469–2473.

Siebeck, O. 1980. Optical orientation of pelagic crustaceans and its consequence in the pelagic and littoral zones. In: Kerfoot, W.C. (Ed.) Evolution and ecology of zooplankton communities, pp. 28–38. The University Press of New England, Hanover.

Siebeck, O. and Böhm, U. 1994. Challenges for an appraisal of UV-B effects upon planktonic crustaceans under natural conditions with a non-migrating (*Daphnia pulex obtusa*) and a migrating cladoceran (*Daphnia galeata*). Arch. Hydrobiol. Beih. Ergebn. Limnol. 43:197–206.

Siebeck, O., Vail, T.L., Williamson, C.E., et al. 1994. Impact of UV-B radiation on zooplankton and fish in pelagic freshwater ecosystems. Arch. Hydrobiol. Beih. Ergebn. Limnol. 43:101–114.

Small, L.F. and Hebard, J.F. 1967. Respiration of a vertically migrating marine crustacean *Euphausia pacifica* Hansen. Limnol. Oceanogr. 12:272–281.

Smith, R.C. and Baker, K.S. 1981. Optical properties of the clearest natural waters (200–00 nm). Appl. Opt. 20:177–184.

Smith, R.C., Tyler, J.E. and Goldman, C.R. 1973. Optical properties and color of Lake Tahor and Crater Lake. Limno. Oceanogr. 18:189–199.

Smith, F.E. and Baylor, E.R. 1953. Color responses in the Cladocera and their ecological significance. Am. Nat. 87:49–55.

Smith, K.C. and Macagno, E.R. 1990. UV receptors in the compound eye of *Daphnia magna* (Crustacea,Branchiopoda). A fourth spectral class in single ommatidia. J. Comp. Physiol. 166:597–606.

Smolander, U. and Arvola, L. 1998. Seasonal variation in diel vertical distribution of the migratory alga *Cryptomonas marssonii* (Cryptophyceae) in a small highly humid lake. Hydrobiologia 161:89–98.

Soeder, C.J. 1967. Tagesperiodische Vertikalwanderung bei begeisselten Planktonalgen. Die Umschau im Wissenschaft und Technik 12:388.

Sokal, R.R. and Rohlf, F.J. 1995. Biometry. W.H. Freeman and Company, New York. Ed.3. pp. 1–887.

Sokolov, E.N. 1960. Neural models of the orienting reflex. In: Brazier, M.A.B. (Ed.) The central nervous system and behavior, pp. 187–276. Josiah Macy, Jr., Foundation, New York (cited in Bullock 1988).

Sokolov, E.N. 1975. Neural mechanisms of the orienting reflex. In: Sokolov, E.N. and Vinogradova, O.S. (Eds.). English Translation edited by Weinberger, N.E., Lawrence Erlbaum Associates, Hillsdale N.J., and John Willey and Sons, New York, pp. 217–238. (cited in Bullock 1988).

Sommer, U. 1985. Differential migration of Cryptophyceae in Lake Constance. In: M.A. Rankin (ed.) Migration: Mechanisms and adaptive significance. Contributions in Marine Science, Suppl. 27:166–175.

Sommer, U., Gliwicz, Z.M., Lampert, W. and Duncan, A. 1995. The PEG-model of seasonal succession of planktonic events in fresh waters. Arch. Hydrobiol. 106:433–471.

Spaak, P. 1994. Genetical ecology of a coexisting *Daphnia* hybrid species complex. Ph. D. Thesis Univ. Utrecht 1–125.

Spaak, P. 1995. Cyclomorphosis as a factor explaining succes of a *Daphnia* hybrid in Tjeukemeer. Hydrobiologia 307:283–289.

Spaak, P. 1996. Temporal changes in genetic structure of the *Daphnia* species complex in Tjeukemeer, with evidence for backcrossing. Heredity 76:539–548.

Spaak, P. and Boersma, M. 2006. Predator coexistence of hybrid and parental taxa. Arch. Hydrobiol. 167:55–76.

Spaak, P. and Hoekstra, J.R. 1993. Clonal structure of the *Daphnia* population lake Maarsseveen: Its implications for diel vertical migration. Arch. Hydrobiol. Beih. Ergebn. Limnol. 39: 157–165.

Spaak, P. and Hoekstra, J.R. 1995. Life history variation and the coexistence of a *Daphnia* hybrid with its parental species. Ecology 76:553–564.

Spaak, P. and Ringelberg, J. 1997. Differential behaviour and shifts in genotype composition during the beginning of a seasonal period of diel vertical migration. Hydrobiologia 360:177–185.

Spitze, K. 1994. *Chaoborus* predation and life-history evolution in *Daphnia pulex*: Temporal pattern of population diversity, fitness, and mean life history. Evolution 45:82–92.

Starkweather, P.L. 1975. Diel patterns of grazing in *Daphnia pulex* Leydig. Verh. Internat. Verein. Limnol. 19:2851–2857.

Starkweather, P. 1978. Diel variations in feeding behavior of *Daphnia pulex*. Influence of food density and nutritional history on mandibular activity. Limnol. Oceanogr. 23:307 317.

Stearns, S.C. 1975. Light responses of *Daphnia pulex*. Limnol. Oceanogr. 20:564–570.

Stearns, D.E. 1983. Control of nocturnal vertical migration in the calanoid copepod *Acartia tonsa* Dana in the Newport River estuary, North Carolina. Ph. D. thesis, Dept. of Zool, Duke Univ., Durham, N. C. 360 pp.

Stearns, D.E. and Forward, Jr., R.B. 1984a. Copepod photobehavior in a simulated natural light environment and its relation to nocturnal vertical migration. Mar. Biol. 82:91–100.

Stearns, D.E. and Forward, Jr., R.B. 1984b. Photosensitivity of the calanoid copepod *Acartia tonsa*. Mar. Biol. 82:85–89.

Sterba, G. 1957. Die neurosekretorischen Zellgruppen einiger Cladoceren (*Daphnia pulex* und *magna, Simocephalus vetulus*). Zool. Jahrbucher Abt. funktionelle Anatomie 76:303–310.

Stibor, H. 1992. Predator induced life-history shifts in a freshwater cladoceran. Oecologia 92: 162–165.

Stibor, H. and Lampert, W. 2000. Components of additivew variance in life-history traits of *Daphnia hyalina*: Seasonal differences in response to predator signals. Oikos 88:129–138.

Stibor, H. and Lüning, J. 1994. Predator-induced phenotypic variation in pattern of growth and reproduction in *Daphnia hyalina* (Crustacea; Cladocera). Funct. Ecol. 8:97–101.

Stibor, H. and Machacek, J. 1998. The influence of fish-exuded chemical signals on the carbon budget of *Daphnia*. Limnol. Oceanogr. 43:997–1000.

Stich, H.B. 1985. Untersuchungen zur tagesperiodischen Vertikalwanderung planktischer Crustaceen im Bodensee. Ph-D. Thesis Albert-Ludwigs-Universitat, Freiburg in Breslau. pp. 1–281.

Stich, H.B. 1989. Seasonal changes of diel vertical migrations of crustacean plankton in Lake Constance. Arch. Hydrobiol. 83:355–405.

Stich, H.-B. and Lampert, W. 1981. Predator evasion as an explanation of diurnal vertical migration. Nature 293:396–398.

Stich, H.-B. and Lampert, W. 1984. Growth and reproduction of migrating and non-migrating *Daphnia* species under simulated food and temperature conditions of diurnal vertical migration. Oecologia (Berlin) 61:192–196.

Strickler, J.R. 1969. Experimentell-oekologische Untersuchungen uber die Vertikalwanderung planktischer Crustaceen. Diss. nr. 4387, Technischen Hochschule, Zurich.

Strickler, J.R. 1970. Uber das Schwimmverhalten von Cyclopoiden bei verminderungen der Bestrahlungsstarke. Schweiz. Z. Hydrol. 32:150–180.

Strickler, J.R. 1975. Intra- and interspecific information flow among planktonic copepods: Receptors. Verh. Internat. Verein. Limnol. 19:2951–2958.

Strickler, J.R. 1982. Calanoid copepods, feeding currents, and the role of gravity. Science 218: 158–160.

Strickler, J.R. 1998. Observing free-swimming copepods mating. Phil. Trans. Roy. Soc. Lond. 353:671–680.

Strogatz, S.H. and Stewart, I. 1993. Coupled oscillators and biological synchronization. Sci. Am. December pp. 68–75.

Stuiver, M., Quay, P.D. and Ostlund, H.G. 1983. Abyssal water carbon-14 distribution and age of the World ocean. Science 220:849–851.

Svetlichny, L.S., Hubareva, E.S., Erkan, F. and Gucu, A.C. 2000. Physiological and behavioral aspects of *Calanus euxinus* females (Copepoda: Calanoida) during vertical migration across temperature and oxygen gradients. Mar. Biol. 137:963–971.

Swain, W.R., Lingeman, R. and Heinis, F. 1987. A characterization and description of the Maarsseveen Lake system. Hydrobiol. Bull. 21:5–16.

Sweat, A.J. and Forward, Jr., R.B. 1985. Diel vertical migration and photoresponse of the chaetognath *Sagitta hispida* Conant. Biol. Bull. 168:18–31.

Swift, M.C. and Forward, Jr., R.B. 1982. Photoresponses of *second-instar Chaoborus* larvae. J. Insect Physiol. 28:183–187.

Swift, M.C. and Forward, Jr., R.B. 1983. Photoresponses of the copepod *Mesocyclops edax*. J. Plankton Res. 5:407–415.

Swift, M.C. and Forward, Jr., R.B. 1988. Absolute light intensity vs. rate of relative change in light intensity: The role of light in the vertical migration of *Chaoborus punctipennis* larvae. Bull. Mar. Sci. 43:604–619.

Taylor, A.H. and Kerr, G.P. 1941. The distribution of energy in the visible spectrum of daylight. J. Opt. Soc. Amer. 31:3–8.

Teraguchi, M. and Northcote, T.G. 1966. Vertical distribution and migration of *Chaoborus flavicans* Larvae in Corbett Lake, British Columbia. Limnol. Oceanogr. 11:164–176.

Tessier, A.J. and Consolatti, N.L. 1989. Variation in offspring size in *Daphnia* and consequences for individual fitness. Oikos 56:269–276.

Tessier, A.J. and Goulden, C.E. 1982. Estimating food limitation in cladoceran populations. Limnol. Oceanogr. 27:707–717.

Tessier, A.J. and Leibold, M.A. 1997. Habitat use and ecological specialization within lake *Daphnia* populations. Oecologia 109:561–570.

Tessier, A.J., Henry, L.L. and Goulden, C.E. 1983. Starvation in *Daphnia*: Energy reserves and reproductive allocation. Limnol. Oceanogr. 28:667–676.

Tilzer, M.M. 1973. Diurnal periodicity in the phytoplankton assemblage of a high mountain lake. Limnol. Oceanogr. 18:15–30.

Tilzer, M.M. 1983. The importance of fractional light absorption by photosynthetic pigments for phytoplankton productivity in Lake Constance. Limnol. Oceanogr. 28:833–846.

Tilzer, M. and Schartz, K. 1976. Seasonal and vertical pattern of phytoplankton, light adaptation in a high mountain lake. Arch. Hydrobiol. 77:488–504.

Tinbergen, N. 1951. The study of instinct. Clarendon Press, Oxford.

Tjossem, S.F. 1990. Effect of fish chemical cues on vertical migration behavior of *Chaoborus*. Limnol. Oceanogr. 37:1456–1468.

Tont, S.A. and Wick, G.L. 1973. Response of a deep scattering layer to the 1972 total solar eclipse. Deep-Sea Res. 20:769–771.

Tyler, J.E. 1975. The in situ quantum efficiency of natural phytoplankton populations. Limnol. Oceanogr. 20:976–980.

Tyler, J.E. and Smith, R.C. 1970. Measurement of spectral irradiance underwater. Gordon and Breach, New York.

Ubrig, U. 1952. Der Einfluss von Sauerstoff und kohlendioxyde auf die taktischen Bewegungen einiger Wassertiere. Z. Vergl. Physiol. 34:471–507.

Uye, S., Huang, C. and Onbe, T. 1990. Ontogenetic diel vertical migration of the planktonic copepod *Calanus sinicus* in the Inland Sea of Japan. Mar. Biol. 104:389–396.

Van der Spoel, S. and Heyman, R.P. 1983. A comparative atlas of zooplankton. Wetenschappelijke Uitgeverij Bunge, Utrecht, The Netherlands.

Van der Stap, I. 2007. Inducible defenses and the dynamics of planktonic food chains. Radboud University, Nijmegen. pp. 1–144.

Van Donk, E., Lurling, M. and Lampert, W. 1997. Consumer-induced changes in phytoplankton: Inducibility, costs, benefits and the impact on grazers. In: Harvell, C.D. and Tollrian, R. (Eds.) Consequences of inducible defences for population biology, Princeton University Press, New York.

Van Duren, L.A. 2000. Moving (in) water, Behavioural kinematics, energetics and hydrodynamics of the calanoid copepod *Temora longicornis* (Müller). Doctoral Thesis, University of Groningen. pp. 1–198.

Van Duren, L.A. and Videler, J.J. 1995. Swimming behaviour of developmental stages of the calanoid copepod *Temore longicornis* at different food concentrations. Mar. Ecol. Progr. Ser 126:154–161.

Van Duren, L.A. and Videler, J.J. 1996. The trade-off between feeding, mate seeking and predator avoidance in copepods: Behavioural responses to chemical cues. J. Plankton Res. 18:805–818.

Van Duren, L.A., Stamhuis, E.J. and Videler, J.J. 1998. Reading the copepod personal ads: Increasing encounter probability with hydromechanical signals. Phil. Trans. Roy. Soc. Lond. B 353:691–700.

Van Gool, E. 1997. Light-induced swimming of *Daphnia:* Can laboratory experiments predict vertical migration? Hydrobiologia 360:161–167.

Van Gool, E. 1998. Diel vertical migration of *Daphnia*: An inquiry into mechanisms of phototactic behaviour. Ph. D. Thesis, Dept.of Aquatic Ecology, Univ. of Amsterdam, pp. 1–117.

Van Gool, E. 1999. Differences in orientation behaviour underlie interclonal differences in phototaxis in *Daphnia magna*. Arch. Hydrobiol. 144:177–194.

Van Gool, E. and Ringelberg, J. 1996. Daphnids respond to algae-associated odours. J. Plankton Res. 18:197–202.

Van Gool, E. and Ringelberg, J. 1997. The effect of accelerations in light increase on the phototactic downward swimming of *Daphnia* and the relevance to diel vertical migration. J. Plankton Res. 19:2041–2050.

Van Gool, E. and Ringelberg, J. 1998. Light-induced migration behaviour of *Daphnia* modified by food and predator kairomones. Anim. Behav. 56:741–747.

Van Gool, E. and Ringelberg, J. 2002. Relation between fish kairomone concentration in a lake and phototactic swimming by *Daphnia*. J. Plankton Res. 24:713–721.

Van Gool, E. and Ringelberg, J. 2003.What goes down must come up: Symmetry in light-induced migration behaviour of *Daphnia*. Hydrobiologia 491:301–307.

Verheijen, F.J. 1958. The mechanism of the trapping effect of artificial light sources upon animals. Arch. Neerl. Zool. 13:1–107.

Verheijen, F.J. 1985. Photopollution: Artificial light optic spatial control systems fail to cope with. Incidents, causations, remedies. Exp. Biol. 44:1–18.

Verheijen, F.J. 1986. The nature and significance of comparative physiology, with optic illustrations, and a sensitive ending. Neth. J. Zool. 36:393–412.

Viaud, G. 1938. Récherches expérimentales sur le Phototropisme des Daphnies. Publ. Fac. Lettres Strassbourg 10e serie 84:1–195.

Viaud, G. 1948. Le phototropism et les deux modes de la photoreception. Experientia 4:81–120.

Viaud, G. 1951. Le phototropism chez les cladocères, les rotiferes et les planaires. Annales de biologique (Paris) 27:365–378.

Vijverberg, J. 1991. Variability and possible adaptive significance of day-time vertical distribution of Leptodora kindtii (Focke) (Cladocera) in a shallow eutrophic lake. Hydrobiol. Bull. 25: 85–91.

Vijverberg, J., Boersma, M., Van Densen, W.L.T., Hoogenboezem, W., Lammens, E.H.R.R. and Mooij, W.M. 1990. Seasonal variation in the interactions between piscivorous fish, planktivorous fish and zooplankton in a shallow eutrophic lake. Hydrobiologia 207:279–286.

Vijverberg, J., Doksæter, A. and Van Donk, E. 2006. Contrasting life history responses to fish released infochemicals of two co-occurring *Daphnia* species that show different migration behaviour. Arch. Hydrobiol. 167:89–114.

Vincente, E., Camacho, A. and Rodrigo, M.A. 1993. Morphometry and physico-chemistry of the cryogenic meromictic Lake El Tobar (Spain). Verh. Internat. Verein. Limnol. 25:698–704.

Vinogradov, M.E., Flint, M.V. and Shushkina, E.A. 1985a. Vertical distribution of mesoplankton in the open Black Sea. Mar. Biol. 89:95–107.

Vinogradov, M.E., Flint, M.V. and Shushkina, E.A. 1985b. Vertical distribution of mesoplankton in the open area of the Black Sea. Mar. Biol. 89:95–107.

Vinogradov, M.E., Arashkevich, E.G. and Ilchenko, S.V. 1992. The ecologuy of *Calanus ponticus* population in the deeper layer of its concentration in the Black Sea. J. Plankton Res. 14: 447–458.

Von Buddenbrock, W. 1931. Untersuchungen über den Schattenreflex. Z. Vergl. Physiol. 13: 164–214.

Von Buddenbrock, W. 1952. Vergleichende Physiologie, Band I. Sinnesphysiologie. Birkhauser Verlag, Basel.

Von Elert, E. and Franck, A. 1999. Colony formation in *Scenedesmus*: Grazer-mediated release and chemical features of the infochemical. J. Plankton Res. 21:789–804.

Von Elert, E. and Loose, C.J. 1996. Predator-induced diel vertical migration in *Daphnia:* Enrichment and preliminary chemical characterization of kairomone exuded by fish. J. Chem. Ecol. 22:885–895.

Von Elert, E. and Pohnert, G. 2000. Predator specificity of kairomones in diel vertical migration of *Daphnia*: A chemical approach. Oikos 88:119–128.

Von Elert, E. and Stibor, H. 2006. Predator-mediated life history shifts in *Daphnia*: Enrichment and preliminary chemical characterisation of a kairomone exuded by fish. Arch. Hydrobiol. 167:21–35.

Von Holst, E. and Mittelstaedt, H. 1950. Das Reafferenzprinzip. Naturwissenschaften 37:464–476.

Von Uexküll, J. 1921. Umwelt und Innenwelt der Tiere. Berlin. Second edition.

Wagner-Döbler, J. 1984. Steuerung der Vertikalwanderung von *Chaoborus flavicans* im Freiland. Dissertation Universität München. pp. 1–266.

Wagner-Döbler, J. 1988. Vertical migration of *Chaoborus flavicans* (Diptera, Chaoboridae): The control of day and night depth by environmental parameters. Arch. Hydrobiol. 114:251–274.

Wanger-Dobier, I. 1990. vertical migration of *Chaoborus flavicans* and migration velocity by environmental stimuli. Arch. Hydrobiol. 117:279–307.

Wahl, C.M., Mills, E.L., McFarland, W.N. and Degisi, J.S. 1993. Ontogenetic changes in prey selection and visual acuity of the yellow perch, *Perca flavescens*. Can. J. Fish. Aquat. Sci. 50:743–749.

Wang, N. 1994. Food and feeding of young perch (*Perca fluviatilis* L.) in Lake Constance. Verh. Internat. Verein. Limnol. 25:2148–2152.

Wang, N. and Eckmann, R. 1994. Distribution of perch (*Perca fluviatilis* L.) during their first year of life in Lake Constance. Hydrobiologia 277:135–143.

Waterman, T.H. 1961. Light sensitivity and vision. In: Waterman, T.H. (Ed.) Physiology of crustacean. Vol. II, pp. 1–64. Academic Press, New York.

Waterman, T.H. 1966a. Interaction of polarized light and turbidity in the orientation of *Daphnia* and *Mysidium*. Z. Vergl. Physiol. 43:149–172.

Waterman, T.H. 1966b. Polarotaxis and primary photoreceptor events in Crustacea. In: Bernhard, C.G. (Ed.) The functional organisation of the compound eye, pp. 493–511. Pergamon Press, New York.

Waterman, T.H. 1974. Underwater light and the orientation of animals. In: Jerlov, N.G. and Steemann Nielsen, E. (Eds.) Optical aspects of oceanography, pp. 415–442. Academic Press, London.

Weber, A. 2001. Interactions between predator kairomone and food level complicate the ecological interpretation of *Daphnia* laboratory results. J. Plankton Res. 23:41–46.

Weber, A. and Declerck, S. 1997. Phenotypic plasticity of *Daphnia* life-history traits in response to predator kairomones: Genetic variability and evolutionary potential. Hydrobiologia 360:89–99.

Weers, P.M.M. and Gulati, R.D. 1997. Effect of the addition of polyunsaturated fatty acids to the diet on the growth and fecundity of *Daphnia galeata*. Freshw. Biol. 38:721–729.

Weetman, D. and Atkinson, D. 2002. Antipredator reaction norms for life history traits in *Daphnia pulex*: Dependence on temperature and food. Oikos 98:299–307.

Wehner, R. 1981. Vision in Arthropods. In: Autrum (Ed.) Comparative physiology and evolution of vision in invertebrates. C invertebrate visual centers and behavior II, Handbook of physiology, p. 378. Springer Verlag, München.

Weider, L.J. 1984. Spatial heterogeneity of *Daphnia* phenotypes: Vertical migration and habitat partitioning. Limnol. Oceanogr. 29:225–236.

Weider, L.J. and Stich, H.B. 1992. Spatial and temporal heterogeneity of *Daphnia* in Lake Constance; intra- and interspecific comparisons. Limnol. Oceanogr. 37:1327–1334.

Weismann, A. 1877. Das Tierleben im Bodensee. Schr. Ver. Gesch. Bodensee Umgebung 7:1–31.

Werner, E.E. and Hall, D.J. 1988. Ontogenetic habitat shifts in bluegill: The foraging rate-predation risk trade-off. Ecology 69:1352–1366.

Wetzel, R.G. 2001. Limnology. Lake and river systems. Academic Press, New York. Ed. 3. 1006 pp.

Whitney, L.V. 1938. Continuous solar radiation measurements in Wisconsin lakes. Trans. Wisc. Acad. Sci. Arts and Letters. 31:175.

Whitney, L.V. 1941. The angular distribution of characteristic diffuse light in natural waters. J. Mar. Res. 4:122–131.

Wichard, T., Poulet, S.A., Halsband Lenk, C., Albaina, A., Harris, R.P. and Dongyan, L. 2005. Survey of the chemical defence potential of diatoms: screening of fifty one species of alpha, beta, gamma, delta-unsaturated aldehydes. J. Chem. Ecol. 31:949–958.

Williams, T.M., Davis, R.W., Fuiman, L.A., et al. 2000. Sink or swim: Strategies for cost-efficient diving by marine mammels. Science 288:133–136.

Williamson, C.E. 1986. The swimming and feeding of *Mesocyclops*. Hydrobiologia 134:11–19.

Williamson, C.E., Neale, P.J., Grad, G., De Lange, H.J. and Hargraeves, B.R. 2001. Beneficial and detrimental effects of UV on aquatic organisms: Implications of spectral variation. Ecol. Appl. 11:1843–1857.

Wiltshire, K., Boersma, M. and Meyer, B. 2003. Grazer-induced changes in the desmid *Staurastrum*. Hydrobiologia 491:255–260.

Winder, M., Spaak, P., Mooij, W. and Burgi, H.R. 2002. Diel vertical migration: Not a predator avoidance strategy in high-mountain lakes? Ph. D. Thesis, ETH, Zurich 107–131.

Winder, M., Spaak, P. and Mooij, W.M. 2004. Trade-offs in *Daphnia* habitat selection. Ecology 85:2027–2036.

Winfree, A.T. 1987. The timing of biological clocks. The Scientific American Library, New York, pp. 1–192.

Wolf, H.G. 1987. Interspecific hybridization between *Daphnia hyalina, D. galeata*, and *D. cucullata* and seasonal abundances of these species and their hybrids. Hydrobiologia 145:213–217.

Wolf, H.G. 1988. Diferences in genetic structure of pond-dwelling and lake-dwelling *Daphnia*. Verh. Internat. Verein. Limnol. 23:2056–2059.

Wolf, H.G. and Hobaek, A. 1986a. Ecological genetics of Norwegian *Daphnia*. Hereditas 104: 193–198.

Wolf, H.G. and Mort, M.A. 1986b. Inter-specific hybridization underlies phenotypic variability in *Daphnia* populations. Oecologia (Berlin) 68:507–511.

Wolken, J.J. 1971. Invertebrate photoreceptors. Academic Press, New York. pp. 74–77.

Wong, C.K. 1988. The swimming behavior of the copepod *Metridia pacifica*. J. Plankton Res. 11:1285–1290.

Wong, C.K. and Sprules, W.G. 1986. The swimming behavior of the freshwater calanoid copepods *Limnocalanus macrurus* Sars, *Senecella calanoides* Juday and *Epischura lacustris* Forbes. J. Plankton Res. 8:79–90.

Wong, C.K., Ramcharan, C.W. and Sprules, W.G. 1986. Behavioral responses of an herbivorous calanoid copepod to the presence of other zooplankton. Can. J. Zool. 64:1422–1425.

Worthington, E.B. 1931. Vertical movements of freshwater macroplankton. Int. Rev. Ges. Hydrobiol. 25:394–436.

Worthington, E.B. and Ricardo, C.K. 1936. Scientific results of the Cambridge expedition to the East African lakes 1930–1931, No. 17. The vertical distribution and movements of the plankton in Lakes Rudolf, Naivasha, Edward and Bunyoni. J. Linn. Soc. Zool. 40:33–69.

Young, S. and Downing, A.C. 1976. The receptive fields of *Daphnia* ommatidia. J. Exp. Biol. 64:185–202.

Zagarese, H.E., Feldman, M. and Williamson, C.E. 1997. UV-B induced damage and photoreactivation in three species of *Boeckellia* (Copepoda, Calanoida). J. Plankton Res. 19:357–367.

Zaret, T.M. 1975. Strategies for existence of zooplankton prey in homogeneous environments. Verh. Internat. Verein. Limnol. 19:1484–1489.

Index

Lightning Source UK Ltd.
Milton Keynes UK
UKOW07n1004190516

274505UK00002B/60/P

9 789048 130924